Complex Variable Methods for Linear Multivariable Feedback Systems

Complex Variable Methods for Linear Multivariable Feedback Systems

Edited by

A. G. J. MACFARLANE

Engineering Department, University of Cambridge

TAYLOR & FRANCIS LTD
10–14 Macklin Street, London WC2B 5NF
1980

First published 1980 by Taylor & Francis Ltd,
10/14 Macklin Street, London WC2B 5NF.

ISBN 0 85066 197 8

Printed and bound in Great Britain by
Taylor & Francis (Printers) Ltd
Rankine Road, Basingstoke, Hampshire RG24 0PR

PREFACE

This collection of reprints from the *International Journal of Control* summarizes an extensive investigation into ways of generalizing the classical Nyquist–Bode–Evans techniques used in the study of single-input, single-output feedback systems to the multivariable case. They are based on work carried out under Science Research Council support by the Control Group of Cambridge University Engineering Department ; this was a continuation of work started in the Control Systems Centre of the University of Manchester Institute of Science and Technology. I am most grateful to the authors of the papers collected here for their agreement to collaborate in producing this book. I would also like to acknowledge my indebtedness to other friends and colleagues whose work played an indispensable role in this investigation, particularly Dr. T. Kontakos, Dr. D. G. Retallack, Dr. J. J. Belletrutti, Dr. N. Karcanias and, especially, Dr. U. Shaked whose investigations of multivariable root loci and zeros were of invaluable assistance.

August 1979

A. G. J. MacFarlane
Cambridge

CONTENTS

I. INTRODUCTION

The work presented in the set of reprints collected here has the following aims :

(i) to extend the concepts underlying the classical work of Nyquist (1932), Bode (1940, 1945) and Evans (1948, 1950, 1953) on feedback systems to the multivariable case ;

(ii) to relate such extensions to state-space model descriptions ; and

(iii) to explore the use of the techniques so developed for the design of multivariable feedback control systems.

Following the early work of Boksenbom and Hood (1949) on non-interacting control systems, the application of frequency-response methods to multivariable control problems was investigated by Golomb and Usdin (1952), Raymond (1953), Kavanagh (1957 a, 1957 b, 1958) and Freeman (1957, 1958). During the hey-day of the ' state-space revolution ' in control theory, the interest in such an approach dropped but began to revive again during the mid-1960s (MacFarlane 1979). Kalman showed that links existed between state-space and frequency-response approaches to control problems by his study of the frequency-domain characterization of optimality (Kalman 1964). Rosenbrock began a systematic attack on the whole problem of developing a frequency-response analysis and design theory for multivariable systems in an important and pioneering paper (Rosenbrock 1966). In his Inverse Nyquist Array (I.N.A.) design method (Rosenbrock 1969, 1974, Rosenbrock and Munro 1978) he developed a means of reducing interaction in a multivariable control system to an amount which would then enable classical single-loop, frequency-response techniques to be deployed to complete the controller design. Rosenbrock's approach was based upon a careful use of a specific criterion of partial interaction, embodied in the concept of diagonal dominance. Following the success of his approach, other investigators have sought to develop ways of reducing a multivariable control problem to a succession of single-loop problems, as in the sequential return difference method developed by Mayne (1973). A discussion of these and other techniques, and a comprehensive survey of the development of frequency-response methods in automatic control has been given by MacFarlane (1979).

The approach adopted in the papers collected here treats the transfer function matrix representation of a linear time-invariant dynamical system as a single object in its own right, and seeks to answer questions of the following sort. How can we extend the key basic concepts associated with the transfer functions which are used in classical single-loop frequency-response feedback investigations to the transfer-function matrices which appear in the multivariable case ? What are the appropriate generalizations to the multivariable case of such specific concepts as pole, zero, Nyquist diagram and root-locus diagram ? How are such generalizations related to state-space models ? How can the classical approach to feedback system design via the manipulation of Nyquist and Bode diagrams be extended to attack the problem of designing multivariable feedback systems ?

An early attempt to extend the use of Nyquist diagrams to the multivariable case was made by Bohn (Bohn 1962, Bohn and Kasvand 1963). A generalization of the Nyquist stability criterion to the multivariable feedback case was put forward by MacFarlane (1970). Following this heuristic treatment, complex-variable-based proofs of a generalized Nyquist stability criterion were supplied by Barman and Katzenelson (1974) and by MacFarlane and Postlethwaite (1977).

The papers reprinted are grouped into three sections : dealing with the use of algebraic functions, with multivariable zeros and multivariable root loci, and with design techniques respectively. In the first section the main topics treated are the complex-variable (as opposed to algebraic) treatment of poles and zeros for transfer-function matrices, the establishment of a generalized Nyquist stability criterion for multivariable feedback systems, the introduction of multivariable root loci and their relationship to generalized Nyquist diagrams, and the use of generalized Nyquist diagrams and multivariable root loci in stability analysis. It is shown how the classical single-loop feedback systems techniques of studying complex gain as a function of complex frequency (the Nyquist–Bode approach) and complex frequency as a function of complex gain can be extended to the multivariable case by associating with transfer function matrices (having the same number of rows as columns) a pair of analytic functions : a characteristic gain function and a characteristic frequency function. These are algebraic functions and each is defined on an appropriate Riemann surface whose construction is described.

The second set of papers deals with the relationships between the parameters of state-space models and multivariable system poles, zeros and root-locus asymptotes. It is shown how both the finite and infinite zeros of the characteristic gain function can be treated from a unified point of view. The asymptotic behaviour of the closed-loop characteristic frequencies of an optimal linear quadratic regulator is also examined.

The development of design techniques for linear multivariable feedback systems is the subject of the final set of papers. Two approaches to the design problem are presented. The first of these is concerned with the manipulation of the frequency-dependent eigenvalues and eigenvectors of a square open-loop transfer-function matrix. The key idea behind this approach is that of an approximately commutative controller, and to enable this to be used algorithms are developed for the approximation of a frame of complex vectors by a frame of real vectors ; these algorithms play a vital role in this design technique. The second method described seeks to produce control schemes which result in a closed-loop frequency response which is as close as possible, in a least-squares sense, to a desired closed-loop response. Design approaches of both these types require for their successful implementation the development of an appropriate set of computer programs for use in computer-aided interactive design with an appropriate form of graphics terminal. Such a set of programs has been developed and described by Edmunds (1978, 1979). This development was supported by the Science Research Council and the resulting package is available from Compeda Limited, Compeda House, Walkern Road, Stevenage, Herts SG1 3QP.

Some further information about the background to the various sets of reprints is given at the beginning of each section and a comprehensive bibliography is given at the end of the book.

References

Barman, J. F., and Katzenelson, J., 1974, A generalized Nyquist-type stability criterion for multivariable feedback systems. *Int. J. Control*, **20,** 593–622.

Bode, H. W., 1940, Relations between attenuation and phase in feedback amplifier design. *Bell Syst. tech. J.*, **19,** 421–454 ; 1945, *Network Analysis and Feedback Amplifier Design* (Princeton, N.J. : Van Nostrand).

Bohn, E. V., 1962, Design and synthesis methods for a class of multivariable feedback control systems based on single-variable methods. *Trans. Am. Inst. elect. Engrs*, **81,** 109–116.

Bohn, E. V., and Kasvand, T., 1963, Use of matrix transformations and system eigenvalues in the design of linear multivariable control systems. *Proc. Instn elect. Engrs*, **110,** 989–996.

Boksenbom, A. S., and Hood, R., 1949, General algebraic method applied to control analysis of complex engine types, National Advisory Committee for Aeronautics, Report NCA-TR-980, Washington, D.C.

Edmunds, J. M., 1978, Cambridge Linear Analysis and Design Programs, Report No. CUED/F-CAMS/TR 170, Cambridge University Engineering Department ; 1979, Cambridge Linear Analysis and Design Programs. *Proceedings of the IFAC Symposium on Computer-Aided Control System Design*, Zurich.

Evans, W. R., 1948. Graphical analysis of control systems. *Trans. Am. Inst. elect. Engrs*, **67,** 547–551 ; 1950, Control system synthesis by root locus method, *Ibid.*, **69,** 1–4 ; 1953, *Control System Dynamics* (New York : McGraw-Hill).

Freeman, H., 1957, A synthesis method for multipole control systems. *Trans. Am. Inst. elect. Engrs*, **76,** 28–31 ; 1958, Stability and physical realizability considerations in the synthesis of multipole control systems. *Ibid.*, Part 2, **77,** 1–15.

Golomb, M., and Usdin, E., 1952, A theory of multidimensional servo systems. *J. Franklin Inst.*, **253(1),** 28–57.

Kalman, R. E., 1964, When is a linear control system optimal ? *J. bas. Engng*, **86,** 51–60.

Kavanagh, R. J., 1957 a, Noninteraction in linear multivariable systems. *Trans. Am. Inst. elect. Engrs*, **76,** 95–100 ; 1957 b, The application of matrix methods to multivariable control systems. *J. Franklin Inst.*, **262,** 349–367 ; 1958, Multivariable control system synthesis. *Trans. Am. Inst. elect. Engrs*, Part 2, **77,** 425–429.

MacFarlane, A. G. J., 1970, The return-difference and return-ratio matrices and their use in the analysis and design of multivariable feedback control systems. *Proc. Instn elect. Engrs*, **117,** 2037–2049.

MacFarlane, A. G. J. (Editor), 1979, *Frequency Response Methods in Control Systems*, I.E.E.E. Reprint Series (New York : I.E.E.E.).

MacFarlane, A. G. J., and Postlethwaite, I., 1977, The generalized Nyquist stability criterion and multivariable root loci. *Int. J. Control*, **25,** 81–127.

Mayne, D. Q., 1973, The design of linear multivariable systems. *Automatica*, **9,** 201–207.

Nyquist, H., 1932, ' Regeneration Theory '. *Bell Syst. tech. J.*, **11,** 126–147.

Raymond, F. H., 1953, Introduction a l'etude des asservissements multiples simultanes. *Bull. Soc. fr. Mec.*, **7,** 18–25.

Rosenbrock, H. H., 1966, On the design of linear multivariable control systems. *Proceedings of the 3rd IFAC Congress*, London, Vol. 1, Book 1, Paper 1A (London : Institution of Mechanical Engineers) ; 1969, Design of multivariable control systems using the inverse Nyquist array. *Proc. Instn elect. Engrs*, **116,** 1929–1936 ; 1974. *Computer-aided Control System Design* (New York : Academic Press).

Rosenbrock, H. H., and Munro, N., 1978, The inverse Nyquist array method. *Proceedings of the International Forum on Alternatives for Linear Multivariable Control* (Chicago : National Engineering Consortium), pp. 101–137.

II. ALGEBRAIC FUNCTION APPROACH TO LINEAR MULTIVARIABLE FEEDBACK SYSTEMS

The classical approaches to single-loop feedback systems analysis involve :

(i) studying open-loop gain as a function of imposed frequency—the Nyquist–Bode approach ; or

(ii) studying closed-loop characteristic frequency as a function of imposed gain—the Evans root-locus approach.

Both of these may be extended to the multivariable case ; furthermore such an extension may be done in a way which shows that there is a fundamental duality between open-loop characteristic gain and closed-loop characteristic frequency. Consider first the determination of closed-loop characteristic frequency as a function of a gain parameter. The closed-loop characteristic frequencies for the feedback configuration shown in Fig. 2.1 are the eigenvalues of the matrix

$$S(g)=A+B(gI_m-D)^{-1}C$$

and are thus found via the equation

$$\nabla(s, g)=\det\,[sI_n-S(g)]=0$$

where $S(g)$ may be called the closed-loop frequency matrix for a system $\mathscr{S}(A, B, C, D)$ having the same number of inputs and outputs. $S(g)$ has a strikingly symmetrical relationship with the open-loop gain matrix

$$G(s)=D+C(sI-A)^{-1}B$$

and this symmetry naturally leads one to seek an open-loop concept which is dual to the concept of closed-loop characteristic frequency. An obvious candidate is one associated with the eigenvalues of $G(s)$. These open-loop characteristic gains are found via the equation

$$\Delta(g, s)=\det\,[gI_m-G(s)]=0$$

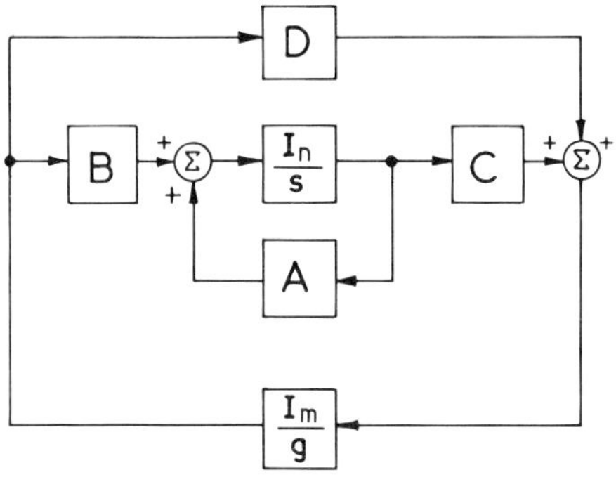

Figure 2.1

and it can be shown that (Postlethwaite and MacFarlane 1979)

$$\det [sI_n - A]\Delta(g, s) = \det [gI_m - D]\nabla(s, g)$$

which implies that, for values of $s \notin \sigma(A)$ and values of $g \notin \sigma(D)$ where $\sigma(A)$ denotes the set of eigenvalues (spectrum) of A :

$$\Delta(g, s) = 0 \Leftrightarrow \nabla(s, g) = 0$$

This shows that a knowledge of characteristic gain as a function of frequency for $G(s)$ is equivalent to a knowledge of characteristic frequency as a function of gain for $S(g)$ and vice versa. These ideas can be used to extend classical Nyquist–Bode and root-locus techniques to the multivariable case.

The use of the frequency-dependent eigenvalues of a matrix transfer function to discuss the stability and performance of a multivariable feedback system in generalized Nyquist terms was discussed in an early paper by Bohn and Kasvand (1963). The systematic use of ' gain spectra ' (the frequency-dependent eigenvalues of the gain matrix $G(s)$) and ' frequency spectra ' (the gain-dependent eigenvalues of the frequency matrix $S(g)$) for the analysis and design of linear multivariable feedback systems was developed in a series of papers by MacFarlane and his collaborators (MacFarlane 1970, MacFarlane and Belletrutti 1973, MacFarlane and Postlethwaite 1977 a, 1977 b, MacFarlane and Kouvaritakis 1977, MacFarlane *et al.* 1978, Postlethwaite and MacFarlane 1979). In these studies there is no essential loss of generality involved in the fact that one needs ' square ' operators (having the same number of inputs and outputs) to have a spectrum. A crucial part of such investigations is concerned with how the state-space parameters of a general linear dynamical system enter into a design procedure when forming the ' square ' operators involved from the ' non-square ' systems around which feedback loops are being closed.

The equations

$$\Delta(g, s) = 0 \quad \text{and} \quad \nabla(s, g) = 0$$

are algebraic equations relating a pair of complex variables s and g. They are used for any dynamical system having the same number of inputs and outputs (such as will arise in considering vector loop transmittances in a multivariable feedback configuration) to define a pair of algebraic functions :

(i) a characteristic gain function $g(s)$ which gives open-loop characteristic gain as a function of frequency ; and

(ii) a characteristic frequency function $s(g)$ which gives closed-loop characteristic frequency as a function of gain (MacFarlane and Postlethwaite 1977 b).

Characteristic gain function

The characteristic gain function $g(s)$ is defined via the equation

$$\Delta(g, s) = \det [gI_m - G(s)] = 0$$

For transfer function matrices arising from practical situations, $\Delta(g, s)$ can be taken to be irreducible over the field of rational function in s. $\Delta(g, s)$ will have the form

$$\Delta(g, s) = g^m + a_1(s)g^{m-1} + \ldots + a_m(s) \tag{2.1}$$

where $\{a_i(s) : i = 1, 2, \ldots, m\}$ are rational functions in s. If $b_0(s)$ is the least common denominator of the $\{a_i(s)\}$ then eqn. (2.1) can be put in the form

$$b_0(s)g^m + b_1(s)g^{m-1} + \ldots + b_m(s) = 0 \tag{2.2}$$

where the coefficients $\{b_i(s) : i = 1, 2, \ldots, m\}$ are now polynomials in s.

A function of a complex variable $g(s)$ defined by an equation of the form (2.2) is called an algebraic function, and is a generalization of the basic concept of a function of a complex variable. There is a well established theory of such functions (Bliss 1933) and its use lies at the heart of the generalization of the classical single-variable feedback system techniques to the multivariable case (MacFarlane and Postlethwaite 1977 a). An algebraic function is said to have a branch point whenever its defining algebraic equation has repeated roots. In every simply connected region of the complex plane which does not contain any branch points, the theory of algebraic functions shows that the m values of the characteristic gain function, obtained as the m roots of the defining algebraic equation, form a set of analytic functions $\{g_i(s) : i = 1, 2, \ldots, m\}$; each of these analytic function is called a branch of $g(s)$ and each branch may be analytically continued into any other branch. A single domain on which $g(s)$ is a single-valued function of position may be constructed out of m copies of the complex plane, suitably joined together along arbitrary cuts between the branch points at which the various branches of $g(s)$ take common values. Such a domain is called a Riemann surface for $g(s)$; the theory underlying the Riemann surface for $g(s)$ and a systematic method for its construction have been described by MacFarlane and Postlethwaite (1977 a) and by Edmunds (1979). The poles of $g(s)$ are defined as the roots of

$$b_0(s) = 0$$

and the zeros of $g(s)$ are defined as the roots of

$$b_m(s) = 0$$

If the Smith–McMillan-form-defined poles and zeros of $G(s)$ (MacFarlane and Karcanias 1976) are the roots of

$$p(s) = 0 \quad \text{and} \quad z(s) = 0$$

respectively then it can be shown that these algebraically defined poles and zeros are related to those defined in terms of the polynomials $b_0(s)$ and $b_m(s)$ via

$$p(s) = e(s)b_0(s) \quad \text{and} \quad z(s) = e(s)b_m(s)$$

where $e(s)$ is the monic polynomial whose zeros are those poles of $G(s)$ which are not poles of $g(s)$.

The poles and zeros of the characteristic gain function $g(s)$ are located on the Riemann surface for $g(s)$. This surface can be thought of as an organized set of all possible closed-loop characteristic frequencies associated with all possible values of the gain parameter g. It is useful to exhibit the nature of $g(s)$ by drawing constant-amplitude and constant-phase contours of $g(s)$ on the Riemann surface.

Characteristic frequency function

The characteristic frequency function $s(g)$ is defined via the equation

$$\nabla(s, g) = \det [sI_n - S(g)] = 0$$

It is also an algebraic function and exactly the same considerations apply as for the characteristic gain function, with the roles of s and g suitably interchanged. It has as domain a Riemann surface formed out of n copies of the complex plane since there are n values of closed-loop characteristic frequency for every given value of g. This Riemann surface can be thought of as the set of all possible open-loop characteristic gains of the open-loop transfer-function matrix $G(s)$ associated with all possible values of open-loop-frequency variable s. Its nature may be exhibited by overlaying it with constant-real-part and constant-imaginary-part or with constant-amplitude and constant-phase contours of $s(g)$. The complex function $s(g)$ of course has poles and zeros but their significance is quite different from those of $g(s)$.

Multivariable Nyquist diagrams: characteristic gain loci and the multivariable Nyquist stability criterion

It is convenient to refer to the Riemann surface for the characteristic gain function $g(s)$ as the frequency-surface (since it is built out of m copies of the frequency-plane or s-plane) and to the Riemann surface for the characteristic frequency function $s(g)$ as the gain-surface (since it is built out of n copies of the gain-plane or g-plane).

Associated with any open loop $m \times m$ transfer function matrix $G(s)$ is (assuming that $\Delta(g, s)$ of eqn. (2.1) is irreducible) a characteristic gain function $g(s)$ having an appropriate frequency-surface as domain. Let D be the standard Nyquist contour in the complex plane, and suppose that a copy of D is placed on each of the m copies of the s-plane out of which the frequency-surface is constructed. Let N be the image set under $g(s)$ on the gain surface of this set of copies of D on the frequency-surface: then this set of complex numbers N is called the characteristic gain locus for $g(s)$ (and hence for $G(s)$). In practice the characteristic gain locus is usually generated as the set of loci in a single copy of the gain plane traced out by the set of m eigenvalues of $G(s)$ as s traverses the Nyquist contour in the s-plane. This set of complex numbers N on the gain-surface corresponds to the $\pm 90°$ constant-phase contours of $s(g)$ on the gain surface. When viewed on the gain-surface (a Riemann surface) the closed image curve N never intersects itself; when represented as a set of loci in a single copy of the complex plane however these loci usually do intersect. These characteristic gain loci are generalizations of the classical single-loop Nyquist diagram and may accordingly be called multivariable Nyquist diagrams (MacFarlane and Postlethwaite 1977 a, 1977 b).

Multivariable Nyquist stability criterion

The multivariable Nyquist stability criterion relates the closed-loop stability of the configuration of Fig. 2.2 to the characteristic gain loci for the loop-gain matrix $G(s)H(s)$. Let the system of Fig. 2.2 have no uncontrollable and/or unobservable modes whose corresponding characteristic frequencies lie in the right-half plane. Then this feedback configuration will be closed-loop stable if and only if the net sum of anti-clockwise encirclements of the critical point $(-1+j0)$ by the set of characteristic gain loci of $G(s)H(s)$ is equal to the total number of right-half-plane poles of $G(s)$ and $H(s)$.

Proofs of multivariable versions of the Nyquist stability criterion were given by Barman and Katzenelson (1974) and MacFarlane and Postlethwaite (1977 a), following earlier heuristic approaches by Bohn and Kasvand (1963) and MacFarlane (1970). The use of the frequency-dependent eigenvalue spectrum of a transfer function matrix played a key role in an important paper by Freedman *et al.* (1969) which generalized the circle criterion for non-linear feedback system stability to a wide class of systems.

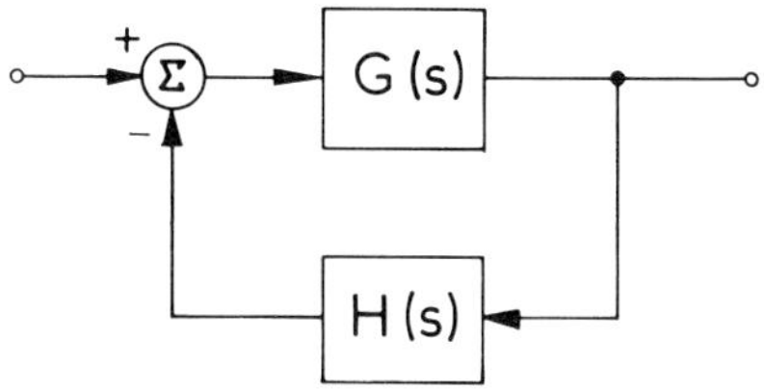

Figure 2.2

Multivariable root loci

The loci of open-loop characteristic gain with frequency are generalizations of classical Nyquist diagrams in the gain-plane. There is a corresponding generalization of the classical Evans root-locus diagrams in the frequency-plane. These are the characteristic frequency loci, which are the 180° phase contours of $g(s)$ on the frequency surface. Let R^- denote the negative real axis in the complex plane, and suppose that a copy of R^- is placed on each of the n copies of the g-plane out of which the gain surface is constructed. Let E be the image set under $s(g)$ on the frequency-surface of this set of copies of R^- on the gain surface ; then this set of complex numbers R is called the characteristic frequency locus for $s(g)$ (and hence for $S(g)$). In practice the characteristic frequency locus is generated as the set of loci in a single copy of the complex frequency plane traced out by the eigenvalues of $S(g)$ as g traverses the negative real axis in the gain-plane. These characteristic frequency loci are generalizations of the classical single-loop root locus diagram and may accordingly be called multivariable root-locus diagrams. In common with the classical root-locus approach, these multivariable root-locus diagrams are usually calibrated in terms of a direct feedback gain parameter $k=g^{-1}$. The use of such multivariable root loci is discussed in papers by Retallack (1970), Kouvaritakis and Shaked (1976), Shaked (1976), MacFarlane *et al.* (1978), Kouvaritakis and Edmunds (1978), Kouvaritakis (1978) and in the book by Owens (1978).

All the papers reprinted in this section deal with this algebraic function approach to generalized Nyquist diagrams for multivariable feedback systems and to the complementary concept of multivariable root loci. The first three are principally concerned with the statement and proof of a generalized Nyquist stability criterion. Relationships between the algebraic functions defining characteristic gain and characteristic frequency functions are then discussed in the next pair of papers. The final paper in this section, by Edmunds, gives a large number of examples of characteristic gain and frequency loci for a wide range of systems.

References

BARMAN, J. F., and KATZENELSON, J., 1974, A generalized Nyquist-type stability criterion for multivariable feedback systems. *Int. J. Control*, **20,** 593–622.

BLISS, G. A., 1933, *Algebraic Functions* (Providence. Rhode Island : American Mathematical Society).

BOHN, E. V., and KASVAND, T., 1963, Use of matrix transformations and system eigenvalues in the design of linear multivariable control systems. *Proc. Instn elect. Engrs*, **110,** 989–996.

EDMUNDS, J. M., 1979, Characteristic gains, characteristic frequencies and stability. *Int. J. Control*, **29,** 669–706.

FREEDMAN, M. I., FALB, P. L., and ZAMES, G., 1969, A Hilbert space stability theory over locally compact Abelian groups. *SIAM J. Control*, **7,** 479–495.

KOUVARITAKIS, B., 1978, Asymptotic multivariable root locus behaviour for non-proper systems. *Int. J. Control*, **28,** 419–440.

KOUVARITAKIS, B., and EDMUNDS, J., 1978, The characteristic frequency and characteristic gain design method for multivariable feedback systems. *Alternatives for Linear Multivariable Control* (Chicago : National Engineering Consortium), pp. 229–246.

KOUVARITAKIS, B., and SHAKED, U., 1976, Asymptotic behaviour of root loci of linear multivariable systems. *Int. J. Control*, **23,** 297–340.

MACFARLANE, A. G. J., 1970, The return-difference and return-ratio matrices and their use in the analysis and design of multivariable feedback control systems. *Proc. Instn elect. Engrs*, **117,** 2037–2049.

MACFARLANE, A. G. J., and BELLETRUTTI, J. J., 1973, The characteristic locus design method. *Automatica*, **9,** 575–588.

MACFARLANE, A. G. J., and KARCANIAS, N., 1976, Poles and zeros of linear multivariable systems : a survey of the algebraic, geometric and complex-variable theory. *Int. J. Control*, **24,** 33–74.

MACFARLANE, A. G. J., and B. KOUVARITAKIS, 1977, A design technique for linear multivariable feedback systems. *Int. J. Control*, **25,** 837–874.

MACFARLANE, A. G. J., KOUVARITAKIS, B., and EDMUNDS, J. M., 1978, Complex variable methods for multivariable feedback systems analysis and design. *Alternatives for Linear Multivariable Control* (Chicago : National Engineering Consortium), pp. 189–228.

MACFARLANE, A. G. J., and POSTLETHWAITE, I., 1977 a, The generalized Nyquist stability criterion and multivariable root loci. *Int. J. Control*, **25,** 81–127 ; 1977 b, Characteristic frequency functions and characteristic gain functions. *Ibid.*, **26,** 265–278.

OWENS, D. H., 1978, *Feedback and Multivariable Systems* (Hitchin : Peter Peregrinus).

POSTLETHWAITE, I., and MACFARLANE, A. G. J., 1979, *A Complex Variable Approach to the Analysis of Linear Multivariable Feedback Systems*, Lecture Notes in Control and Information Sciences, Vol. 12 (Berlin : Springer-Verlag).

RETALLACK, D. G., 1970, Extended root-locus technique for design of linear multivariable feedback systems. *Proc. Instn elect. Engrs*, **117,** 618–622.

SHAKED, U., 1976, The angles of departure and approach of the root loci in linear multivariable systems. *Int. J. Control*, **23,** 445–457.

The generalized Nyquist stability criterion and multivariable root loci

A. G. J. MacFARLANE and I. POSTLETHWAITE

A comprehensive discussion is given of the background to the generalized Nyquist stability criterion for linear multivariable feedback systems. This leads to a proof based on the use of the Principle of the Argument applied to an algebraic function defined on an appropriate Riemann surface. It is shown how the matrix-valued functions of a complex variable which define the loop transmittance, return-ratio and return-difference matrices of feedback systems analysis may be associated with a set of characteristic algebraic functions, each associated with a Riemann surface. These characteristic functions enable the characteristic loci, which featured in previous heuristic treatments of the generalized Nyquist stability criterion, to be put on a sound basis. The relationship between the algebraic structure of the matrix-valued functions and the appropriate complex-variable theory is carefully discussed. These extensions of the complex-variable concepts underlying the Nyquist criterion are then related to an appropriate generalization of the root locus concept. It is shown that multivariable root loci are the 180° phase loci of the characteristic functions of a return-ratio matrix on an appropriate Riemann surface, plus some possibly degenerate loci each consisting of a single point.

1. Introduction

The Nyquist stability criterion (Nyquist 1932) is one of the most fundamental results in the theory of linear feedback systems and its generalization to the multivariable case is of great interest. Such a generalization was put forward (MacFarlane 1970 a) and used as part of a technique called the Characteristic Locus Method (MacFarlane and Belletrutti 1973) for feedback systems analysis and design, but no satisfactory proof was supplied. The proof of a generalized Nyquist stability theorem was undertaken by Barman and Katzenelson (1973, 1974) but their approach ignored certain key algebraic properties of the quantities involved ; it also leaned very heavily on the use of cuts in the complex plane ; this made their treatment technically complicated, and obscured the essential simplicity of the result. The purpose of this paper is to give a comprehensive discussion of the background to the generalized criterion and a rigorous proof based on a fundamental result in complex variable theory : the Principle of the Argument applied to an algebraic function (Bliss 1966) defined on an appropriate Riemann surface (Springer 1957).

The feedback configuration considered is shown in Fig. 1 (a). S_1 and S_2 are linear multivariable dynamical systems described by appropriate sets of state-space equations. Suppose all the feedback loops are broken at the point shown in Fig. 1 (b). Let $Z(s)$ be the loop-transmission transfer-function matrix relating signals injected at the point α to those returned at the point α'.

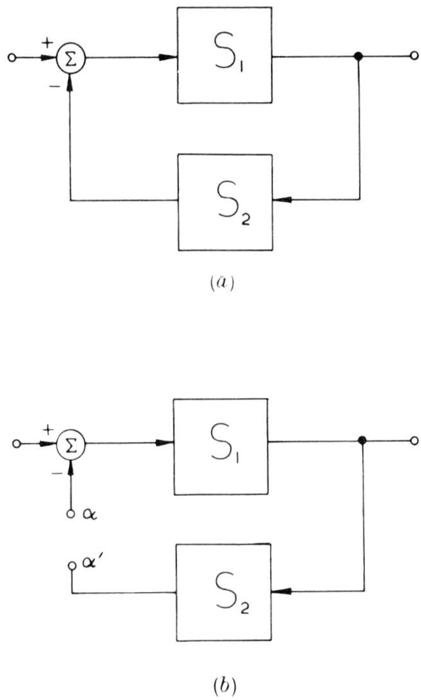

Figure 1. Multivariable feedback configuration. (*a*) Closed loop: (*b*) open loop.

Then the corresponding return-difference matrix (MacFarlane 1970 a) for this break point is

$$F(s)=I-Z(s)=I+L(s)$$

where $L(s)$ is defined as the system return-ratio matrix (MacFarlane 1970 a) and I is a unit matrix of appropriate order. A return-difference operator generates the difference between injected and returned signals from the injected signals. It plays a key role in feedback theory, since the essence of forging a feedback link is making two sets of signals identically equal and thus making the difference between them identically zero. Both $L(s)$ and $F(s)$ are matrix-valued functions of a complex variable s. The relevant basic properties of such entities are discussed in §§ 3, 4 and 5 from both the algebraic and the complex-variable point of view. The key concepts in the proof of the generalized criterion revolve around the properties of the eigenvalues of a matrix-valued function of a complex variable. It is shown in § 3 that a matrix-valued function of a complex variable is associated with a set of algebraic functions which may be called its characteristic functions. These characteristic functions accurately define the eigenvalues of the matrix-valued function as functions of a complex variable. It is also shown that the sets of characteristic functions have sets of poles and zeros. These sets of poles and zeros are related in § 4 to a set of poles and zeros for the matrix-valued function defined in algebraic terms.

There is a well-established theory of algebraic functions (Bliss 1966). This theory therefore provides the rigorous foundation on which the proof of the generalized stability criterion may be built. Each characteristic function is an algebraic function, which is a generalization of the elementary function of a complex variable. It is therefore associated with a Riemann surface (Springer 1957), on which domain of definition it becomes a single-valued function analytic everywhere except at a set of critical points. Many standard properties of analytic function theory hold in this extended domain. In particular the Principle of the Argument (Hille 1959) can be applied and this, together with the relationships between the poles and zeros of the set of characteristic functions of a return-difference matrix and the poles and zeros of the return-difference matrix itself, forms the basis of the proof.

Together with the Nyquist criterion (Nyquist 1932), and the related concept of Nyquist diagrams, the Root Locus method (Evans 1954) provides the basis for most single-loop linear feedback design techniques. It is therefore of interest and importance to relate these extensions of the Nyquist approach to root-locus concepts ; this is done in § 9, where it is shown that the multivariable root loci are the 180° phase loci of the characteristic functions of $L(s)$ on appropriate Riemann surfaces, plus some possibly degenerate loci each consisting of a single point.

2. Relationship between open-loop and closed-loop characteristic polynomials for basic feedback configuration

Consider the multivariable feedback arrangement shown in Fig. 1 (*a*). Let the state-space equation sets for the systems S_1 and S_2 be

$$\frac{dx_i}{dt} = A_i x_i + B_i u_i$$

$$y_i = C_i x_i + D_i u_i \quad i = 1, 2 \tag{2.1}$$

Let the transfer-function matrices for S_1 and S_2 be $G_1(s)$ and $G_2(s)$ respectively, and let their characteristic polynomials be $\Delta_1(s)$ and $\Delta_2(s)$ respectively. Then

$$G_i(s) = C_i(sI - A_i)^{-1}B_i + D_i \tag{2.2}$$

$$\Delta_i(s) = \det [sI - A_i] \qquad i = 1, 2 \tag{2.3}$$

$$G_i(\infty) = D_i \tag{2.4}$$

Further assume that

$$\det [I + G_2(\infty)G_1(\infty)] \neq 0 \tag{2.5}$$

Let the overall system, as shown in Fig. 2, have input vector u, state vector x and output vector y. Then breaking the feedback loops at y_2 puts the two

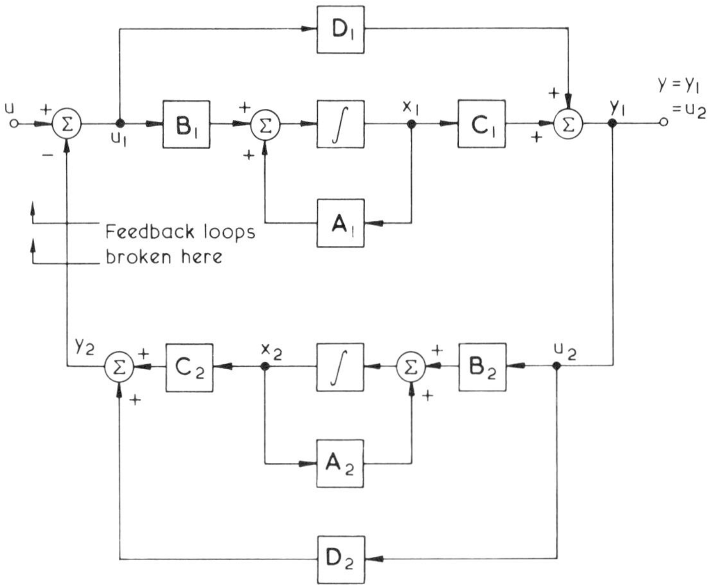

Figure 2. State-space representation of multivariable feedback system.

systems S_1 and S_2 in cascade, resulting in an open-loop system whose state-space description is given by :

$$\dot{x}_1 = A_1 x_1 + B_1 u_1 \tag{2.6}$$

$$\begin{aligned} \dot{x}_2 &= A_2 x_2 + B_2 u_2 \\ &= A_2 x_2 + B_2 C_1 x_1 + B_2 D_1 u_1 \end{aligned} \tag{2.7}$$

$$\begin{aligned} y_2 &= C_2 x_2 + D_2 u_2 \\ &= C_2 x_2 + D_2 C_1 x_1 + D_2 D_1 u_1 \end{aligned} \tag{2.8}$$

Combining these expressions we have that this open-loop system has a set of state-space equations

$$\dot{x} = Ax + Bu_1 \tag{2.9}$$

$$y_2 = Cx + Du_1 \tag{2.10}$$

where

$$x = \begin{bmatrix} x_1 \\ x_2 \end{bmatrix}$$

$$\left.\begin{aligned} A &= \begin{bmatrix} A_1 & 0 \\ B_2 C_1 & A_2 \end{bmatrix}, \quad B = \begin{bmatrix} B_1 \\ B_2 D_1 \end{bmatrix} \\ C &= [D_2 C_1 \quad C_2], \quad D = D_2 D_1 \end{aligned}\right\} \tag{2.11}$$

Now suppose the loops are all reconnected at y_2 so that

$$u_1 = u - y_2 \tag{2.12}$$

Substituting this into (2.9) and (2.10) we get

$$\dot{x} = Ax + B(u - y_2) \tag{2.13}$$

$$y_2 = Cx + D(u - y_2) \tag{2.14}$$

from which we obtain the set of closed-loop state-space equations

$$\begin{aligned} \dot{x} &= Ax + Bu - B(I+D)^{-1}Cx - B(I+D)^{-1}Du \\ &= [A - B(I+D)^{-1}C]x + [B - B(I+D)^{-1}D]u \end{aligned} \tag{2.15}$$

$$y_2 = (I+D)^{-1}Cx + (I+D)^{-1}Du \tag{2.16}$$

If we write the first of these closed-loop equations in the form

$$\dot{x} = A_c x + B_c u \tag{2.17}$$

then we have that

$$\begin{aligned} A_c &= A - B(I+D)^{-1}C \\ &= \begin{bmatrix} A_1 & 0 \\ B_2C_1 & A_2 \end{bmatrix} - \begin{bmatrix} B_1 \\ B_2D_1 \end{bmatrix} J\,[D_2C_1 \quad C_2] \\ &= \begin{bmatrix} A_1 - B_1JD_2C_1 & -B_1JC_2 \\ B_2C_1 - B_2D_1JD_2C_1 & A_2 - B_2D_1JC_2 \end{bmatrix} \end{aligned} \tag{2.18}$$

where

$$J = (I + D_2D_1)^{-1} \tag{2.19}$$

Let $F(s)$ be the return difference operator (MacFarlane 1970 a) for the set of opened loops. Then, applying Schur's formula (Gantmacher 1959) for the evaluation of partitioned determinants we have that

$$\begin{aligned} \det F(s) &= \det\,[C(sI-A)^{-1}B + D + I] \\ &= \det \begin{bmatrix} sI - A & B \\ -C & I + D \end{bmatrix} \div [\det\,(sI - A)] \\ &= \det \begin{bmatrix} sI - A_1 & 0 & B_1 \\ -B_2C_1 & sI - A_2 & B_2D_1 \\ -D_2C_1 & -C_2 & J^{-1} \end{bmatrix} \div \Delta_1(s)\Delta_2(s) \end{aligned} \tag{2.20}$$

Finally we may apply elementary column operations to the right-hand side of eqn. (2.20) in order to replace $-D_2C_1$ and $-C_2$ by zero blocks in the following way:

$$\det F(s) = \det F(s) \det \begin{bmatrix} I & 0 & 0 \\ 0 & I & 0 \\ JD_2C_1 & JC_2 & I \end{bmatrix}$$

$$= \det \begin{bmatrix} sI - A_1 + B_1JD_2C_1 & B_1JC_2 & B_1 \\ -B_2C_1 + B_2D_1JD_2C_1 & sI - A_2 + B_2D_1JC_2 & B_2D_1 \\ 0 & 0 & J^{-1} \end{bmatrix} \div \Delta_1(s)\Delta_2(s) \tag{2.21}$$

Then using eqns. (2.11), (2.18) and (2.19) we have

$$\det F(s) = \det \begin{bmatrix} sI - A_c & B \\ 0 & J^{-1} \end{bmatrix} \div \Delta_1(s)\Delta_2(s)$$

$$= \det [sI - A_c] \det [J^{-1}] \div \Delta_1(s)\Delta_2(s) \tag{2.22}$$

so that

$$\frac{\det F(s)}{\det F(\infty)} = \frac{\det [sI - A_c]}{\det [sI - A]} = \frac{CLCP(s)}{OLCP(s)} \tag{2.23}$$

where $CLCP(s)$ and $OLCP(s)$ are the system closed-loop and open-loop characteristic polynomials respectively, and

$$OLCP(s) = \det [sI - A] = \det [sI - A_1] \det [sI - A_2] \tag{2.24}$$

The relationship has been derived for a restricted class of non-proper systems, that is one in which D is a finite constant matrix which ensures that $\det F(\infty)$ will be non-zero. This type of system is considered throughout the paper.

3. Characteristic functions

Let $Q(s)$ be an $m \times m$ square matrix-valued function of a complex variable s whose elements are rational functions in s. For any specific value of complex frequency, $\bar{s}$ say, the corresponding matrix $Q(\bar{s})$ will be a matrix of complex numbers and will have a set of eigenvalues $\{q_i(\bar{s}) \mid i = 1, 2, \ldots, m\}$ which are a set of complex numbers. In some sense therefore (which we hope to shortly make precise) the eigenvalues of $Q(s)$ are functions of a complex variable. However, if we form the relevant characteristic equation

$$\det [q(s)I_m - Q(s)] \triangleq \Delta(q, s) = 0 \tag{3.1}$$

it is obvious that the polynomial in q, $\Delta(q, s)$, will not in general be expressible as a product of factors linear in q. Hence the matrix $Q(s)$ will not normally have eigenvalues which are rational functions of s. Or, to put it in another way, although the elements of $Q(s)$ are drawn from a field (the field of rational functions) the eigenvalues of $Q(s)$ do not all necessarily lie in that field. In general $\Delta(q, s)$ will be reducible to a form

$$\Delta(q, s)=\Delta_1(q, s)\Delta_2(q, s)\ \dots\ \Delta_k(q, s) \tag{3.2}$$

where the factors $\{\Delta_i(q, s) : i=1, 2, \dots, k\}$ are polynomials which are irreducible over the field of rational functions in s. Let the irreducible factors $\Delta_i(q, s)$ have the form

$$\Delta_i(q, s)=q_i^{t_i}(s)+a_{i1}q_i^{t_i-1}(s)+\dots+a_{it_i}(s) \quad i=1, 2, \dots, k \tag{3.3}$$

where t_i is the degree of the ith irreducible polynomial and the coefficients $\{a_{ij} : i=1, 2, \dots, k ; j=1, 2, \dots, t_i\}$ are rational functions in s.

If $b_{i0}(s)$ is the least common denominator of the coefficients $\{a_{ij}(s) : j=1, 2, \dots, t\}$, eqn. (3.3) can be put in the form

$$b_{i0}(s)q_i^{t_i}+b_{i1}(s)q_i^{t_i-1}+\dots+b_{it_i}(s)=0 \quad i=1, 2, \dots, k \tag{3.4}$$

where the coefficients $\{b_{ij}(s) : i=1, 2, \dots, k ; j=1, 2, \dots, t_i\}$ are now polynomials in s. The function of a complex variable $q_i(s)$ which is defined by eqn. (3.4) is called an algebraic function (Bliss 1966). Thus a square matrix-valued function of a complex variable $Q(s)$ is associated with a set of algebraic functions $\{q_i(s) : i=1, 2, \dots, k\}$ which play a role for $Q(s)$ related to its set of eigenvalues. The precise sense in which the set of eigenvalues of $Q(s)$ are functions of a complex variable is that they are the values of a set of algebraic functions ; and an algebraic function is a generalisation of the concept of a function of a complex variable. The *characteristic functions* of $Q(s)$ are defined to be the set of algebraic functions $\{q_i(s) : i=1, 2, \dots, k\}$.

The problems of finding the characteristic functions of the matrix $Q(s)$ is closely linked to the problem of finding an appropriate canonical form. If $\Delta(q, s)$ were reducible to factors linear in q then $Q(s)$ could be put into Jordan form (Cohn 1974). In general this will not be the case and a suitable canonical form is defined as follows.

Let

$$C(\Delta_i) \triangleq \begin{bmatrix} 0 & 0 \dots 0 & -a_{i,t_i}(s) \\ 1 & 0 \dots 0 & -a_{i,t_i-1}(s) \\ 0 & 1 \dots 0 & -a_{i,t_i-2}(s) \\ \vdots & & \\ 0 & \quad\ 1 & -a_{i1}(s) \end{bmatrix} \tag{3.5}$$

for $t_i > 1$ with

$$C(\Delta_i) \triangleq [-a_{i1}] \quad \text{if} \quad t_i=1 \tag{3.6}$$

then a transformation matrix $T(s)$ exists such that

$$Q(s)=T(s)R(s)T^{-1}(s) \tag{3.7}$$

where $R(s)$ is a unique matrix, which we will define to be the *irreducible rational canonical form* of $Q(s)$ and given by

$$R(s) \triangleq \text{diag}\,[C(\Delta_1), C(\Delta_2), \ldots, C(\Delta_k)] \tag{3.8}$$

It is clear that, given $R(s)$, the irreducible factors $\Delta_i(q, s)$ can easily be obtained. A proposed method for finding $R(s)$ for any given $Q(s)$ is given in Appendix 1.

3.1. *Riemann surface of a characteristic function*

Each characteristic function $q(s)$ is defined by an irreducible equation of the form

$$b_0(s)q^t + b_1(s)q^{t-1} + \ldots + b_t(s) = 0 \tag{3.9}$$

having in general t distinct roots. An exception occurs only if

(*a*) $b_0(s) = 0$, because the degree of the equation is then lowered, or if

(*b*) the equation has multiple roots.

This last situation can occur if, and only if, an expression, called the discriminant of the equation, vanishes. The discriminant (Barnett 1971) is an entire rational function of the equation coefficients ; it will be denoted by $D_q(s)$, and is discussed in Appendix 2†.

Ordinary points of the characteristic function

An ordinary point (Bliss 1966) of the characteristic function $q(s)$ is any point of the complex plane such that

$$b_0(s) \neq 0 \quad \text{and} \quad D_q(s) \neq 0$$

Critical points of the characteristic function

A critical point (Bliss 1966) of $q(s)$ is any point of the complex plane at which either

$$b_0(s) = 0 \quad \text{or} \quad D_q(s) = 0$$

or both.

Branch points of the characteristic function

Solutions of

$$D_q(s) = 0 \tag{3.10}$$

are called *branch points* of the characteristic function.

At every ordinary point the eqn. (3.9) defining the characteristic function has t distinct roots, since the discriminant does not vanish. The theory of algebraic functions (Bliss 1966) then shows that in a simply connected region of the complex plane punctured by the exclusion of the critical points the values of the characteristic function $q(s)$ form a set of analytic functions ; each of these analytic functions is called a *branch* of the characteristic function $q(s)$. Arguments based on standard techniques of analytic continuation, together with the properties of algebraic equations, show that the various branches can be organized into a single entity : the corresponding algebraic function.

† The point $s = \infty$ requires special discussion ; this is given in Appendix 7.

This is summarized in the following basic theorem of algebraic function theory: an irreducible algebraic equation of the form (3.9) defines precisely one t-valued regular function $q(s)$ in the punctured plane (Knopp 1947).

Functions defined in this way are called *algebraic functions*, and can be regarded as natural generalizations of the familiar elementary functions of a complex variable. An elementary function of a complex variable has the set of complex numbers $\mathbf{C}$ as both its domain and its range. An algebraic function has the complex number set $\mathbf{C}$ as its range but has a new and appropriately defined domain $\mathbf{R}$ which is called its *Riemann Surface* (Springer 1957). Since the Riemann surface of a characteristic function plays a crucial role in the proof of the generalized Nyquist stability criterion presented later, it is important to have an intuitive grasp of the ideas underlying its definition and formation, which is therefore now briefly considered.

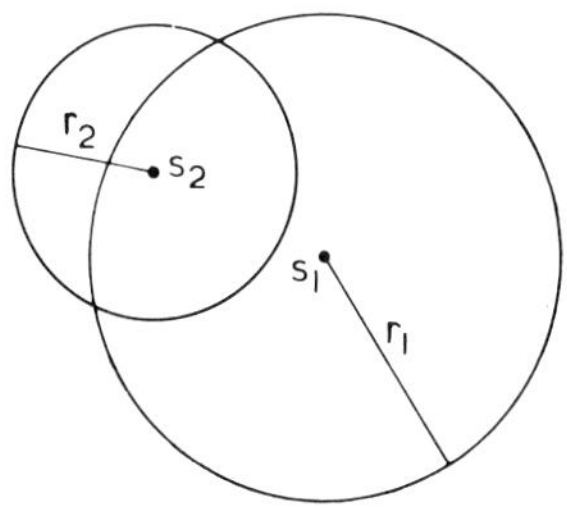

Figure 3. Analytic continuation.

Suppose we have a representation of part of one branch of an algebraic function in the form of a power series; such a representation is usually called a functional element. Imagine its circle of convergence to be cut out of paper and that the individual points of the paper disc are made bearers of the unique functional values of the elements. If now this initial element is analytically continued by means of a second power series, another circle of convergence can be thought of as being cut out and pasted partly over the first, as illustrated by Fig. 3. The parts pasted together are made bearers of the same functional values and are accordingly treated as a single region covered once with values. If a further analytic continuation is carried out, a further disc is similarly pasted onto the preceding one. Now suppose that, after repeated analytic continuations, one of the discs lies over another disc, not associated with an immediately preceding analytic continuation, as shown in Fig. 4. Such an overlapping disc is pasted together with the one it overlaps if and only if both are bearers of the same functional values. If however they bear different functional values they are allowed to overlap but remain disconnected. Thus two sheets, which are bearers of different functional values, become superimposed on this part of the complex plane.

Continuing this process for as long as possible, a surface-like configuration is obtained covering t ‘ sheets ’ of the complex plane, where t is the degree of the algebraic function. To form the Riemann surface these sheets can be joined together in the most varied of ways. This may involve connecting together two sheets which are separated by several other sheets lying between them. Although such a construction cannot be carried out in a three-dimensional space it is not difficult to give a perfectly satisfactory topological description of the process required. This surface-like configuration is called the *Riemann surface* of the multiple-valued algebraic function. On the Riemann surface the entire domain of values of the algebraic function is spread out in a completely single-valued manner so that, on every one of the t copies of the complex plane involved, every point is the bearer of one and only one value of the function.

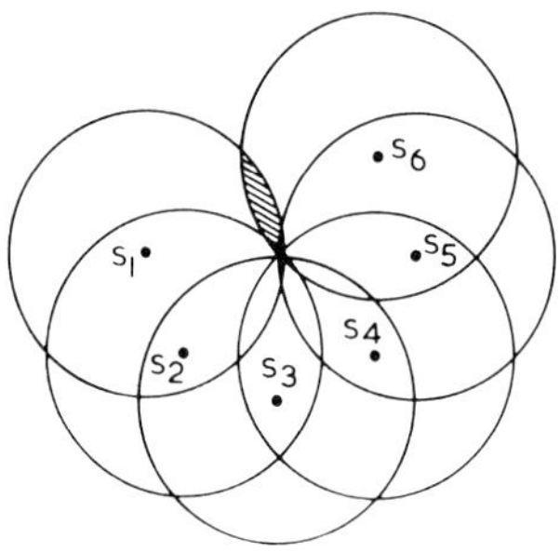

Figure 4. Repeated analytic continuation.

A method for building Riemann surfaces is given in Appendix 3. This involves the use of cuts in the complex plane and it may be helpful to say a word about them at this point. Let an algebraic function $q(s)$ have r critical points $\{a_1, a_2, \ldots, a_r\}$. Suppose them to be joined to one another and then to the point at infinity by a line L. Any line joining critical points will be called a *cut*. Let $\mathbf{L}$ denote the set of complex numbers defined by the line L. We then have that the solutions of eqn. (3.9) define a set of t ‘ distinct ’ analytical functions $\{q_1(s), q_2(s), \ldots, q_t(s)\}$ in the cut plane $\mathbf{C}-\mathbf{L}$. Each of these functions can be analytically continued, by standard procedures, across the cut L. Now it follows from the fundamental principles of analytic continuation that if an analytic function satisfies an algebraic equation in one part of its domain of definition, it must satisfy that equation in every region into which it is analytically continued. We must therefore have that

(i) there are only t ‘ distinct ’ analytic functions which satisfy the defining algebraic equation in the cut plane $\mathbf{C}-\mathbf{L}$:

(ii) each analytic continuation of any of these analytic functions $\{q_i(s) : i = 1, 2, \ldots, t\}$ gives rise to an analytic function which also satisfies the defining algebraic equation. It follows from this that the set of analytic functions associated with one side of the cut L must be a simple permutation of the set of analytic functions associated with the other side of the cut. Therefore by identifying and suitably matching up corresponding analytic functions (via their sets of computed values) on opposite sides of the cut L, one can produce an appropriate domain on which a single analytic function may be specified which defines a continuous single-valued mapping from this domain into the complex plane. This function is of course the algebraic function, conceived of as a single entity, and the domain so constructed is its Riemann surface.

It is sufficient for the purposes of understanding this paper for the reader to know that a Riemann surface can be constructed for any given characteristic function, on which its values form a single-valued function of position. Many standard relationships and properties of analytic function theory generalize, using the Riemann surface concept, to the algebraic function case and, in particular the Principle of the Argument holds on the Riemann surface (Behnke *et al.* 1974).

3.2. *Poles and zeros of characteristic functions*

Consider the defining equation for a characteristic function $q(s)$:

$$b_0(s)q^t + b_1(s)q^{t-1} + \ldots + b_t(s) = 0$$

We will take both

$$b_0(s) \not\equiv 0 \quad \text{and} \quad b_t(s) \not\equiv 0$$

since, if either or both of these polynomial coefficients were to vanish identically, we could find a reduced-order equation such that both the coefficients of the highest and zeroth powers of $q(s)$ were non-zero ; this reduced-order equation would then be taken as defining an appropriate new algebraic function for whose defining equation the supposition would be true.

It may happen however that $b_0(s)$ and $b_t(s)$ share a common factor and thus both vanish together at some specific set of values of s. Before looking at the effect of this, consider the situation when $b_0(s)$ and $b_t(s)$ do not share a common factor. The algebraic function will obviously be zero when

$$b_t(s) = 0 \tag{3.11}$$

and will tend to infinity as

$$b_0(s) \to 0 \tag{3.12}$$

For this reason those values of s which satisfy eqn. (3.11) are defined to be the *zeros* of the algebraic function $q(s)$, and those values of s which satisfy the equation

$$b_0(s) = 0 \tag{3.13}$$

are defined to be the *poles* of the algebraic function $q(s)$†.

---

† The point $s = \infty$ requires special discussion ; this is given in Appendix 7.

In order to be able to take eqns. (3.11) and (3.13) as defining the zeros and poles of $q(s)$ in the general case, we must show that they remain appropriate when $b_0(s)$ and $b_t(s)$ share a common factor. Let us first dispose of the trivial case when all the coefficients $\{b_i(s) : i=0, 2, \ldots, t\}$ share a common factor by saying that such a common factor would simply be divided out to get a new defining equation for an appropriate algebraic function. Suppose then that $b_0(s)$ and $b_t(s)$ have a common factor, but that some non-empty set of coefficients $\{b_k(s), \ldots, b_l(s)\}$ do not share this common factor. Then dividing through the left-hand side of eqn. (3.9) by $b_0(s)$ we get

$$q^t+\frac{b_1(s)}{b_0(s)}q^{t-1}+\ldots+\frac{b_k(s)}{b_0(s)}q^{t-k}+\ldots+\frac{b_l(s)}{b_0(s)}q^{t-l}+\ldots+\frac{b_t(s)}{b_0(s)}=0 \tag{3.14}$$

Then, as $s\to\bar{s}$, where $\bar{s}$ is a zero of the common factor of $b_0(s)$ and $b_t(s)$, the moduli of the coefficient set

$$\left\{\frac{b_k(s)}{b_0(s)}, \ldots, \frac{b_l(s)}{b_0(s)}\right\}$$

all become arbitrarily large, and it is obvious that $q(s)$ will have a pole at $s=\bar{s}$.

Again, suppose that $b_0(s)$ and $b_t(s)$ have a common factor but that some non-empty set of coefficients $\{b_j(s), \ldots, b_m(s)\}$ do not. Then as $s\to\bar{s}$, where $\bar{s}$ is a zero of the common factor, the algebraic eqn. (3.9) may be replaced by

$$b_j(\bar{s})q^{t-j}(\bar{s})+\ldots+b_m(\bar{s})q^{t-m}(\bar{s})=0 \tag{3.15}$$

where

$$b_j(\bar{s})\neq 0, \ldots, b_m(\bar{s})\neq 0$$

so that we must have

$$q(\bar{s})=0$$

showing that $\bar{s}$ is indeed a zero of the algebraic function $q(s)$.

We thus conclude that eqns. (3.11) and (3.13) may be taken as defining the zeros and poles of the algebraic function $q(s)$, and that use of these definitions enables us to cope with the existence of coincident poles and zeros.

3.3. *The eigenvector functions for a matrix-valued function of a complex variable*

We have seen how a matrix-valued function of a complex variable $Q(s)$ is associated with a set of algebraic functions $q_i(s)$ whose values are the eigenvalues of $Q(s)$. It is shown in Appendix 4 that associated with each algebraic function $q_i(s)$ is a vector-valued function of a complex variable $w_i(s)$, whose elements are algebraic functions having the same Riemann surface as $q_i(s)$, and whose vector values are the eigenvectors of $Q(s)$.

4. Algebraic definition of poles and zeros of a transfer-function matrix

Let $G(s)$ be an $m\times l$ rational matrix-valued function of the complex variable s. Then there exists a canonical form for $G(s)$, the Smith-McMillan form (Rosenbrock 1970), such that

$$G(s)=H(s)M(s)J(s) \tag{4.1}$$

where the $m \times m$ matrix $H(s)$ and the $l \times l$ matrix $J(s)$ are both unimodular (that is having a constant value for their determinants, independent of s). If r is the normal rank of $G(s)$ (that is $G(s)$ has rank r for almost all values of s) then $M(s)$ has the form

$$M(s) = \begin{bmatrix} M^*(s)_{rr} & 0_{r,\, l-r} \\ 0_{m-r,\, r} & 0_{m-r,\, m-r} \end{bmatrix} \tag{4.2}$$

with

$$M^*(s) = \operatorname{diag}\left[\frac{\epsilon_1(s)}{\psi_1(s)}, \frac{\epsilon_2(s)}{\psi_2(s)}, \ldots, \frac{\epsilon_r(s)}{\psi_r(s)}\right] \tag{4.3}$$

where

(i) each $\epsilon_i(s)$ divides all $\epsilon_{i+j}(s)$ and
(ii) each $\psi_i(s)$ divides all $\psi_{i-j}(s)$.

With an appropriate partitioning of H, M and J we therefore have

$$G = [H_1 \;\; H_2] \begin{bmatrix} M^* & 0 \\ 0 & 0 \end{bmatrix} \begin{bmatrix} J_1 \\ J_2 \end{bmatrix}$$
$$= H_1 M^* J_1 \tag{4.4}$$

where M^* is as defined in eqn. (4.3).

Thus $G(s)$ may be expressed in the form

$$G(s) = H_1(s)\left[\operatorname{diag}\frac{\epsilon_i(s)}{\psi_i(s)}\right] J_1(s)$$
$$= \sum_{i=1}^{r} h_i(s)\frac{\epsilon_i(s)}{\psi_i(s)} j_i{}^t(s) \tag{4.5}$$

where

(i) $\{h_i(s) : i = 1, 2, \ldots, r\}$ are the columns of the matrix $H_1(s)$;
(ii) $\{j_i{}^t(s) : i = 1, 2, \ldots, r\}$ are the rows of the matrix $J_1(s)$.

We know that

$$r \leqslant \min(l, m)$$

and that $H(s)$ and $J(s)$ are unimodular matrices of full rank m and l respectively for all s. Suppose $G(s)$ is the transfer-function matrix for a system with input transform vector $u(s)$ and output transform vector $y(s)$. Then any input vector $u(s)$ is turned into an output vector $y(s)$ by

$$y(s) = \sum_{i=1}^{r} h_i(s)\frac{\epsilon_i(s)}{\psi_i(s)}[j_i{}^t(s)u(s)] \tag{4.6}$$

For the single-input single-output case where

$$y(s) = \frac{k\epsilon(s)}{\psi(s)} u(s) \tag{4.7}$$

with k a constant, the transfer function

$$g(s) = \frac{k\epsilon(s)}{\psi(s)} \tag{4.8}$$

is defined as having zeros at those values of s where $\epsilon(s)$ vanishes and poles at those values of s where $\psi(s)$ vanishes. Thus for a non-zero $u(s)$ the modulus of $y(s)$ vanishes when s is a zero of $g(s)$, and becomes arbitrarily large when s is a pole of $g(s)$. A natural way therefore to characterize the zeros and poles of $G(s)$ is in terms of those values of s for which $\|y(s)\|$ becomes zero for non-zero $\|u(s)\|$, and arbitrarily large for finite $\|u(s)\|$ where $\|\cdot\|$ denotes the standard vector norm. This natural extension of scalar case ideas leads directly to definitions of zeros and poles of $G(s)$ in terms of the Smith–McMillan form quantities

$$\left\{\frac{\epsilon_i(s)}{\psi_i(s)}\right\}$$

because of the following pair of simple results.

Zero lemma. $\|y(s)\|$ vanishes for $\|u(s)\| \neq 0$ and s finite if and only if some $\epsilon_i(s)$ is zero.

Pole lemma. $\|y(s)\| \to \infty$ for $\|u(s)\| < \infty$ if and only if some $\psi_i(s) \to 0$.

These considerations lead naturally to the following definitions (Rosenbrock 1970).

Poles of $G(s)$. The poles of $G(s)$ are defined to be the set of all zeros of the set of polynomials $\{\psi_i(s) : i = 1, 2, \ldots, r\}$. In what follows we will usually denote the poles of $G(s)$ by $\{p_1, p_2, \ldots, p_n\}$ and put

$$p(s) = (s - p_1)(s - p_2) \ldots (s - p_n) \tag{4.9}$$

where $p(s)$ is conveniently referred to as the pole polynomial of $G(s)$ and is given by

$$p(s) = \prod_{i=1}^{r} \psi_i(s) \tag{4.10}$$

Zeros of $G(s)$. The zeros of $G(s)$ are defined to be the set of all zeros of the set of polynomials $\{\epsilon_i(s) : i = 1, 2, \ldots, r\}$. We will normally denote the zeros of $G(s)$ by $\{z_1, z_2, \ldots, z_q\}$ and put

$$z(s) = (s - z_1)(s - z_2) \ldots (s - z_q) \tag{4.11}$$

where $z(s)$ is conveniently referred to as the zero polynomial of $G(s)$ and is given by

$$z(s) = \prod_{i=1}^{r} \epsilon_i(s) \tag{4.12}$$

It is important to remember that $z(s)$ and $p(s)$ are not necessarily relatively prime ; for this reason it is wrong to simply define $z(s)$ and $p(s)$ for a square matrix $G(s)$ as the numerator and denominator polynomials of det $G(s)$.

4.1. *Rules for calculating pole polynomials and zero polynomials*

The route via the Smith–McMillan form is not always convenient for the determination of the poles and zeros of $G(s)$, particularly if the calculation is being done by hand. The following rules (Kontakos 1973) can be shown to give the same results as the Smith–McMillan definitions.

Pole polynomial rule. $p(s)$ is the least common denominator of all non-zero minors of all orders of $G(s)$.

Zero polynomial rule. $z(s)$ is the greatest common divisor of the numerators of all minors of $G(s)$ of order r (r being the normal rank of $G(s)$) *which minors have all been adjusted to have* $p(s)$ *as their common denominator.*

4.2. *Example of pole and zero calculation* (*Kontakos* 1973)

In what follows, the expression

$$G^{i_1, i_2, \ldots, i_r}_{j_1, j_2, \ldots, j_r}$$

denotes the minor of $G(s)$ of order r formed by taking the determinant of the matrix obtained from $G(s)$ by deleting all rows except rows $i_1, i_2, \ldots, i_r$ and all columns except $j_1, j_2, \ldots, j_r$.

Let

$$G(s) = \frac{1}{(s+1)(s+2)(s-1)} \begin{bmatrix} (s-1)(s+2) & 0 & (s-1)^2 \\ -(s+1)(s+2) & (s-1)(s+1) & (s-1)(s+1) \end{bmatrix}$$

The minors of order 2 are

$$G^{1,\,2}_{1,\,2} = \frac{1}{(s+1)(s+2)} \qquad G^{1,\,2}_{1,\,3} = \frac{2}{(s+1)(s+2)} \qquad G^{1,\,2}_{2,\,3} = \frac{-(s-1)}{(s+1)(s+2)^2}$$

so that, considering the minors of all orders we get

$$p(s) = (s+1)(s+2)^2(s-1)$$

On adjusting the denominators of all minors of order 2 to be $p(s)$ we have

$$\bar{G}^{1,\,2}_{1,\,2} = \frac{(s-1)(s+2)}{p(s)} \qquad \bar{G}^{1,\,2}_{1,\,3} = \frac{2(s-1)(s+2)}{p(s)} \qquad \bar{G}^{1,\,2}_{2,\,3} = \frac{-(s-1)^2}{p(s)}$$

and so

$$z(s) = s - 1$$

The Smith–McMillan form of $G(s)$ can be calculated to be

$$M(s) = \begin{bmatrix} \dfrac{1}{(s+1)(s+2)(s-1)} & 0 & 0 \\ 0 & \dfrac{(s-1)}{(s+2)} & 0 \end{bmatrix}$$

which verifies the result.

5. Relationship between algebraically defined poles and zeros and the poles and zeros of the set of characteristic functions†

As a key step in the establishment of a generalized Nyquist stability criterion, it is crucially important to relate the poles and zeros defined by algebraic means to complex variable theory, and thus to the poles and zeros of the set of characteristic functions.

The coefficients $a_i(s)$ in the expansion

$$\det [qI_m - Q(s)] = q^m + a_1(s)q^{m-1} + a_2(s)q^{m-2} + \ldots + a_m(s) \tag{5.1}$$

are all appropriate sums of minors of $Q(s)$ since it can be shown that

$$\det [qI_m - Q(s)] = q^m - [\text{trace } Q(s)]q^{m-1} + [\textstyle\sum \text{principal minors of } Q(s) \text{ of order } 2]q^{m-2} - \ldots + (-1)^m \det Q(s) \tag{5.2}$$

and thus the term $b_0(s)$ in eqn. (3.9) is the least common denominator of all non-zero *principal* minors of all orders of $Q(s)$.

Now the pole polynomial $p(s)$ of a square matrix $Q(s)$ is the least common denominator of all non-zero minors of all orders of $Q(s)$. Therefore, if $e(s)$ is the least common denominator of all non-zero *non-principal* minors, with all factors common to $b_0(s)$ removed, we have that

$$p(s) = e(s)b_0(s) \tag{5.3}$$

Furthermore, since

$$a_m(s) = \det Q(s) \tag{5.4}$$

and

$$a_m(s) = \frac{b_m(s)}{b_0(s)} = \frac{b_m(s)e(s)}{p(s)} \tag{5.5}$$

and since the zero polynomial of a square matrix $Q(s)$ is the numerator of the rational function obtained from det $Q(s)$ after adjusting det $Q(s)$ in such a way as to obtain an equivalent rational function with $p(s)$ as its denominator, we must have that

$$z(s) = e(s)b_m(s) \tag{5.6}$$

In many cases the least common denominator of the non-zero non-principal minors of $Q(s)$ will divide $b_0(s)$, in which case $e(s)$ will be unity and the pole and zero polynomials for $Q(s)$ will be $b_0(s)$ and $b_m(s)$ respectively. In general a square-matrix-valued function of a complex variable $Q(s)$ will have a set of k irreducible characteristic functions in the form specified by eqn. (3.3) and the general form for the pole and zero polynomials can be written as

$$p(s) = e(s) \prod_{i=1}^{k} b_{i0}(s) \tag{5.7}$$

and

$$z(s) = e(s) \prod_{i=1}^{k} b_{i,\, t_i}(s) \tag{5.8}$$

where the pole and zero polynomials for the jth characteristic function $q_j(s)$ are $b_{j0}(s)$ and $b_{j,\, t_j}(s)$ respectively.

---

† Some minor errors appear in this section ; please refer to the corrections given on p. 56.

5.1. *Example demonstrating the pole-zero relationships*

Let

$$Q(s)=\frac{1}{(s+1)(s+2)(s-1)}\begin{bmatrix}(s-1)(s+2) & 0\\ -(s+1)(s+2) & (s-1)(s+1)\end{bmatrix}$$

$$=\begin{bmatrix}\dfrac{1}{s+1} & 0\\ \dfrac{-1}{s-1} & \dfrac{1}{s+2}\end{bmatrix}$$

The pole polynomial for $Q(s)$ is obviously

$$p(s)=(s+1)(s+2)(s-1)$$

and consequently the zero polynomial is

$$z(s)=(s-1)$$

The characteristic equation for $Q(s)$ is

$$\det\,[qI-Q(s)]=\left(q-\frac{1}{s+1}\right)\left(q-\frac{1}{s+2}\right)=0$$

so that the irreducible characteristic equations are

$$\Delta_1(s,\,q)=q-\frac{1}{s+1}=0$$

and

$$\Delta_2(s,\,q)=q-\frac{1}{s+2}=0$$

which may be written as

$$(s+1)q-1=0$$

and

$$(s+2)q-1=0$$

Therefore the pole and zero polynomials for the characteristic functions $q_1(s)$ and $q_2(s)$ are

$$p_{q_1}(s)=b_{10}(s)=(s+1)\quad z_{q_1}(s)=b_{11}(s)=-1$$
$$p_{q_2}(s)=b_{20}(s)=(s+2)\quad z_{q_2}(s)=b_{21}(s)=-1$$

Now for $Q(s)$ the least common denominator of all non-zero non-principal minors with all factors common to $b_0(s)$ $(=b_{10}(s)b_{20}(s))$ removed is given by

$$e(s)=(s-1)$$

which verifies the relationships

$$p(s)=e(s)\prod_{i=1}^{2}b_{i0}(s)$$

and

$$z(s)=e(s)\prod_{i=1}^{2}b_{i1}(s)$$

6. Extended principle of the argument

For each characteristic function $q(s)$ there is an appropriate Riemann surface $\mathbf{R}_q$ on which it becomes single-valued, and mappings from the Riemann surface to a q-plane of values are one-to-one and continuous. Suppose a characteristic function $q(s)$ has a defining algebraic equation of order t_j; then $\mathbf{R}_q$ is formed by piecing together t_j copies of the complex plane $\mathbf{C}$. Suppose now that a sub-set of $\mathbf{C}$ defining a closed Jordan contour $\mathbf{J}$ is drawn on every one of these copies of $\mathbf{C}$ before they are pieced together to form $\mathbf{R}_q$. Then, when the Riemann surface is formed, the set of copies of $\mathbf{J}$ will be joined into a set of closed contours on $\mathbf{R}_q$. The exact number of closed contours formed on $\mathbf{R}_q$ will lie between 1 and t_j and will depend on the location of the branch points of the characteristic function with respect to $\mathbf{J}$. Associated with each closed contour on $\mathbf{R}_q$ will be a set of values of the algebraic function $q(s)$ which will define a corresponding set of closed curves in the q-plane ($\mathbf{C}$).

Suppose that $q(s)$ has $^{\#}p$ poles and $^{\#}z$ zeros located within the set of closed contours on $\mathbf{R}_q$. Suppose that when $\mathbf{R}_q$ is formed the t_j copies of $\mathbf{J}$ join into a set of closed contours $\mathbf{A}_1, \mathbf{A}_2, \ldots, \mathbf{A}_l$ on $\mathbf{R}_q$ and that $\mathbf{A}_i$ encloses $^{\#}p_i$ poles and $^{\#}z_i$ zeros of $q(s)$. Let $\mathbf{\Gamma}_1, \mathbf{\Gamma}_2, \ldots, \mathbf{\Gamma}_l$ be the closed image curves in the q-plane corresponding to the clockwise traversal of the closed curves $\mathbf{A}_1, \mathbf{A}_2, \ldots, \mathbf{A}_l$ on $\mathbf{R}_q$. For each closed curve on $\mathbf{R}_q$ the Principle of the Argument will apply and thus :

$$\sum \text{clockwise encirclements of origin by } \mathbf{\Gamma}_i = {}^{\#}z_i - {}^{\#}p_i$$

Thus

$$\text{Total} \sum \text{clockwise encirclements of origin by } \mathbf{\Gamma}_1, \ldots, \mathbf{\Gamma}_l$$

$$= \sum_{i=1}^{l} {}^{\#}z_i - \sum_{i=1}^{l} {}^{\#}p_i = {}^{\#}z - {}^{\#}p$$

7. Characteristic loci

Let $Q(s)$ be an $m \times m$ matrix-valued function of a complex variable. Associated with $Q(s)$ is a set of characteristic functions $\{q_1(s), \ldots, q_k(s)\}$ each having associated with it a corresponding Riemann surface $\mathbf{R}_1, \ldots, \mathbf{R}_k$. Let $\mathbf{D}$ denote the standard Nyquist contour in $\mathbf{C}$, as shown in Fig. 5, and suppose that there is a copy of $\mathbf{D}$ on each of the replicas of $\mathbf{C}$ out of which $\mathbf{R}_1, \ldots, \mathbf{R}_k$ are constructed. Then, associated with the standard Nyquist contour in $\mathbf{C}$ are sets of closed contours on each of the surfaces. Let $\mathbf{N}_1, \ldots, \mathbf{N}_k$ be the

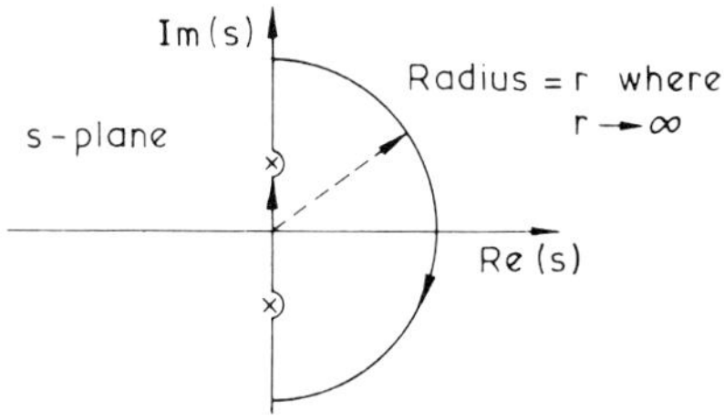

Figure 5. Standard Nyquist contour.

sets of closed image curves in the q-plane corresponding to the sets of closed contours on $\mathbf{R}_1, \ldots, \mathbf{R}_k$; then these subsets $\mathbf{N}_1, \ldots, \mathbf{N}_k$ of $\mathbf{C}$ are defined to be the *characteristic loci* of $Q(s)$ (MacFarlane and Belletrutti 1973).

In practice the characteristic loci are generated as the loci in the complex plane traced out by the eigenvalues of the matrix $Q(s)$ as s traverses the Nyquist contour in the standard (clockwise) direction. Suppose that we consider a portion of the imaginary axis. We can then compute a set of loci corresponding to the eigenvalues $q_1(j\omega), \ldots, q_m(j\omega)$ in the following way :

(i) Select a value of angular frequency, say ω_a.

(ii) Compute the complex matrix $Q(j\omega_a)$.

(iii) Use a standard computer algorithm to compute the eigenvalues of $Q(j\omega_a)$, which are a set of complex numbers denoted by $\{q_i(j\omega_a)\}$.

(iv) Plot the numbers $\{q_i(j\omega_a)\}$ in the complex plane.

(v) Repeat with further values of angular frequency $\omega_b, \omega_c, \ldots$, etc. and join the resulting plots up into continuous loci using a sorting routine based on the continuity of the various branches of the characteristic functions involved.

An example of the characteristic loci for a second order return-ratio matrix is given in Fig. 8 and discussed in § 10.

8. The generalized Nyquist stability criterion

We are now in a position to state and prove the generalized Nyquist stability criterion. The feedback system considered is shown in Fig. 1, and S_1 and S_2 are a pair of linear multivariable dynamical systems whose behaviour is governed by the state-space equation set (2.1) of § 2.

S_1 has m inputs and S_2 has m outputs and the set of feedback loops is broken at the point α. Let $L(s)$ be the system return-ratio matrix (MacFarlane 1970 a) for signals injected at point α and returned at point α'. Let $L(s)$ have a set of characteristic functions $\{l_1(s), l_2(s), \ldots, l_k(s)\}$ with characteristic loci $\mathbf{N}_1, \mathbf{N}_2, \ldots, \mathbf{N}_k$ in the complex plane as s traverses the standard Nyquist contour of Fig. 5 in a clockwise direction. Then we may state the generalized Nyquist criterion in the following form.

Generalized Nyquist stability criterion

Let the multivariable feedback system shown in Fig. 1 have no open-loop uncontrollable and/or unobservable modes whose corresponding characteristic frequencies lie in the right-half plane. Then this configuration will be closed-loop stable if and only if the net sum of anti-clockwise encirclements of the critical point $(-1+j0)$ by the set of characteristic loci of $L(s)$ is equal to the total number of right-half plane poles of $G_1(s)$ and $G_2(s)$.

Since uncontrollable and/or unobservable modes are not influenced by feedback, their possible presence adds a further necessary condition for stability : that all their corresponding characteristic frequencies must be in the left-half plane. In such a case the encirclement criterion above remains necessary but is no longer sufficient for closed-loop stability.

Since we are dealing with a return-ratio matrix, we must be careful, as pointed out by Desoer and Chan (1975), to consider the possibility of pole-zero cancellations between the cascade systems which make up the loop transmittance. The generalized Nyquist stability criterion statement given above, and the proof which follows below, both cope with this situation.

8.1. *Proof of generalized Nyquist criterion*

For the break point shown in Fig. 1 the corresponding return-difference matrix (MacFarlane 1970 a) is

$$F(s)=I_m+L(s) \tag{8.1}$$

Let

$OLCP(s)$ = open-loop characteristic polynomial of complete system with all loops broken at point α, as shown in Fig. 1 (*b*)

$CLCP(s)$ = closed-loop characteristic polynomial of complete system with all feedback loops closed, as shown in Fig. 1 (*a*)

Then, as shown in § 2 we have that

$$\frac{\det F(s)}{\det F(\infty)}=\frac{CLCP(s)}{OLCP(s)} \tag{8.2}$$

The eigenvalues of the return-difference matrix $F(s)$ are given by

$$\det [f(s)I_m-F(s)]=0 \tag{8.3}$$

which in general, as explained in § 3, can be expressed as a product of irreducible characteristic equations of the form

$$b_{i0}(s)f_i^{\,t_i}+b_{i1}(s)f_i^{\,t_i-1}+\ldots+b_{it_i}(s)=0 \quad i=1, 2, \ldots, k \tag{8.4}$$

where the coefficients are polynomials in s. Therefore from § 5 we have that the pole and zero polynomials for $F(s)$ are

$$p(s)=e(s)\prod_{i=1}^{k} b_{i0}(s) \tag{8.5}$$

and

$$z(s)=e(s)\prod_{i=1}^{k} b_{it_i}(s) \tag{8.6}$$

respectively, where $e(s)$ is the least common denominator of all non-zero non-principal minors of $F(s)$, with all factors common to $\prod_{i=1}^{k} b_{i0}(s)$ removed†.

By definition, the open-loop characteristic polynomial of the system of Fig. 1 is given by

$$OLCP(s)=\det [sI-A]=\det [sI-A_1]\det [sI-A_2] \tag{8.7}$$

where A, A_1 and A_2 are as defined in § 2. It is easily shown (MacFarlane and Karcanias 1976) that

$$\det [sI-A_i]=p_{G_i}(s)p_{d_i}(s) \quad i=1, 2 \tag{8.8}$$

† See correction given on p. 56.

where $p_{G_i}(s)$ is the pole polynomial for the transfer function $G_i(s)$ and the polynomial $p_{d_i}(s)$ has as its zeros the decoupling zeros of the system associated with that set of characteristic frequencies (eigenvalues) of A_i which correspond to modes of the system which are uncontrollable and/or unobservable.

Let $p_l(s)$ be the pole polynomial for the return-ratio matrix $L(s)$, then

$$p_{G_1}(s)p_{G_2}(s)=p_l(s)p_x(s) \tag{8.9}$$

where $p_x(s)$ has as its zeros those poles of $G_1(s)$ and $G_2(s)$ which are lost when $L(s)$ is formed (Desoer and Chan 1975). Thus we have that

$$OLCP(s)=p_l(s)p_x(s)p_{d_1}(s)p_{d_2}(s)=p_l(s)p_x(s)p_d(s) \tag{8.10}$$

where

$$p_d(s)=p_{d_1}(s)p_{d_2}(s) \tag{8.11}$$

It is obvious, from an inspection of eqn. (8.1), that the return-ratio matrix $L(s)$ and the corresponding return-difference matrix $F(s)$ have the same poles, and therefore

$$OLCP(s)=p(s)p_x(s)p_d(s) \tag{8.12}$$

giving

$$OLCP(s)=p_x(s)p_d(s)e(s)\prod_{i=1}^{k} b_{i0}(s) \tag{8.13}$$

It then follows, using the fact that

$$\det F(s)=\frac{z(s)}{p(s)} \tag{8.14}$$

and using eqns. (8.2) and (8.6) that the closed-loop characteristic polynomial is given by

$$CLCP(s)=\frac{p_x(s)p_d(s)e(s)\prod_{i=1}^{k} b_{it_i}(s)}{\det F(\infty)} \tag{8.15}$$

An example of a situation in which the polynomial $p_x(s)\neq 1$ is given in Appendix 6.

For the set of irreducible characteristic equations associated with the return-difference operator $F(s)$, there is a corresponding set of Riemann surfaces on which the appropriate characteristic algebraic functions become single-valued, and mappings from these surfaces on to a corresponding f-plane are one-to-one and continuous. Consider the jth equation of the set defined in (8.4). The degree of the equation is t_j and therefore the corresponding Riemann surface $\mathbf{R}_j$ is formed by piecing together t_j copies of the complex s-plane $\mathbf{C}$. Suppose now that a standard Nyquist contour $\mathbf{D}$, as shown in Fig. 5, is drawn on every copy of $\mathbf{C}$ before they are pieced together to form $\mathbf{R}_j$. Then, when $\mathbf{R}_j$ is formed, the set of Nyquist contours will be joined into a set of closed contours on $\mathbf{R}_j$. Since the Principle of the Argument will apply to each closed contour on $\mathbf{R}_j$, the number of clockwise encirclements of the origin in $\mathbf{C}$, by the appropriate image curve, will be the difference between the number of zeros and poles of the characteristic function $f_j(s)$ which are enclosed by the domain curve on $\mathbf{R}_j$. Therefore, considering all the closed contours formed on $\mathbf{R}_j$ from the original Nyquist contour set, and applying the Argument Principle to each gives that

$$n_{f_j} = z_{f_j} - p_{f_j} \tag{8.16}$$

where :

(i) n_{f_j} is the net sum of clockwise encirclements of the origin in $\mathbf{C}$ by the set of image curves under f_j of the set of curves formed on $\mathbf{R}_j$ by piecing together the appropriate set of Nyquist contours when forming $\mathbf{R}_j$;

(ii) z_{f_j} is the number of right-half plane zeros of $f_j(s)$; and

(iii) p_{f_j} is the number of right-half plane poles of $f_j(s)$.

From eqn. (8.15) we have that necessary and sufficient conditions for the closed-loop stability of the feedback configuration shown in Fig. 1 are that

(*a*) $e(s)=0$ has no roots in the right-half plane ;

(*b*) $\prod_{i=1}^{k} b_{it_i}(s)=0$ has no roots in the right-half plane ;

(*c*) $p_x(s)=0$ has no roots in the right-half plane ; and

(*d*) $p_d(s)=0$ has no roots in the right-half plane.

Condition (*b*) is equivalent to saying that there are no zeros of

$$\{f_i(s) : i=1, 2, \ldots, k\}$$

in the right-half plane, and can therefore be replaced by

$$z_{f_i}=0 \qquad i=1, 2, \ldots, k \tag{8.17}$$

or

$$n_{f_i}=-p_{f_i} \quad i=1, 2, \ldots, k \tag{8.18}$$

which imply, and are implied by,

$$\sum_{i=1}^{k} n_{f_i} = -\sum_{i=1}^{k} p_{f_i} \tag{8.19}$$

We therefore conclude that an equivalent set of necessary and sufficient conditions for closed-loop stability are that

(*a*) $e(s)=0$ has no roots in the right-half plane ;

(*b*) $\sum_{i=1}^{k} n_{f_i} = -\sum_{i=1}^{k} p_{f_i}$;

(*c*) $p_x(s)=0$ has no roots in the right-half plane ; and

(*d*) $p_d(s)=0$ has no roots in the right-half plane.

Now, as shown by eqn. (8.5), $e(s)$ together with the set of pole polynomials for the set of characteristic functions $\{f_i(s) : i=1, 2, \ldots, k\}$, make up the pole polynomial for $F(s)$.

This leads us to consider combining the three conditions (*a*), (*b*) and (*c*) into the single equivalent condition

$$\sum_{i=1}^{k} n_{f_i} = -p \tag{8.20}$$

where p is the total number of right-half plane poles of $G_1(s)$ and $G_2(s)$.

Now the uncontrollable and/or unobservable parts of the system will not be affected by the feedback. We therefore confine our attention to the case where $p_d(s)=0$ has no roots in the right-half plane.

Necessity and sufficiency of condition (8.20) *for closed-loop stability when* $p_d(s)=0$ *has no roots in the right-half plane*

We have that

$$\sum_{i=1}^{k} n_{f_i} = \sum_{i=1}^{k} z_{f_i} - \sum_{i=1}^{k} p_{f_i} \tag{8.21}$$

and

$$p = \sum_{i=1}^{k} p_{f_i} + e + p_x \tag{8.22}$$

where e is the number of right-half plane roots of

$$e(s) = 0 \tag{8.23}$$

and p_x is the number of right-half plane roots of

$$p_x(s) = 0 \tag{8.24}$$

Combining the two relationships (8.21) and (8.22) we thus have

$$\sum_{i=1}^{k} n_{f_i} = \sum_{i=1}^{k} z_{f_i} + e + p_x - p \tag{8.25}$$

where $\sum_{i=1}^{k} z_{f_i}$, e, p_x and p are all positive integers or zero.

To establish the necessity of (8.20) suppose that

$$\sum_{i=1}^{k} n_{f_i} \neq -p$$

Then, from (8.25), this implies that

$$\sum_{i=1}^{k} z_{f_i} \neq 0, \quad \text{or} \quad e \neq 0, \quad \text{or} \quad p_x \neq 0$$

or any combination of these, and thus that the system will be closed-loop unstable. Hence we conclude that (8.20) is necessary for closed-loop stability.

For sufficiency suppose that

$$\sum_{i=1}^{k} n_{f_i} = -p$$

Then, from (8.25) we must have that

$$p_x = e = \sum_{i=1}^{k} z_{f_i} = 0$$

and hence that the system is closed-loop stable. Thus the sufficiency of condition (8.20) is established.

We have therefore shown that, when $p_d(s)=0$ has no roots in the right-half plane,

$$\sum_{i=1}^{k} n_{f_i} = -p$$

is a necessary and sufficiency condition for closed-loop stability. Now, using the eigenvalue shift theorem (MacFarlane 1970 b) we have that the characteristic functions $\{f_i(s) : i=1, 2, \ldots, k\}$ of $F(s)$ and $\{l_i(s) : i=1, 2, \ldots, k\}$ of $L(s)$ are related by

$$f_i(s) = 1 + l_i(s), \quad i = 1, 2, \ldots, k \tag{8.26}$$

The image curve sets in **C** of the Nyquist contour sets mapped under $f_i(s)$ and $l_i(s)$ are therefore simply related by a unit horizontal shift in **C**. Hence if n_{l_j} is defined as the net sum of clockwise encirclements of the critical point $(-1+j0)$ in **C** by the set of characteristic loci of $L(s)$ corresponding to the jth characteristic function, then the necessary and sufficient conditions for closed-loop stability in the case where $p_d(s)=0$ has no roots in the right-half plane simply become

$$\sum_{i=1}^{k} n_{l_i} = -p \tag{8.27}$$

Adding the obvious necessary qualifications regarding the role of uncontrollable and/or unobservable unstable open-loop modes then completes the proof of the generalized Nyquist stability criterion.

In the previous treatments of the generalized Nyquist stability criterion (MacFarlane and Belletrutti 1973) the necessary conditions on the uncontrollable and unobservable unstable modes are incorporated in the encirclement theorem : the closed-loop system is stable if and only if the net sum of anti-clockwise encirclements of the critical point $(-1+j0)$ by the set of characteristic loci of $L(s)$ is equal to the number of right-half plane zeros of $OLCP(s)$. The equivalence of this theorem with the stability criterion presented in this paper is established in Appendix 5 where a rigorous proof is given for the early statement of the generalized Nyquist stability criterion.

9. Multivariable root loci and their associated Riemann surfaces

Consider the feedback system shown in Fig. 6, where the linear dynamical system S_1 has m inputs and S_2 has m outputs ; k is a scalar gain multiplier. Let $L(s)$ be the system return-ratio matrix for signals injected at point α and returned at point α' with $k=1$.

The return-difference for this configuration is given by

$$F(s, k) = I_m + kL(s) \tag{9.1}$$

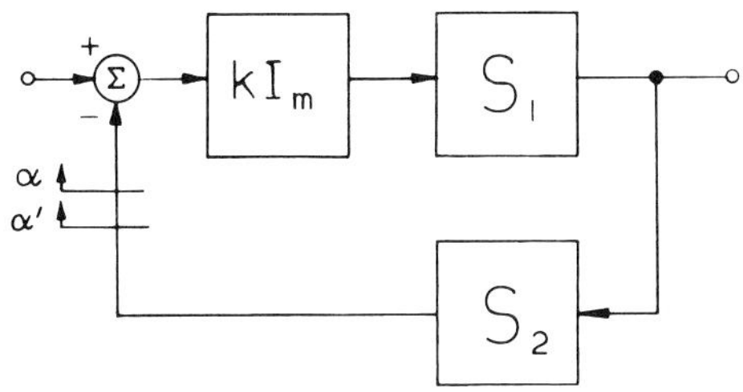

Figure 6. Multivariable feedback configuration with scalar gain k.

and the set of characteristic functions associated with this return-difference matrix are defined by

$$\det\,[f(s,k)I_m - F(s,k)] = 0 \tag{9.2}$$

Assume, for simplicity of exposition, that there is just one irreducible characteristic equation of the form

$$b_0(s)f^m(s,k) + b_1(s,k)f^{m-1}(s,k) + \dots + b_m(s,k) = 0 \tag{9.3}$$

Denoting the system's open-loop and closed-loop characteristic polynomials by $OLCP(s)$ and $CLCP(s,k)$ respectively we will have that

$$\frac{\det F(s,k)}{\det F(\infty,k)} = \frac{CLCP(s,k)}{OLCP(s)} = \frac{b_m(s,k)}{b_0(s)} \tag{9.4}$$

and also that

$$OLCP(s) = p_x(s)p_d(s)e(s)b_0(s) \tag{9.5}$$

where $e(s)$ is here defined as the least common denominator of all non-zero non-principal minors of $L(s)$ with all factors common to $b_0(s)$ removed, $p_d(s)$ has as its zeros the decoupling zeros of the system, and $p_x(s)$ has as its zeros those poles of $G_1(s)$ and $G_2(s)$ which are lost when $L(s)$ is formed. In § 8, $e(s)$ was defined in terms of $F(s)$. It is shown in Appendix 6 that it is immaterial whether $e(s)$ is defined for $F(s)$ or $L(s)$ since the same polynomial results in either case.

It therefore follows that

$$CLCP(s,k) = \frac{p_x(s)p_d(s)e(s)b_m(s,k)}{\det F(\infty,k)} \tag{9.6}$$

so that the solution of the equation

$$p_x(s)p_d(s)e(s)b_m(s,k) = 0 \tag{9.7}$$

for s in terms of k will determine the dependence of the closed-loop poles on the scalar feedback gain parameter k. The graphical exhibition of this dependence in the complex plane is what constitutes the multivariable root loci of this system.

Note that for any value of k the solutions of

$$p_x(s)=0 \tag{9.8}$$

$$p_d(s)=0 \tag{9.9}$$

and

$$e(s)=0 \tag{9.10}$$

are solutions of eqn. (9.7). Thus for $e(s)\neq 1$ (an unusual condition) and/or $p_d(s)\neq 1$ (indicating unobservable and/or uncontrollable modes) and/or $p_x(s)\neq 1$ (indicating pole cancellations) some of the root loci degenerate to single points, these being the solutions of eqns. (9.8), (9.9) and (9.10). The other solutions of eqn. (9.7), which satisfy

$$b_m(s, k)=0 \tag{9.11}$$

are equivalent to finding the zeros of the characteristic function $f(s, k)$. If $l(s, k)$ is the characteristic function of $kL(s)$ (assumed irreducible), then the eigenvalue shift theorem gives us that

$$f(s, k)=1+l(s, k) \tag{9.12}$$

and thus finding the zeros of $f(s, k)$ is equivalent to finding those values of s for which

$$l(s, k)=-1 \tag{9.13}$$

The multivariable root loci for the feedback configuration of Fig. 6 are therefore defined by the solutions of

$$p_x(s)=0 \tag{9.14}$$

$$p_d(s)=0 \tag{9.15}$$

$$e(s)=0 \tag{9.16}$$

and

$$l(s, k)=-1 \tag{9.17}$$

for all values of k.

Now the characteristic equation for $L(s)$, the system return-ratio matrix with $k=1$, is

$$\det\,[l(s)I_m-L(s)]=0 \tag{9.18}$$

from which we can obtain the equation

$$c_0(s)l^m(s)+c_1(s)l^{m-1}(s)+\ldots+c_m(s)=0 \tag{9.19}$$

where the coefficients $\{c_i(s):\ i=1, 2, \ldots, m\}$ are polynomials in s. The characteristic equation for $kL(s)$ is given by

$$c_0(s)l^m(s)+c_1(s)kl^{m-1}(s)+\ldots+c_m(s)k^m=0 \tag{9.20}$$

Thus the introduction of the scalar gain multiplier k is equivalent to multiplying the characteristic function $l(s)$ by k. Thus the solutions of eqn. (9.17) are equivalent to those of

$$l(s)=-\frac{1}{k} \tag{9.21}$$

for all values of k, which is obviously a direct generalization of the defining equations for the single-loop root locus.

It is clear from this discussion that the complex loci defined by eqn. (9.21) are simply the 180° phase loci of the characteristic function $l(s)$ on its associated Riemann surface. The fact that multivariable root loci ‘live’ on a Riemann surface explains their complicated behaviour (Kouvaritakis and Shaked 1976) as compared with the single-input single-output case, where the root loci lie in a simple complex plane. In the general case, there will be several irreducible characteristic equations for a given $L(s)$ and thus associated with a given $L(s)$ there will normally be several sets of root loci each on a different Riemann surface. It is also interesting to note that, when $e(s) \neq 1$, or $p_d(s) \neq 1$, or $p_x(s) \neq 1$ there will also be degenerate loci each consisting of a single point, a feature not present in the single-input single-output case. Note that the single point loci associated with $e(s)$ and $p_x(s)$ can be removed with feedback whereas those associated with $p_d(s)$ are invariant.

We have that

$$CLCP(s) = \frac{p_x(s)p_d(s)e(s) \prod_{i=1}^{k} b_{i0}(s)}{\det F(\infty)} \tag{9.22}$$

Now the solutions of

$$\prod_{i=1}^{k} b_{i0}(s) = 0 \tag{9.23}$$

are the zeros of $f_i(s)$ and are also values of s at which

$$l_i(s) = -1 \tag{9.24}$$

Consequently those points on the sheets of the Riemann surface at which the 0 dB contours cut the 180° phase contours are closed-loop poles of the system with direct unit-gain negative feedback. Note that, due to the presence of the factors $p_d(s)$, $p_x(s)$ and $e(s)$ in eqn. (9.22), not all the closed-loop poles appear on the Riemann surface.

10. An example of the generalized Nyquist stability criterion with Riemann surface construction and multivariable root loci illustration

As an illustrative example (Kontakos 1973) consider the standard multivariable feedback configuration of Fig. 7, where

$$G(s) = \frac{1}{1{\cdot}25(s+1)(s+2)} \begin{bmatrix} (s-1) & s \\ -6 & (s-2) \end{bmatrix}, \quad K(s) = kI_2 \quad \text{and} \quad H(s) = I_2$$

The pole polynomial for $G_1(s)$ $[=kG(s)]$ is

$$p_{G_1}(s) = (s+1)(s+2)$$

and for $G_2(s)[=H(s)]$ is

$$P_{G_2}(s) = 1$$

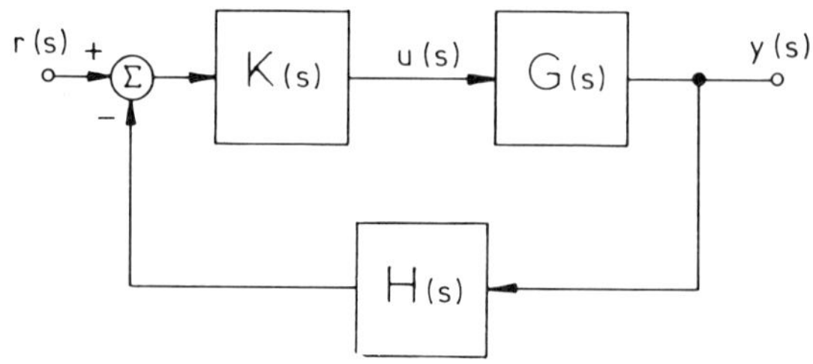

Figure 7. Standard multivariable feedback configuration.

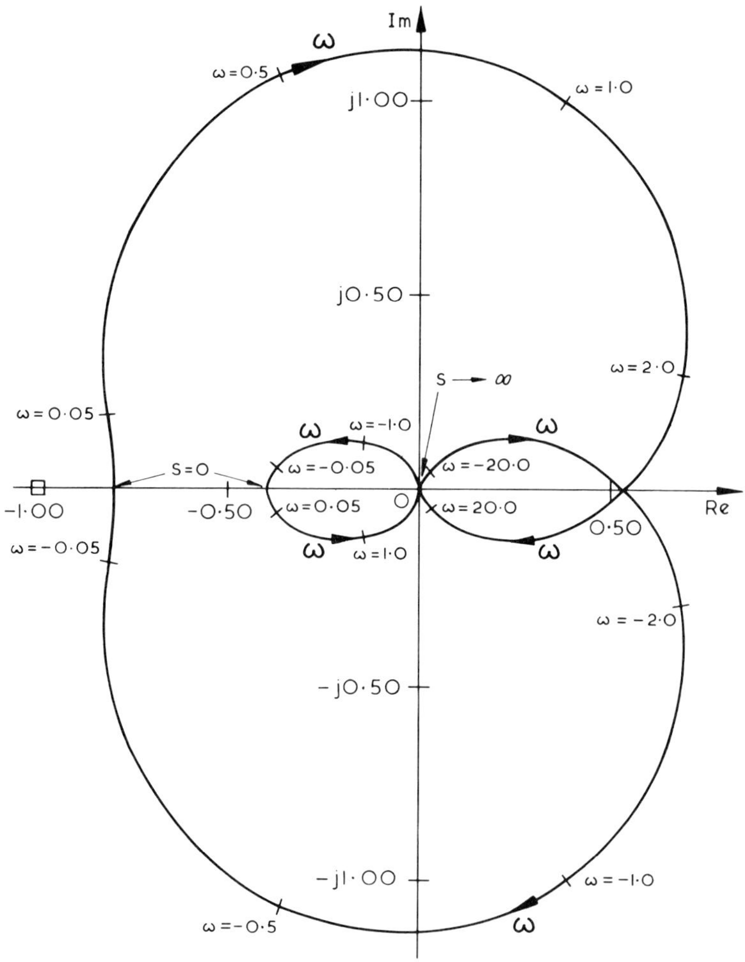

Figure 8. Example (§ 10) of characteristic loci.

and therefore the total number of poles in the right-half plane for $G_1(s)$ and $G_2(s)$ is $p=0$ and closed-loop stability will be ensured if

$$\sum_{i=1}^{k'} n_{l_i}=0$$

The characteristic loci of $L(s)$ $[=G(s)$ for $k=1]$ are shown in Fig. 8.

Taking the critical point as $(-1/k,\ 0)$ the following stability conditions are obtained.

(i) For $-\infty < -1/k < -0{\cdot}8$ there are no encirclements of the critical point

$$\sum_{i=1}^{k'} n_{l_i}=0$$

and thus the closed loop system is stable for $0 \leqslant k < 1{\cdot}25$.

(ii) For $-0{\cdot}8 \leqslant -1/k \leqslant 0{\cdot}4$ there is one encirclement of the critical point, and therefore the system is closed-loop unstable for $1{\cdot}25 \leqslant k \leqslant 2{\cdot}5$.

(iii) For $-0{\cdot}4 < -1/k \leqslant 0$ the net sum of clockwise encirclements is zero,

$$\sum_{i=1}^{k'} n_{l_i}=0$$

and thus the closed-loop system is stable for $2{\cdot}5 < k < \infty$.

(iv) For $0 \leqslant -1/k \leqslant 0{\cdot}533$ there are two clockwise encirclements of the critical point and therefore the system is closed-loop unstable for $-\infty < k \leqslant -1{\cdot}875$.

(v) For $0{\cdot}533 < -1/k < \infty$ there are no encirclements of the critical point,

$$\sum_{i=1}^{k'} n_{l_i}=0$$

and thus the closed-loop system is stable for $-1{\cdot}875 < k \leqslant 0$. Note that conditions (iv) and (v) correspond to positive feedback.

The system exhibits the conditionally stable characteristic familiar in scalar feedback control systems. On increasing the gain (equally in both loops) the closed-loop system experiences stability, instability, and stability again. This phenomenon is linked with the presence of a branch point in the right-half plane (at $s=1/24$). The role of the branch point is best illustrated by looking at the Riemann surface corresponding to $L(s)$.

The two sheets of the surface are shown in Figs. 9 and 10, with constant phase and magnitude contours of $l(s)$ shown to emphasise the position of the cuts. The 180° phase contours are obviously the multivariable root loci and imply possible instability by being partly in the right-half plane on both sheets. The root locus branch starting at the -1 pole on sheet one moves with increasing gain into the right-half plane, implying instability ; passes through the branch point to sheet two, still implying instability ; and then moves into the stable left-half plane and off to infinity. The crossings of the imaginary axis into the right-half plane of sheet one, and out of the right-half plane of sheet two, are for $k=1{\cdot}25$ and $k=2{\cdot}5$ respectively. Therefore the system is closed-loop unstable for $1{\cdot}25 \leqslant k \leqslant 2{\cdot}5$ which is the condition obtained from the characteristic loci of $L(s)$.

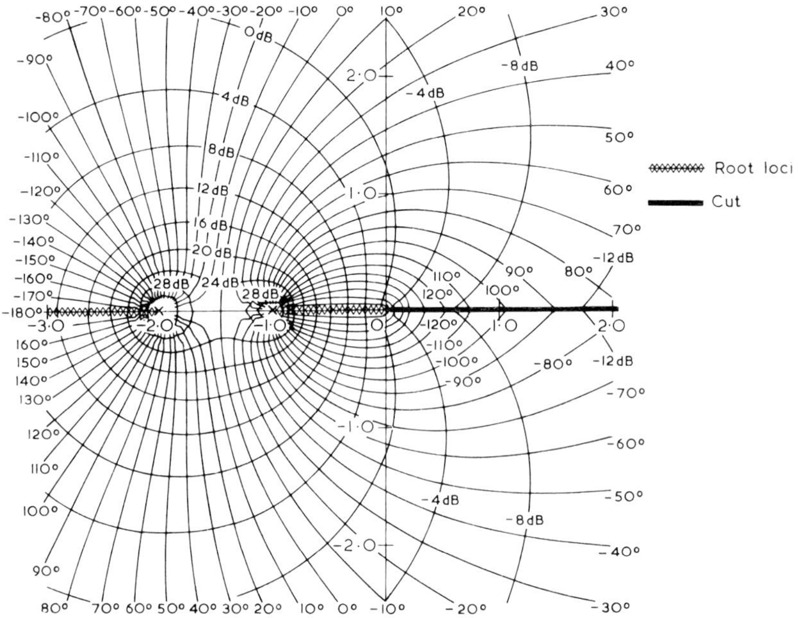

Figure 9. Sheet 1 of the Riemann surface for the example in § 10.

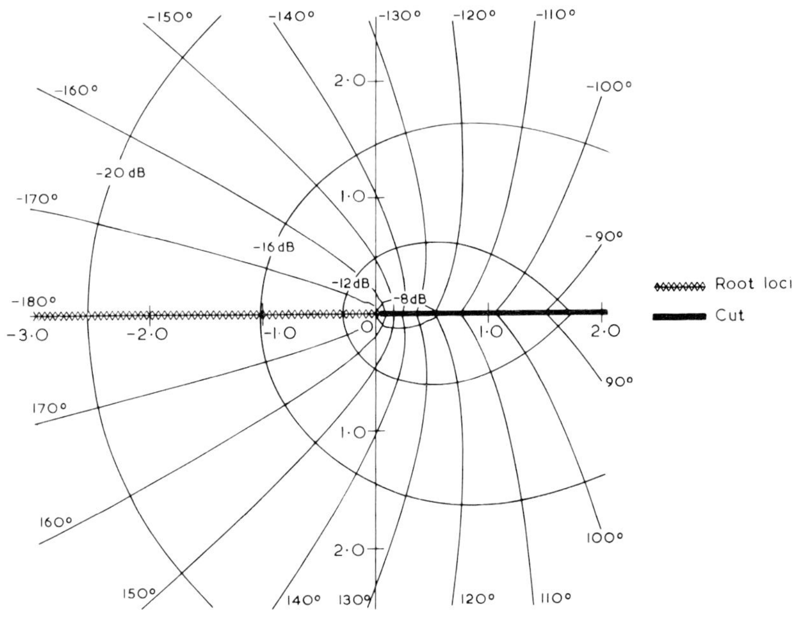

Figure 10. Sheet 2 of the Riemann surface for the example in § 10.

The movement of the root loci from poles to ' zeros at infinity ' is in agreement with work carried out by Kouvaritakis and Shaked (1976) on the asymptotic behaviour of the root loci of linear multivariable systems. Kouvaritakis and Shaked (1976) also pointed out that contrary to the scalar case the multivariable root loci can not only cross each other but two or more loci may even coincide in part. This is the case when the root loci are projected onto a single complex plane but when viewed on the Riemann surface the root loci can never cut each other, as demonstrated in Figs. 9 and 10.

A characteristic equation defining a single characteristic function $l(s)$ can equally well be viewed as an equation defining a single algebraic function $s(l)$ (assuming irreducibility). A Riemann surface, which is a domain for $s(l)$, can then be conceived and used to display constant phase and magnitude contours of $s(l)$. If $l(s)$ corresponds to a return-ratio matrix $L(s)$, then the $90°$ and $-90°$ phase contours of $s(l)$ obviously represent the characteristic loci for $L(s)$. Therefore viewing the characteristic loci on appropriate surfaces they appear never to cut each other, as with the root loci on their corresponding surfaces.

For an open-loop stable system (i.e. $p=0$), it is necessary for closed-loop stability that the characteristic loci do not encircle the critical point $(-1+j0)$. If the characteristic loci are considered as subsets of points on an appropriate Riemann surface punctured by removal of the $-1+j0$ points, then it is sufficient for closed-loop stability that on this punctured surface the image of the Nyquist D-contour bounded by the characteristic loci can be continuously deformed into a point. This topological description of the criterion is intuitively more satisfying than considering an encirclement criterion for loci in $\mathbf{C}$. It arose out of a discussion between one of the authors (I.P.) and Professor R. Saeks of his generalized approach to Nyquist-type criteria (Saeks 1975) which uses such topological arguments.

Appendix 1

A reduction to the irreducible rational canonical form

In this Appendix a proposed method is given for reducing any m-square matrix to its *irreducible* rational canonical form. The method used is a variation on that given by Ayres (1965) for finding the rational canonical form of a square matrix. The rational canonical form of a square matrix Q is similar to the irreducible form except that its diagonal blocks correspond to the invariant factors of $qI-Q$, rather than the irreducible factors. The procedure given by Ayres (1965) for finding the rational canonical form is outlined below along with some necessary definitions.

Definitions

If the vectors

$$X, QX, Q^2X, \ldots, Q^{t-1}X \tag{A 1}$$

are linearly independent but

$$X, QX, Q^2X, \ldots, Q^{t-1}X, Q^tX \tag{A 2}$$

are not, then (A 1) is called a *chain* of length t having X as its *leader*.

Procedure

For a given m-square matrix Q over any field F:

(i) let X_m be the leader of a chain C_m of *maximum* length for all m-vectors over F;

(ii) let X_{m-1} be the leader of a chain C_{m-1} of *maximum* length (any member of which is linearly independent of the preceding members and those of C_m) for all m-vectors over F which are linearly independent of the vectors of C_m;

(iii) let X_{m-2} be the leader of a chain C_{m-2} of *maximum* length (any member of which is linearly independent of the preceding members and those of C_m and C_{m-1}) for all m-vectors over F which are linearly independent of the vectors C_m and C_{m-1}; and so on. Then, for

$$T = [X_j, QX_j, \ldots, Q^{t_j-1}X_j;\ X_{j+1}, QX_{j+1}, \ldots, Q^{t_{j+1}-1}X_{j+1};\ \ldots;\ X_m, QX_m, \ldots, Q^{t_m-1}X_m]$$

we have that $T^{-1}QT$ is the rational canonical form of Q.

In this approach the chains of maximum length are used to pick out the invariant factors. Now the invariant factors are made up from products of the irreducible characteristic equations which are required in the irreducible rational canonical form. Therefore, if instead of using chains of maximum length those of *minimum* length are found, the transformation matrix T so formed is that required to give the irreducible rational canonical form R.

The problem with both of these methods is that no indication is given which enables one to know when a chain of maximum or minimum length has been obtained, except that in the contrary case there will appear a chain of longer or shorter length.

It is interesting to note that the chains of maximum and minimum length form bases for the Q-invariant subspaces of maximum and minimum dimensions respectively. Therefore the problem of finding the chains of minimum length is equivalent to that of finding the invariant subspaces of minimum dimension. It also happens that just as the 1-dimensional invariant subspace (eigenvector) picks out an eigenvalue, a t-dimensional invariant subspace (from the set of those of minimum dimensions) picks out an irreducible equation of degree t, defining t eigenvalues.

Example

To find the irreducible rational canonical form of

$$Q(s) = \begin{bmatrix} \dfrac{(s+3)}{(s+1)^2} & \dfrac{(s+2)}{(s+1)^2} & 0 \\[2ex] \dfrac{(s-3)}{(s+1)^2} & \dfrac{-2}{(s+1)^2} & 0 \\[2ex] \dfrac{-1}{(s+1)^2} & \dfrac{-(s+2)}{2(s+1)^2} & \dfrac{1}{s+1} \end{bmatrix}$$

Let $X=(0\ 0\ 1)^{\mathrm{t}}$, where ' t ' denotes the transpose, then

$$QX=\frac{1}{(s+1)}X$$

and X is a chain of minimum length.

Let

$$Y=(0\ 1\ 0)^{\mathrm{t}}$$

then

$$QY=\left[\frac{s+2}{(s+1)^2}\quad\frac{-2}{(s+1)^2}\quad\frac{-(s+2)}{2(s+1)^2}\right]^{\mathrm{t}}$$

and

$$Q^2Y=\left[\frac{s+2}{(s+1)^3}\quad\frac{s-2}{(s+1)^3}\quad\frac{-(s+2)}{2(s+1)^3}\right]^{\mathrm{t}}=\frac{1}{(s+1)}QY+\frac{s}{(s+1)^3}Y$$

No chain of smaller length can be found and therefore Y, QY completes the set of minimum length chains. The transformation matrix $T(s)$ is therefore given by

$$T(s)=[X\ \ Y\ \ QY]=\begin{bmatrix} 0 & 0 & \dfrac{s+2}{(s+1)^2} \\ 0 & 1 & \dfrac{-2}{(s+1)^2} \\ 1 & 0 & \dfrac{-(s+2)}{2(s+1)^2} \end{bmatrix}$$

and

$$R(s)=T^{-1}(s)Q(s)T(s)$$

$$=\begin{bmatrix} \dfrac{1}{s+1} & 0 & 0 \\ 0 & 0 & \dfrac{s}{(s+1)^3} \\ 0 & 1 & \dfrac{1}{(s+1)} \end{bmatrix}$$

which implies that the irreducible characteristic equations are

$$\Delta_1(s,g)=g-\frac{1}{s+1}$$

$$\Delta_2(s,g)=g^2-\frac{1}{(s+1)}g-\frac{s}{(s+1)^3}$$

Note that these are also obvious from the dependence relations obtained when finding the chains of minimum length.

Appendix 2

The discriminant

In this Appendix two methods are given for finding the discriminant of an equation of the form

$$b(s, q) = b_0(s)q^t + b_1(s)q^{t-1} + \dots + b_t(s)$$

Method 1 (Barnett 1971)

The resultant $R[a(q), c(q)]$ of two polynomials $a(q)$ and $c(q)$ given by

$$a(q) = a_0 q^n + a_1 q^{n-1} + \dots + a_n$$
$$c(q) = c_0 q^m + c_1 q^{m-1} + \dots + c_m$$

where $a_i, c_i \in \mathbf{C}$ is the determinant

$$R[a(q), c(q)] = \left| \begin{array}{cccccccccc} a_0 & a_1 & a_2 \dots & & & a_n & 0 & & & \\ 0 & a_0 & a_1 \dots & & & a_{n-1} & a_n & & & \\ \cdot & \cdot & \cdot \dots & & & & & & & \\ \cdot & \cdot & \cdot \dots & & & & a_{n-1} & a_n & & \\ \cdot & \cdot & \cdot \dots & & c_0 & c_1 & c_{m-1} & c_m & & \\ \cdot & \cdot & \cdot \dots & c_0 & c_1 & c_2 & c_m & 0 & & \\ \cdot & \cdot & \cdot \dots & & & & & & & \\ c_0 & c_1 & c_2 \dots & & & & & & & \end{array} \right| \begin{array}{l} \uparrow \\ m \text{ rows} \\ \downarrow \\ \uparrow \\ n \text{ rows} \\ \downarrow \end{array}$$

The polynomials $a(q)$ and $c(q)$ have a common factor (of degree greater than zero) if and only if the $(n+m)$-order determinant $R[a(q), c(q)]$ is zero, provided that a_0 and c_0 are not both zero.

Let the derivative with respect to q of the polynomial $a(q)$ be denoted by $a'(q)$. Then the discriminant of the polynomial $a(q)$ is the determinant $D_q(a_0, a_1, \dots, a_n)$ defined by

$$D_q(a_0, \dots, a_n) = \frac{1}{a_0} R[a(q), a'(q)]$$

The polynomial $a(q)$ has a repeated factor if and only if the discriminant $D_q(a_0, \dots, a_n)$ is zero.

Now consider a polynomial of the type

$$b(s, q) = b_0(s)q^t + b_1(s)q^{t-1} + \dots + b_t(s), \quad t > 0$$

where the coefficients $\{b_i(s) : i = 1, 2, \dots, t\}$ are all polynomials in s. The discriminant of this polynomial in q is found from $D_q(b_0, b_1, \dots, b_t)$ as defined above by replacing $b_0, \dots, b_t$ by $b_0(s), \dots, b_t(s)$ respectively. Thus there is a function $D_q(s)$, again called the discriminant and defined by

$$D_q(s) = \frac{1}{b_0(s)} R[b(s, q), b'(s, q)]$$

where $b'(s, q)$ is the derivative with respect to q of $b(s, q)$.

Method 2 (Sansone and Gerretsen 1969)

Consider the polynomial

$$a(q) = a_0 q^n + a_1 q^{n-1} + \dots + a_n, \quad n > 0$$

where $a_i \in \mathbf{C}$; then the discriminant of $a(q)$ is given by the expression

$$D_q(a_0, \ldots, a_n) = a_0^{2n-2} P$$

where P is a determinant given by

$$P = \begin{vmatrix} \sigma_0 & \sigma_1 \ldots & \sigma_{n-1} \\ \sigma_1 & \sigma_2 \ldots & \sigma_n \\ . & . \quad . & . \\ \sigma_{n-1} & \sigma_n \ldots & \sigma_{2n-2} \end{vmatrix}$$

and the elements $\{\sigma_i : i = 1, 2, \ldots, 2n-2\}$ are functions of the coefficients $\{a_i : i = 0, 1, \ldots, n\}$, and

$$\sigma_0 = n$$

The elements $\sigma_1, \ldots, \sigma_{n-1}$ can be found from

$$\begin{aligned} a_1 + a_0\sigma_1 &= 0 \\ 2a_2 + a_1\sigma_1 + a_0\sigma_2 &= 0 \\ \ldots\ldots\ldots\ldots\ldots\ldots \\ (n-1)a_{n-1} + a_{n-2}\sigma_1 \ldots + a_0\sigma_{n-1} &= 0 \end{aligned}$$

and $\sigma_n, \ldots, \sigma_{2n-2}$ from

$$\begin{aligned} a_n\sigma_0 + a_{n-1}\sigma_1 + \ldots + a_0\sigma_n &= 0 \\ a_n\sigma_1 + a_{n-1}\sigma_2 + \ldots + a_0\sigma_{n+1} &= 0 \\ \ldots\ldots\ldots\ldots\ldots\ldots\ldots\ldots \\ a_n\sigma_m + a_{n-1}\sigma_{m+1} + \ldots + a_0\sigma_{n+m} &= 0 \\ \ldots\ldots\ldots\ldots\ldots\ldots\ldots\ldots \end{aligned}$$

Consider now the polynomial $b(s, q)$ given by

$$b(s, q) = b_0(s)q^t + b_1(s)q^{t-1} + \ldots + b_t(s)$$

The discriminant of this polynomial in q is found from $D_q(b_0, \ldots, b_t)$ as defined above by replacing $b_0, \ldots, b_t$ by $b_0(s), \ldots, b_t(s)$ respectively. The polynomial $b(s, q)$ therefore has repeated factors if and only if

$$D_q(s) = b_0^{2t-2}(s)P$$

is zero, where P is the determinant of a matrix whose elements are functions of the coefficients $\{b_i(s) : i = 0, 1, 2, \ldots, t\}$.

Appendix 3

A method for constructing the Riemann surfaces of the characteristic functions corresponding to a transfer-function matrix

In § 3 it was shown that for a characteristic function of degree t the corresponding Riemann surface is made up from t sheets of the complex s-plane stitched together along cuts made between branch points and infinity. Although the cuts are in some sense arbitrary (i.e. there is no unique set of cuts), it is still a problem to choose a set that is consistent, and then to be able to identify in what order the sheets are connected together. In this Appendix a systematic method is given for solving this problem. The method is quite elegant in that the resulting cuts are symmetrical about the real axis and are always parallel to the imaginary axis except for possible cuts along the real axis. Also, the approach uses the transfer function matrix directly, so that there is no need to find the characteristic equations, and the resulting non-connected sets of connected sheets represent the Riemann surfaces for the irreducible characteristic equations.

The method is based on finding the eigenvalues of the matrix for a grid of values covering the s-plane and then sorting the eigenvalues in a continuous form along certain lines of the grid. If the transfer function is of order $m \times m$, m arrays which will effectively represent the m sheets of the Riemann surfaces are needed to store the eigenvalues. The process is analogous to the analytic continuation procedure described in § 3 where individual points of a circular disc are made bearers of unique functional values. Here individual points of the s-plane sheets (arrays) are being made the bearers of functional values.

The first line along which the eigenvalues are calculated and sorted is the real axis, and then the calculation and sorting are carried out along lines parallel to the imaginary axis and emanating from the real axis, as shown in Fig. 11. The calculation is only necessary in the upper-half plane since the eigenvalues in the lower-half plane are the complex conjugate of those in the upper half. Continuing this process the s-plane is covered, and m arrays of eigenvalues are obtained which are continuous along the real axis and along lines parallel to the imaginary axis but not necessarily crossing the real axis.

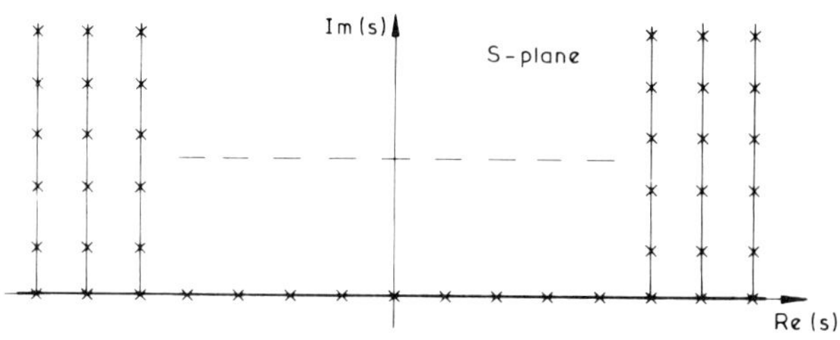

Figure 11. Lines along which eigenvalues are calculated and sorted. ×——× indicates continuity between eigenvalues.

By observation of each array or sheet the necessary cuts are obvious. If parallel to the imaginary axis a line of eigenvalues is not continuous with an adjacent line then these must be separated by a cut starting at a branch point and ending at a branch point or infinity as shown in Figs. 12 and 13. Eigenvalues corresponding to values of s in the upper-half plane are complex conjugate to those in the lower-half plane and therefore continuity of eigenvalues across the real axis is impossible if the eigenvalues situated on the real axis are complex. Therefore a cut along the real axis is necessary whenever the real axis eigenvalues are complex. Again all the cuts start at a branch point and end at a branch point or infinity, as shown in Fig. 14.

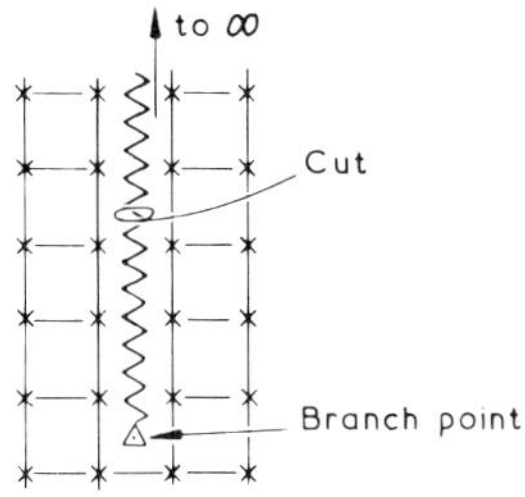

Figure 12. Infinite cut parallel to the imaginary axis. ×——× indicates continuity between eigenvalues.

The stitching together of the sheets also becomes obvious by matching eigenvalues on one side of a cut on one sheet to those on another. When the matching process is complete, in general, there will be sets of connected sheets each set defining a Riemann surface.

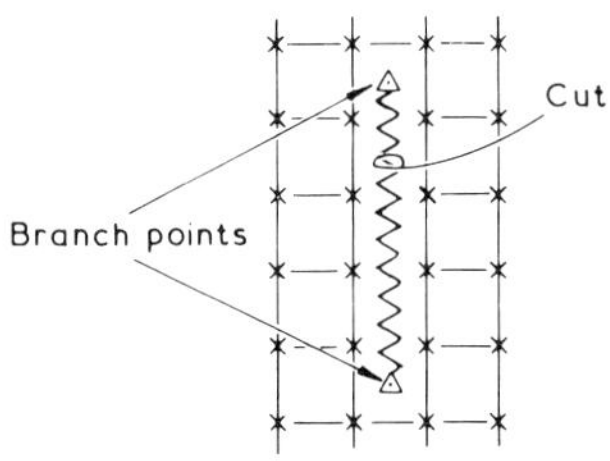

Figure 13. Finite cut parallel to the imaginary axis. ×——× indicates continuity between eigenvalues.

To facilitate the identification of cuts and the matching of sheets it is very useful to draw the constant phase and constant magnitude contours on each sheet; this is demonstrated in the example given in § 10 of the paper.

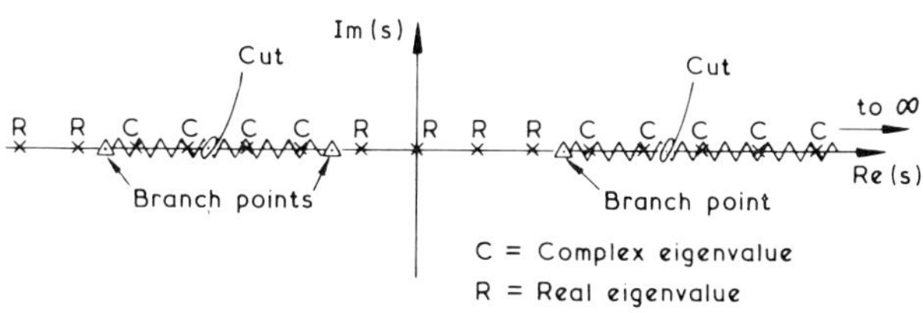

Figure 14. Cuts along the real axis.

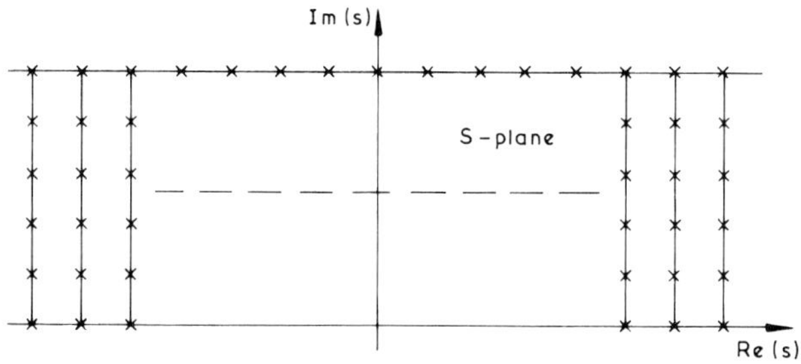

Figure 15. Lines along which eigenvalues are calculated and sorted (modified method). ×——× indicates continuity between eigenvalues.

Computationally sorting the eigenvalues along the real axis can be difficult because of the likelihood of real axis poles. This problem is overcome if the first line of calculation and sorting is changed to be a line parallel to, and a ' large ' distance away from, the real axis. The rest of the calculations and sorting are then carried out along lines parallel to the imaginary axis starting at this new line and finishing at the real axis ; as shown in Fig. 15. With this new approach the only possible cuts are either along the real axis, as before (see Fig. 14), or between complex conjugate branch points as shown in Fig. 16.

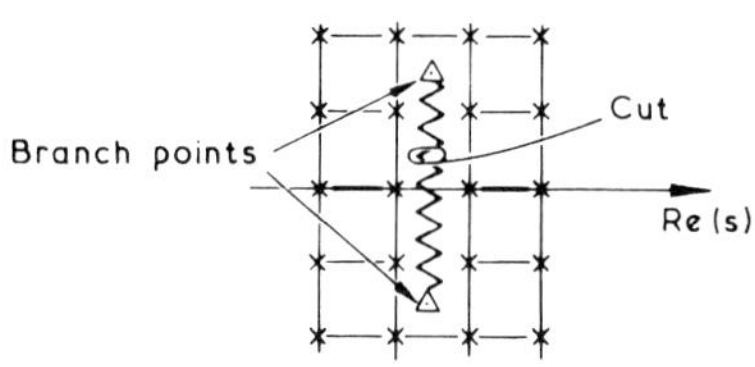

Figure 16. Cut parallel to the imaginary axis (modified method). ×——× indicates continuity between eigenvalues.

A Riemann surface construction for

$$L(s)=\frac{1}{1\cdot 25(s+1)(s+2)}\begin{bmatrix} 5s-2 & 2s-1 \\ 3s-18 & s-8 \end{bmatrix}$$

is shown using the original method in Figs. 17 and 18, and using the modified method in Figs. 19 and 20, and illustrates the arbitrariness of the cuts.

Appendix 4

The eigenvector functions of a matrix-valued function of a complex variable

The eigenvalue functions of a matrix-valued function of a complex variable have been shown in § 3 to be associated with a set of algebraic functions called

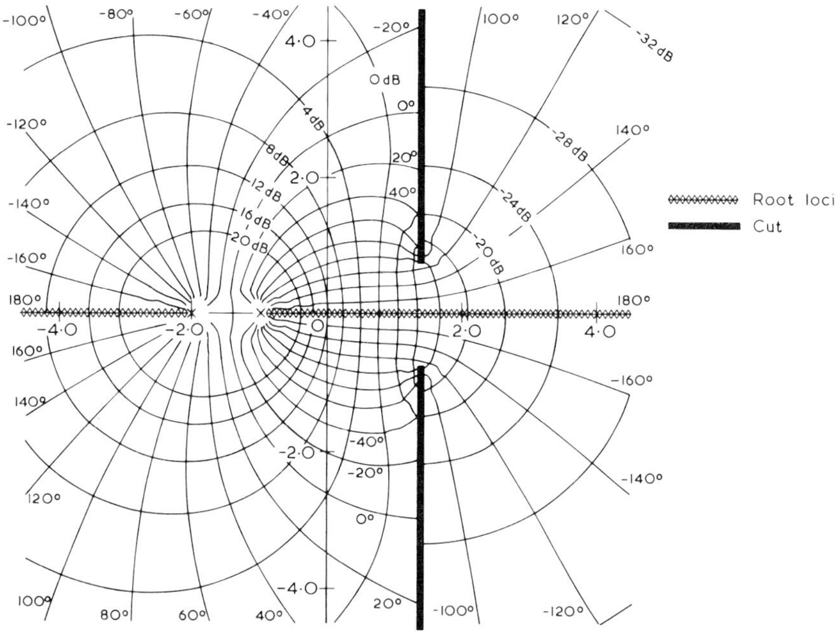

Figure 17. Sheet 1 of the Riemann surface for the example in Appendix 3.

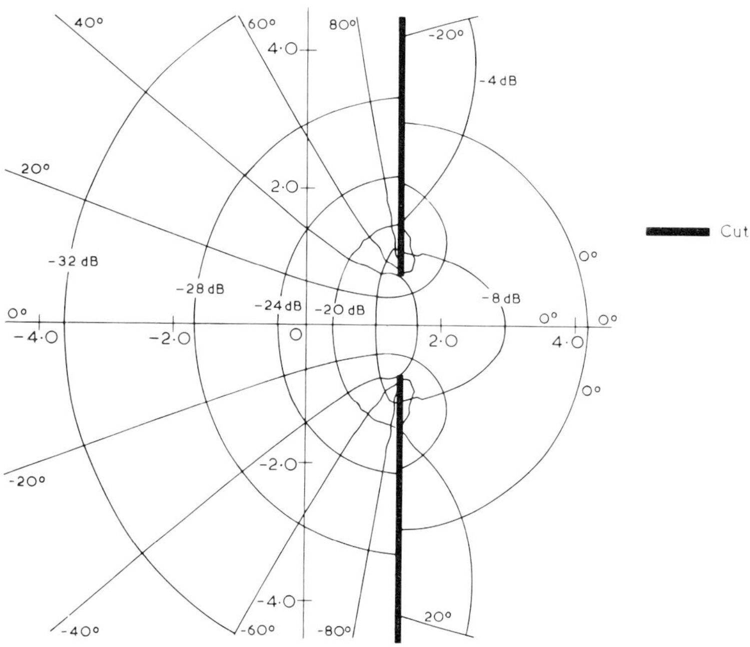

Figure 18. Sheet 2 of the Riemann surface for the example in Appendix 3.

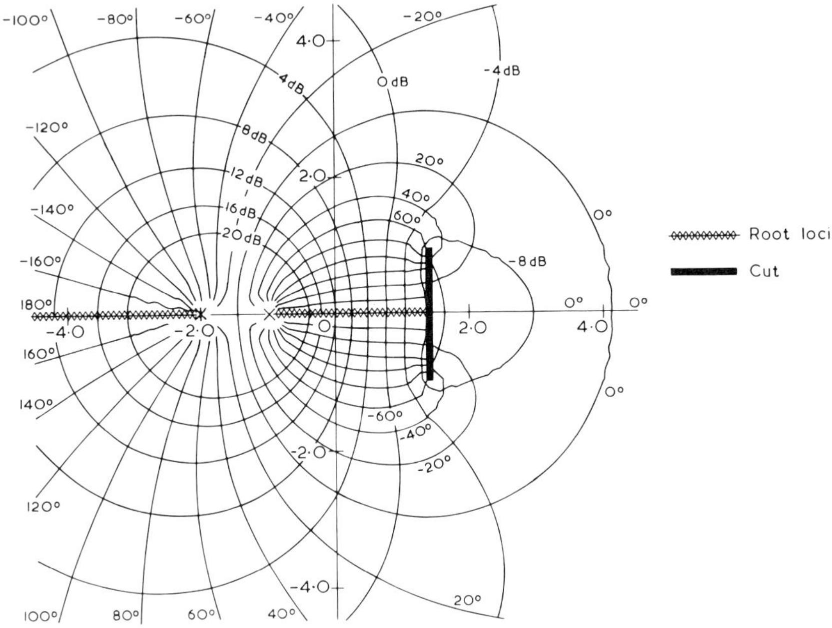

Figure 19. Sheet 1 of the Riemann surface for the example in Appendix 3 (modified method of construction).

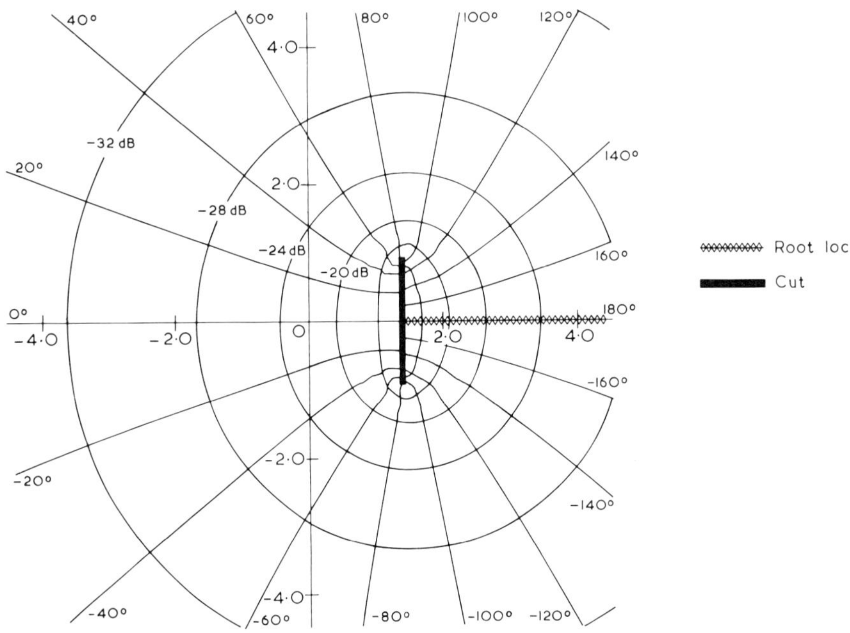

Figure 20. Sheet 2 of the Riemann surface for the example in Appendix 3 (modified method of construction).

the characteristic functions of the matrix. In order to complete the description of the relationships between the eigenstructure of a matrix-valued function of a complex variable and Riemann surfaces, we must consider the corresponding set of eigenvector functions.

For any given matrix-valued function of a complex variable $Q(s)$ there is a corresponding set of Riemann surfaces $\mathbf{R}_{qi}$ which are the domain sets for its characteristic functions $\{q_i : i=1, 2, \ldots, k\}$. It is shown in standard textbooks on complex function theory that the set of all algebraic functions having a given Riemann surface $\mathbf{R}$ as domain set form a field $K_{\mathbf{R}}$ (Behnke *et al.* 1974). For any such field $K_{\mathbf{R}}$ the set of all elementary rational functions are members, since such elementary functions are defined on the domain $\mathbf{C}$ and therefore on any domain formed out of multiple copies of $\mathbf{C}$. From an algebraic point of view each algebraic function defines a particular *extension* of the basic field of elementary rational functions. Now consider the equation defining the eigenvector function $w(s)$ corresponding to an eigenvalue function $q(s)$:

$$Q(s)w(s)=q(s)w(s)$$

The elements of $w(s)$ are found by solving linear sets of equations involving rational functions (the elements of $Q(s)$) and the algebraic function $q(s)$. They must therefore belong to the field of algebraic functions having $\mathbf{R}_q$, the Riemann surface for $q(s)$, as domain.

Thus, corresponding to every characteristic algebraic function $q(s)$ of a matrix-valued function of a complex variable $Q(s)$, we have a vector-valued function of a complex variable $w(s)$ whose elements are algebraic functions having the same Riemann surface domain as the characteristic function $q(s)$.

Appendix 5

Proof of an early statement of the generalized Nyquist stability criterion (MacFarlane and Belletrutti 1973)

Theorem

The closed-loop system is stable if and only if the net sum of anti-clockwise encirclements of the critical point $(-1+j0)$ by the set of characteristic loci of $L(s)$ is equal to the number of right-half plane zeros of $OLCP(s)$ where, as defined in § 2, $OLCP(s)=\det[sI-A_1]\det[sI-A_2]$.

Proof

From § 8 we have that the necessary and sufficient conditions for closed loop stability are that

(*a*) $e(s)=0$ has no roots in the right-half plane ;

(*b*) $\sum_{i=1}^{k} n_{f_i} = -\sum_{i=1}^{k} p_{f_i}$;

(*c*) $p_x(s)=0$ has no roots in the right-half plane ; and

(*d*) $p_d(s)=0$ has no roots in the right-half plane.

Also, from eqn. (8.13) the open loop characteristic polynomial is given by

$$OLCP(s)=p_x(s)p_d(s)e(s)\prod_{i=1}^{k} b_{i0}(s)$$

so that if p_0 is defined as the number of right-half plane zeros of $OLCP(s)$ we have that

$$p_0 = p_x + p_d + e + \sum_{i=1}^{k} p_{f_i}$$

where p_d is the number of right-half plane zeros of $p_d(s)$ and e and p_x are as defined in § 8.

This leads us to consider combining the four conditions (*a*), (*b*), (*c*) and (*d*) into the single equivalent condition

$$\sum_{i=1}^{k} n_{f_i} = -p_0 \tag{A 3}$$

Necessity and sufficiency of condition (A 3) for closed-loop stability

We have that

$$\sum_{i=1}^{k} n_{f_i} = \sum_{i=1}^{k} z_{f_i} - \sum_{i=1}^{k} p_{f_i}$$

and

$$p_0 = p_x + p_d + e + \sum_{i=1}^{k} p_{f_i}$$

and combining the two relationships we thus have

$$\sum_{i=1}^{k} n_{f_i} = \sum_{i=1}^{k} z_{f_i} + p_d + e + p_x - p_0 \tag{A 4}$$

where $\sum_{i=1}^{k} z_{f_i}$, p_d, e, p_x and p_0 are all positive integers or zero.

To establish the necessity of (A 3) suppose that

$$\sum_{i=1}^{k} n_{f_i} \neq -p_0$$

Then, from (A 4), this implies that

$$\sum_{i=1}^{k} z_{f_i} \neq 0, \quad \text{or} \quad p_d \neq 0, \quad \text{or} \quad e \neq 0, \quad \text{or} \quad p_x \neq 0$$

or any combination of these, and thus that the system will be closed-loop unstable.

Hence we conclude that (A 3) is necessary for closed-loop stability.

For sufficiency suppose that

$$\sum_{i=1}^{k} n_{f_i} = -p_0$$

Then, from (A 4) we must have that

$$p_x = p_d = e = \sum_{i=1}^{k} z_{f_i} = 0$$

and hence the system is closed-loop stable. Thus the sufficiency of condition (A 3) is established.

We have therefore shown that

$$\sum_{i=1}^{k} n_{f_i} = -p_0$$

is a necessary and sufficient condition for closed-loop stability. Using the eigenvalue shift theorem (MacFarlane 1970 b) and continuing as in § 8 it then follows that the necessary and sufficient condition for closed-loop stability becomes

$$\sum_{i=1}^{k} n_{l_i} = -p_0$$

where $\sum_{i=1}^{k} n_{l_i}$ is the net sum of clockwise encirclements of the critical point $(-1+j0)$ by the characteristic loci of $L(s)$. This then completes the proof of the early statement of the generalized Nyquist stability criterion.

Appendix 6

A note on the polynomials $e(s)$ and $p_x(s)$

In § 8 a polynomial $e(s)$ is defined in terms of the matrix $F(s)$ and in § 10 it is defined in terms of the matrix $L(s)$. The purpose of the first part of this brief Appendix is to show that the same polynomial $e(s)$ is obtained regardless of whether $F(s)$ or $L(s)$ is involved in the definition. The second part gives an example of the occurrence of a polynomial $p_x(s) \neq 1$.

We have that the pole polynomial for $F(s)$ is given by

$$p_f(s) = e_f(s) \prod_{i=1}^{k} b_{fio}(s)$$

where $e_f(s)$ is the least common denominator of all non-zero non-principal minors of $F(s)$ with all factors common to $\prod_{i=1}^{k} b_{fio}(a)$ removed. Also we have that the pole polynomial for $L(s)$ is given by

$$p_l(s) = e_l(s) \prod_{i=1}^{k} b_{lio}(s)$$

with an obvious definition of $e_l(s)$.

Furthermore :

$$\text{pole polynomial for } f_i(s) = b_{fio}(s)$$

and

$$\text{pole polynomial for } l_i(s) = b_{lio}(s)$$

But

$$f_i(s) = 1 + l_i(s)$$

so that

$$b_{fio}(s) = b_{lio}(s)$$

and thus

$$\prod_{i=1}^{k} b_{fio}(s) = \prod_{i=1}^{k} b_{lio}(s)$$

Finally, since

$$F(s) = I_m + L(s)$$

we must have that

$$p_f(s) = p_l(s)$$

and so

$$e_f(s) = e_l(s) = e(s)$$

To see how a polynomial $p_x(s) \neq 1$ can arise suppose we have

$$G_1(s) = G_2(s) = \begin{bmatrix} \dfrac{2}{(s+1)} & \dfrac{1}{(s-1)} \\ 0 & \dfrac{2}{(s+1)} \end{bmatrix}$$

Then we have

$$L(s) = G_2(s)G_1(s) = \begin{bmatrix} \dfrac{4}{(s+1)^2} & \dfrac{4}{(s+1)(s-1)} \\ 0 & \dfrac{4}{(s+1)^2} \end{bmatrix}$$

so that

$$p_l(s) = (s+1)^4(s-1)$$

and

$$p_{G_1}(s) = p_{G_2}(s) = (s+1)^2(s-1)$$

Thus

$$p_x(s) = \frac{p_{G_1}(s)p_{G_2}(s)}{p_l(s)} = \frac{(s+1)^2(s-1)(s+1)^2(s-1)}{(s+1)^4(s-1)}$$

$$= (s-1).$$

Appendix 7.

Characterization of the point at infinity for the characteristic function $q(s)$

The characteristic function $q(s)$ is defined by an irreducible equation of the form

$$\Delta(q, s) = b_0(s)q^t + b_1(s)q^{t-1} + \dots + b_t(s) = 0$$

For the purpose of considering $q(s)$ at the point $s = \infty$ one may put

$$s = \frac{1}{z}$$

Then

$$\Delta(q, s) = \Delta(q, z^{-1}) = z^{-n}\Psi(q, z) \tag{A 5}$$

and in any neighbourhood of the value $z = 0$ (the point $z = 0$ itself being excluded from it) the equation $\Delta(q, s) = 0$ is equivalent to the equation $\Psi(q, z) = 0$. Considering the equation

$$\Psi(q, z) = \phi_0(z)q^t + \phi_1(z)q^{t-1} + \dots + \phi_t(z) = 0 \tag{A 6}$$

it therefore follows that:

(i) $s=\infty$ is a pole of the characteristic function $q(s)$ if and only if $\phi_0(0)=0$.
(ii) $s=\infty$ is a zero of the characteristic function $q(s)$ if and only if $\phi_t(0)=0$.
(iii) $s=\infty$ is a branch point of the characteristic function $q(s)$ if and only if $D(0)=0$, where $D(z)$ is the discriminant of $\Psi(q, z)$.

For a transfer function matrix describing a physically-realizable system it is not possible to have poles at infinity, and therefore they have not been considered when deriving the pole polynomial for $q(s)$ in § 3.2, or when determining the poles of a transfer function matrix $G(s)$ in § 4. Zeros at infinity are also neglected, but the reason for this is not so obvious since for any practical transfer function matrix they will always exist, and the relationship

$$z(s)=e(s)\prod_{i=1}^{k} b_{it_i}(s)$$

of § 5 only relates the finite zeros of a transfer-function matrix to the finite zeros of its set of characteristic functions. The infinite zeros need not be considered in this paper because the proof of the generalized Nyquist stability criterion is based on the return-difference matrix $F(s)$ which almost always has no infinite zeros. This can be shown very easily as follows.

The return-difference matrix $F(s)$ is defined as

$$F(s)=I+L(s)$$

where $L(s)$ is a transfer function matrix known as the system return-ratio matrix; the corresponding characteristic functions of $F(s)$ and $L(s)$ are related by

$$f(s)=1+l(s)$$

Therefore providing $l(\infty)\neq -1$ the characteristic function $f(s)$ has no zeros at infinity, and it also follows that $F(s)$ will have no zeros at infinity.

For the restricted class of non-proper systems considered in this paper the condition $l(\infty)=-1$ is not possible because of the conditions imposed by eqn. (2.5), i.e. $\det[F(\infty)]\neq 0$. For a proper system, that is one in which $D_1=D_2=0$, we have $l(\infty)=0$ and hence $f(\infty)=1$. Consequently a return-difference matrix $F(s)$ almost always has no zeros at infinity, and thus this possibility has not been considered in formulating the proof of the generalized Nyquist criterion.

Example

$$G(s)=\frac{1}{1\cdot 25(s+1)(s+2)}\begin{bmatrix} s-1 & s \\ -6 & s-2 \end{bmatrix}$$

$$\det[gI-Gs]=\left(g-\frac{(s-1)}{1\cdot 25(s+1)(s+2)}\right)\left(g-\frac{(s-2)}{1\cdot 25(s+1)(s+2)}\right)+\frac{6s}{1\cdot 25^2(s+1)^2(s+2)^2}$$

$$=g^2-\frac{(2s-3)}{1\cdot 25(s+1)(s+2)}\,g+\frac{1}{1\cdot 25^2(s+1)(s+2)}$$

therefore

$$\Delta(g, s) = 1{\cdot}25^2(s+1)(s+2)g^2 - 1{\cdot}25(2s-3)g + 1 = 0$$

To consider $g(s)$ as $s \to \infty$ put $s = 1/z$

$$\Delta\left(g, \frac{1}{z}\right) = 1{\cdot}25^2\left(\frac{1}{z}+1\right)\left(\frac{1}{z}+2\right)g^2 - 1{\cdot}25\left(\frac{2}{z}-3\right)g + 1$$

$$= z^{-2}\Psi(g, z)$$

$$= 0$$

where

$$\Psi(g, z) = 1{\cdot}25^2(1+z)(1+2z)g^2 - 1{\cdot}25z(2-3z)g + z^2$$

Thus the coefficient $\phi_2(z)$ of eqn. (A 6) is given by

$$\phi_2(z) = z^2$$

$$\phi_2(0) = 0$$

which shows that there are two zeros located at infinity. Furthermore the discriminant of $\Psi(q, z)$ is given by

$$D(z) = 1{\cdot}25^2z^2(2-3z)^2 - 4(1{\cdot}25)^2(1+z)(1+2z)z^2$$

so that

$$D(0) = 0$$

which shows that we have a branch point at infinity.

Corrections

The leading coefficient $b_0(s)$ in the algebraic eqn. (3.9) formed from

$$\det\,[qI_m - Q(s)] = 0$$

is *not* in general, as stated in § 5 of this paper, the least common denominator of all non-zero principal minors of all orders of $Q(s)$. A correct statement is as follows :

$b_0(s)$ is the least common denominator of all the non-zero *sums* of the principal minors of the same order for $Q(s)$.

This in turn requires a correction to the definition given in § 5 and used in § 8 for the polynomial $e(s)$ in terms of non-principal minors. The polynomial $e(s)$ is perhaps best and most simply thought of as the monic polynomial whose zeros are the poles of $Q(s)$ which are not poles of $q(s)$. If $Q(s)$ is a return-ratio matrix for the standard multivariable configuration of Fig. 1, with a scalar gain k common to all the feedback loops, then the zeros of $e(s)$ are those poles of $Q(s)$ which remain poles of the closed-loop system as k varies.

ACKNOWLEDGMENTS

We are grateful to Dr. B. Kouvaritakis for the derivation of the relationship between open-loop and closed-loop characteristic polynomials used in § 2, and to Dr. J. Edmunds for the development of the computer algorithm, described in Appendix 3, used to construct the Riemann surfaces of Figs. 9, 10, 17, 18, 19 and 20.

References

AYRES, F. M., 1965, *Theory and Problems of Modern Algebra* (New York : Schaum).

BARMAN, J. F., and KATZENELSON, J., 1973, Memorandum ERL–383, Electronics Research Laboratory, College of Engineering, University of California, Berkeley ; 1974, *Int. J. Control*, **20,** 593.

BARNETT, S., 1971, *Matrices in Control Theory* (London : Van Nostrand-Reinhold).

BEHNKE, H., BACHMANN, F., FLADT, K., and SUSS, W., 1974, *Fundamentals of Mathematics*, Vol. 3 *Analysis* (Cambridge, Mass. : MIT Press).

BLISS, G. A., 1966 (reprint of 1933 original), *Algebraic Functions* (New York : Dover).

COHN, P. M., 1974, *Algebra*, Vol. 1 (London : Wiley).

DESOER, C. A., and CHAN, W. S., 1975, *J. Franklin Inst.*, **300,** 335.

EVANS, W. R., 1954, *Control System Dynamics* (New York : McGraw-Hill).

GANTMACHER, F. R., 1959, *Theory of Matrices*, Vol. 1 (New York : Chelsea).

HILLE, E., 1959, *Analytic Function Theory*, Vol. 1 (Waltham, Mass. : Blaisdell).

KNOPP, K., 1947, *Theory of Functions*, Part 2 (New York : Dover).

KONTAKOS, T., 1973, Ph.D. Thesis, University of Manchester.

KOUVARITAKIS, B., and SHAKED, U., 1976, *Int. J. Control*, **23,** 297.

MACFARLANE, A. G. J., 1970 a, *Proc. Inst. elect. Engrs*, **117,** 2037 ; 1970 b, *Dynamical System Models* (London : Harrap).

MACFARLANE, A. G. J., and BELLETRUTTI, J. J., 1973, *Automatica*, **9,** 575.

MACFARLANE, A. G. J., and KARCANIAS, N., 1976, *Int. J. Control*, **24,** 33.

NYQUIST, H., 1932, *Bell Syst. tech. J.*, **11,** 126.

ROSENBROCK, H. H., 1970, *State Space and Multivariable Theory* (London : Nelson).

SAEKS, R., 1975, *Trans. I.E.E.E. Circuits Systems*, **22,** 780.

SANSONE, G., and GERRETSEN, J., 1969, *Lectures on the Theory of Functions of a Complex Variable* Vol. 2 (Groningen : Wolters-Nordhoff).

SPRINGER, G., 1957, *Introduction to Riemann Surfaces* (Reading, Mass. : Addison-Wesley).

A generalized inverse Nyquist stability criterion

I. POSTLETHWAITE

In this paper the inverse Nyquist stability criterion, for single-input single-output linear feedback systems, is generalized to the multivariable case. This is based on an association of a transfer function matrix with a set of inverse Nyquist diagrams, which will be termed the inverse characteristic gain loci. A simple example is given to illustrate the stability tests, and also to show how the inverse characteristic gain loci may be used in feedback system design.

1. Introduction

In recent papers (MacFarlane and Postlethwaite 1977 a,b ; Postlethwaite 1977) the theory of algebraic functions (Bliss 1966) has been used to extend the basic classical techniques of single-loop feedback systems to the multivariable case. The main feature of these generalizations has been the association of a set of algebraic functions with a square transfer function matrix $G(s)$ by means of the characteristic equation

$$\Delta(g, s) = \det [gI_m - G(s)] = 0 \tag{1.1}$$

where I_m is a unit matrix of order m, and $G(s)$ is $m \times m$. For simplicity of exposition, in the following, it is assumed that $\Delta(g, s)$ is irreducible over the field of rational functions in the complex variable s.

If $\Delta(g, s)$ is regarded as a polynomial in g with coefficients which are rational functions of s then eqn. (1.1) defines an algebraic function $g(s)$ which is called the *characteristic gain function* (MacFarlane and Postlethwaite 1977 b). Similarly if we assume that $G(s)$ has normal rank m, then we can form the characteristic equation

$$\Gamma(\hat{g}, s) = \det [\hat{g}I_m - G(s)^{-1}] = 0 \tag{1.2}$$

where $[\,.\,]^{-1}$ denotes the inverse. If $\Gamma(\hat{g}, s)$ is then regarded as a polynomial in $\hat{g}$, with coefficients which are rational functions of s, eqn. (1.2) defines an algebraic function $\hat{g}(s)$ which will be called the *inverse characteristic gain function*. This function forms the basis for the generalization of the inverse Nyquist stability criterion which is presented in this paper.

The system description is given in § 2, and in §§ 3, 4 and 5 relationships are developed between the poles and zeros of $G(s)$, its inverse $G(s)^{-1}$, and their corresponding characteristic functions $g(s)$ and $\hat{g}(s)$. In § 6 the inverse characteristic gain loci, which are a natural generalization of the inverse Nyquist diagram (Rosenbrock 1974), are defined. Two statements of the generalized inverse Nyquist stability criterion are then given in § 7 and proved in § 8. The extension of the criterion to a more complicated feedback configuration is discussed in § 9. Finally in § 10 a simple example is given to illustrate the stability tests, and also to show how the inverse characteristic gain loci may be used in feedback system design.

2. System description

The feedback configuration is shown in Fig. 1. S is a linear multivariable dynamical system described in state space terms by the triple of matrices $\{A, B, C\}$ and in frequency response terms by $G(s)$ which is a proper $m \times m$ matrix of rational functions in s, with normal rank m.

It is also assumed, for simplicity of exposition, that the characteristic equation for $G(s)$ is irreducible over the field of rational functions in s, and hence that the characteristic equation for $G(s)^{-1}$ is also irreducible over the same field.

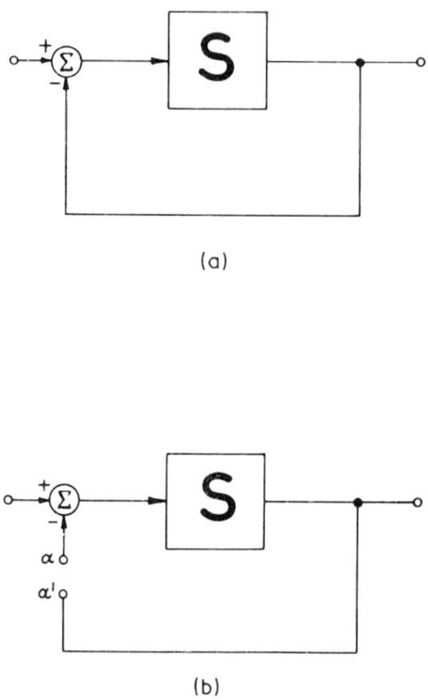

Figure 1. Multivariable feedback configuration. (a) Closed-loop ; (b) open-loop.

3. Poles and zeros of the characteristic gain and inverse characteristic gain functions

Let the characteristic equation for $G(s)$ have the form

$$\Delta(g, s) = \det [gI_m - G(s)] = g^m + a_1(s)g^{m-1} + \ldots + a_m(s) = 0 \tag{3.1}$$

and let $b_0(s)$ be the least common denominator of the rational functions $\{a_i(s) : i = 1, 2, \ldots, m\}$. On multiplying through by $b_0(s)$ eqn. (3.1) becomes

$$b_0(s)g^m + b_1(s)g^{m-1} + \ldots + b_m(s) = 0 \tag{3.2}$$

where the coefficients $\{b_i(s) : i = 1, 2, \ldots, m\}$ are now polynomials in s. As shown by MacFarlane and Postlethwaite (1977 a) the poles of the characteristic gain function $g(s)$ are then defined as zeros of the pole polynomial

$$p_g(s) \triangleq b_0(s) \tag{3.3}$$

and the zeros of $g(s)$ are defined as the zeros of the zero polynomial

$$z_g(s) \triangleq b_m(s) \tag{3.4}$$

Similarly, by considering the characteristic eqn. (1.2) for $G(s)^{-1}$, we can find expressions for the pole and zero polynomials of the inverse characteristic gain function $\hat{g}(s)$. However, since the eigenvalues of a matrix are the reciprocals of the eigenvalues of the inverse of the matrix we have that the pole polynomial for $\hat{g}(s)$ is identical to the zero polynomial for $g(s)$, that is

$$p_{\hat{g}}(s) = z_g(s) \tag{3.5}$$

and the zero polynomial for $\hat{g}(s)$ is the same as the pole polynomial for $g(s)$, that is

$$z_{\hat{g}}(s) = p_g(s) \tag{3.6}$$

where $p_{\hat{g}}(s)$ and $z_{\hat{g}}(s)$ are the pole and zero polynomials of $\hat{g}(s)$, respectively.

4. Poles and zeros of a square transfer function matrix and its inverse

It is well known that there exists for the $m \times m$ transfer function matrix $G(s)$ a canonical form, the Smith–McMillan form (Rosenbrock 1970), such that

$$G(s) = H(s)M(s)J(s) \tag{4.1}$$

where $H(s)$ and $J(s)$ are both $m \times m$ unimodular matrices (that is having a constant determinant independent of s), and

$$M(s) = \text{diag}\left[\frac{\epsilon_1(s)}{\psi_1(s)}, \frac{\epsilon_2(s)}{\psi_2(s)}, \ldots, \frac{\epsilon_m(s)}{\psi_m(s)}\right] \tag{4.2}$$

where

(i) each $\epsilon_i(s)$ divides all $\epsilon_{i+j}(s)$; and
(ii) each $\psi_i(s)$ divides all $\psi_{i-j}(s)$.

Physical considerations, as outlined by McFarlane and Postlethwaite (1977 a), then lead to the following definitions (Rosenbrock 1970) for the poles and zeros of $G(s)$.

(*a*) *Poles of* $G(s)$: The poles of $G(s)$ are defined to be the set of all zeros of the set of polynomials $\{\psi_i(s) : i = 1, 2, \ldots, m\}$.
(*b*) *Zeros of* $G(s)$: The zeros of $G(s)$ are defined to be the set of all zeros of the set of polynomials $\{\epsilon_i(s) : i = 1, 2, \ldots, m\}$.

In the following we shall denote the pole and zero polynomials of $G(s)$ by $p_G(s)$ and $z_G(s)$, respectively, so that

$$p_G(s) = \prod_{i=i}^{m} \psi_i(s) \tag{4.3}$$

and

$$z_G(s) = \prod_{i=i}^{m} \epsilon_i(s) \tag{4.4}$$

If we now consider inverting eqn. (4.1) we find that

$$G(s)^{-1} = J(s)^{-1}M(s)^{-1}H(s)^{-1} \tag{4.5}$$

If we also introduce an elementary transformation matrix E, of order m, and given as

$$E \triangleq \begin{bmatrix} 0 & 0 & \dots & 0 & 1 \\ 0 & 0 & \dots & 1 & 0 \\ \vdots & \vdots & \iddots & \vdots & \vdots \\ 0 & 1 & \dots & 0 & 0 \\ 1 & 0 & \dots & 0 & 0 \end{bmatrix} = E^{-1} \tag{4.6}$$

then eqn. (4.5) can be rewritten as

$$\begin{aligned} G(s)^{-1} &= J(s)^{-1}EEM(s)^{-1}EEH(s)^{-1} \\ &= J(s)^{-1}EQ(s)EH(s)^{-1} \end{aligned} \tag{4.7}$$

The matrix

$$\begin{aligned} Q(s) &= EM(s)^{-1}E \\ &= \operatorname{diag}\left[\frac{\psi_m(s)}{\epsilon_m(s)}, \frac{\psi_{m-1}(s)}{\epsilon_{m-1}(s)}, \dots, \frac{\psi_1(s)}{\epsilon_1(s)}\right] \end{aligned} \tag{4.8}$$

and is the Smith–McMillan form for $G(s)^{-1}$. Therefore applying the pole-zero definitions for a transfer function matrix we see that the pole polynomial for $G(s)^{-1}$ is the zero polynomial for $G(s)$, and the zero polynomial for $G(s)^{-1}$ is the pole polynomial for $G(s)$. In the following we will denote the pole and zero polynomials of $G(s)^{-1}$ by $p_{\hat{G}}(s)$ and $z_{\hat{G}}(s)$, respectively, so that

$$p_{\hat{G}}(s) = z_G(s) \tag{4.9}$$

and

$$z_{\hat{G}}(s) = p_G(s) \tag{4.10}$$

5. Pole-zero relationships

MacFarlane and Postlethwaite (1977 a) have shown that the algebraically defined poles and zeros of a transfer function matrix $G(s)$ are related to the poles and zeros of $g(s)$, defined via complex variable theory, by the relationships

$$p_G(s) = e(s)p_g(s) \tag{5.1}$$

and

$$z_G(s) = e(s)z_g(s) \tag{5.2}$$

where $e(s)$ is a polynomial whose zeros are the poles of $G(s)$ which remain poles of the closed-loop system as k varies, where k is a scalar gain common to all the feedback loops.

If we now combine the results of §§ 3 and 4 with eqns. (5.1) and (5.2), we obtain the following pole-zero relationships.

$$\text{(i)}\quad p_G(s) = z_{\hat{G}}(s) = e(s)p_g(s) = e(s)z_{\hat{g}}(s) \tag{5.3}$$

$$\text{(ii)}\quad z_G(s) = p_{\hat{G}}(s) = e(s)z_g(s) = e(s)p_{\hat{g}}(s) \tag{5.4}$$

6. Inverse characteristic gain loci

The inverse characteristic gain function $\hat{g}(s)$, like $g(s)$, is an algebraic function which has as its domain an appropriate Riemann surface (Springer 1957) formed out of m copies of the complex plane suitably joined together. The theory underlying the Riemann surface of an algebraic function, and a suitable method for its construction, is described in a paper by MacFarlane and Postlethwaite (1977 a). An algebraic function has branch points wherever its defining equation has repeated roots. Therefore, because the eigenvalues of a matrix are the reciprocals of the eigenvalues of the inverse matrix, the branch points of $g(s)$ and $\hat{g}(s)$ are the same. Consequently, the characteristic gain and inverse characteristic gain functions share the same Riemann surface.

Let $\mathfrak{D}$ denote the Nyquist D-contour in $\mathfrak{C}$ (the complex s-plane) as shown in Fig. 2. The indentations in the D-contour, to avoid poles, zeros and branch points, are necessary for the use of the extended Argument Principle (MacFarlane and Postlethwaite 1977 c). Suppose that a copy of $\mathfrak{D}$ is placed on each of the m sheets of the s-plane out of which the frequency surface $\mathfrak{R}$ for $\hat{g}(s)$ is constructed. Let $\hat{\mathfrak{R}}$ be the image set on the gain plane of the set of copies of $\mathfrak{D}$ on the frequency surface ; then the set of complex numbers $\hat{\mathfrak{R}}$ is called the inverse characteristic gain loci set for $G(s)$.

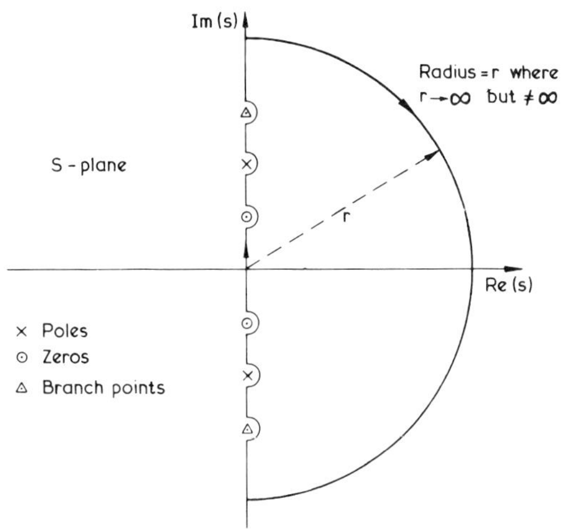

Figure 2. Standard Nyquist D-contour.

In practice the inverse characteristic gain loci are generated as the loci in the complex plane traced out by the reciprocal of each of the eigenvalues of the matrix $G(s)$ as s traverses the Nyquist D-contour in the standard (clockwise) direction. Suppose that we consider a portion of the imaginary axis. We can then compute a set of loci corresponding to the reciprocal of each of the eigenvalues $g_1(j\omega), \ldots, g_m(j\omega)$ in the following way :

(i) Select a value of angular frequency ω_a.
(ii) Compute the complex matrix $G(j\omega_a)$.

(iii) Use a standard computer algorithm to compute the eigenvalues of $G(j\omega_a)$, which are a set of complex numbers denoted by $\{g_i(j\omega_a)\}$.

(iv) Plot the reciprocal of each of the numbers $\{g_i(j\omega_a)\}$ in the complex gain plane.

(v) Repeat with further values of angular frequency ω_b, ω_c, ..., etc., and join the resulting plots up into continuous loci using a sorting routine based on the continuity of the various branches of the inverse characteristic function involved.

7. The generalized inverse Nyquist stability criterion

Let the multivariable feedback configuration shown in Fig. 1 have no open-loop uncontrollable and or unobservable modes whose characteristic frequencies lie in the right-half plane. Then the stability criterion can be stated in two ways.

(1) The configuration will be closed-loop stable if and only if the net sum of anticlockwise encirclements of the critical point $(-1+j0)$ by the set of inverse characteristic gain loci of $G(s)$, minus the net sum of anticlockwise encirclements of the origin by the set of inverse characteristic gain loci of $G(s)$, is equal to the number of right-half plane poles of $G(s)$.

(2) Alternatively, the configuration will be closed-loop stable if and only if the net sum of anticlockwise encirclements of the critical point $(-1+j0)$ by the set of inverse characteristic gain loci of $G(s)$ is equal to the number of right-half plane zeros of $G(s)$.

For proper systems the inverse characteristic gain loci approach infinity as s approaches infinity and therefore $s=\infty$ is a pole of the inverse function. (Essentially the D-contour excludes the point at infinity from the right-half plane, and this is the same as excluding a singularity from the imaginary axis when viewed on a Riemann sphere.) Therefore, in practice, it is necessary to traverse the whole of the D-contour, rather than just the imaginary axis, in order to obtain closed curves. In this way the net sum of encirclements of the critical point and the origin by the inverse characteristic gain loci can be obtained. As an example, the inverse characteristic gain loci for

$$G(s)=\frac{1}{(s+1)^3} \tag{7.1}$$

are shown in Fig. 3.

The traversal of the D-contour off the imaginary axis, however, is not necessary if we are using statement (1) of the stability criterion. This is because the extra loci, corresponding to the circular part of the D-contour, encircle the origin and the critical point the same number of times, and thereby cancel each other. A useful rule, when using statement (1) of the stability criterion, is therefore to join up the corresponding loose ends of the loci via a large semicircle, and to forget any extra encirclements that may actually exist. Whether the ends are joined via the right-half plane or left-half plane is fixed by the fact that the mapping under $\hat{g}(s)$ from the $\mathfrak{D}$-contours to the $\hat{g}$-plane is conformal. This means that if we imagine two people, X and Y, X walking

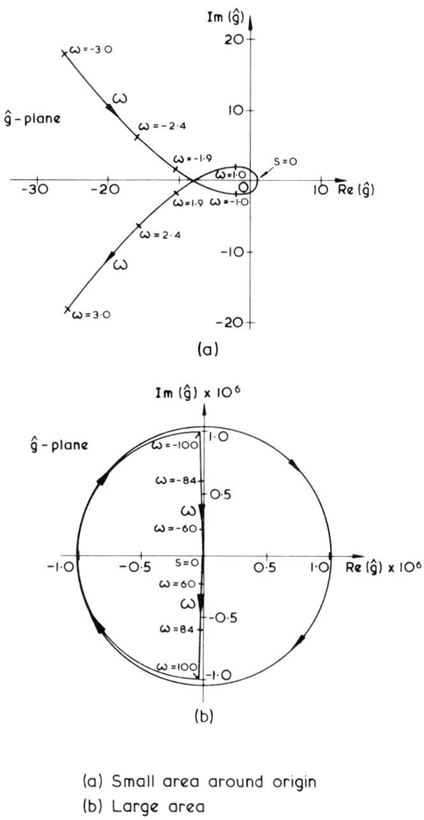

Figure 3. Inverse characteristic gain loci for $G(s)=1/(s+1)^3$. (*a*) Small area around origin ; (*b*) large area.

along the domain curve, and Y walking along the corresponding image curve, such that at each step X defines a point on the domain curve and Y defines the corresponding image point, then if X turns to his right, Y will turn to *his* right.

8. Proof of the generalized inverse Nyquist stability criterion

For the break point shown in Fig. 1 (*b*), the corresponding return difference matrix (MacFarlane 1970 a) is

$$F(s)=I_m+G(s) \tag{8.1}$$

Let

OLCP(s) = open-loop characteristic polynomial of the complete system with all loops broken at the point α, as shown in Fig. 1 (*b*).

CLCP(s) = closed-loop characteristic polynomial of the complete system with all feedback loops closed as shown in Fig. 1 (*a*).

Then as shown by MacFarlane and Postlethwaite (1977 a) we have the fundamental equation

$$\frac{\det F(s)}{\det F(\infty)}=\frac{\text{CLCP}(s)}{\text{OLCP}(s)} \tag{8.2}$$

relating the open-loop and closed-loop system behaviour.

The closed-loop transfer function matrix $R(s)$ is given by

$$R(s)=G(s)[I_m+G(s)]^{-1} \tag{8.3}$$

and therefore

$$R(s)^{-1}G(s)=I_m+G(s)=F(s) \tag{8.4}$$

the return difference matrix. From eqn. (8.2) and (8.4) we therefore have that

$$\det F(s)=\frac{\det [R(s)^{-1}]}{\det [G(s)^{-1}]}=\frac{\text{CLCP}(s)}{\text{OLCP}(s)} \tag{8.5}$$

Note that for a proper system, $G(\infty)=0$, and hence $\det F(\infty)=1$.

If we post-multiply eqn. (8.4) by $G(s)^{-1}$ we have that

$$R(s)^{-1}=G(s)^{-1}+I_m \tag{8.6}$$

and by the eigenvalue shift theorem (MacFarlane 1970 b)

$$\hat{r}(s)=\hat{g}(s)+1 \tag{8.7}$$

where $\hat{r}(s)$ is the algebraic function defining the eigenvalues of $R(s)^{-1}$. We will denote the pole and zero polynomials of $\hat{r}(s)$ by $p_{\hat{r}}(s)$ and $z_{\hat{r}}(s)$, respectively. From the pole-zero definitions for an algebraic function it is obvious that

$$\det [R(s)^{-1}]=\frac{z_{\hat{r}}(s)}{p_{\hat{r}}(s)} \tag{8.8}$$

and

$$\det [G(s)^{-1}]=\frac{z_{\hat{g}}(s)}{p_{\hat{g}}(s)} \tag{8.9}$$

and therefore

$$\frac{\det [R(s)^{-1}]}{\det [G(s)^{-1}]}=\frac{z_{\hat{r}}(s)p_{\hat{g}}(s)}{p_{\hat{r}}(s)z_{\hat{g}}(s)} \tag{8.10}$$

However, from eqn. (8.7) it is clear that

$$p_{\hat{r}}(s)=p_{\hat{g}}(s) \tag{8.11}$$

and hence

$$\frac{\det [R(s)^{-1}]}{\det [G(s)^{-1}]}=\frac{z_{\hat{r}}(s)}{z_{\hat{g}}(s)}=\frac{\text{CLCP}(s)}{\text{OLCP}(s)} \tag{8.12}$$

MacFarlane and Postlethwaite (1977 a) have shown that the open-loop characteristic polynomial is given by

$$\text{OLCP}(s)=p_d(s)e(s)p_g(s) \tag{8.13}$$

where $p_d(s)$ has as its zeros the decoupling zeros of the system associated with that set of characteristic frequencies (eigenvalues) of A which correspond to modes of the system which are uncontrollable and/or unobservable. If we relate the poles of $g(s)$ to the zeros of $\hat{g}(s)$, eqn. (3.6), we have from eqn. (8.13) that

$$\mathrm{OLCP}(s) = p_d(s)e(s)z_{\hat{g}}(s) \tag{8.14}$$

and hence from eqn. (8.12) that the closed-loop characteristic polynomial is given by

$$\mathrm{CLCP}(s) = p_d(s)e(s)z_{\hat{r}}(s) \tag{8.15}$$

This implies that the following conditions are necessary and sufficient for closed-loop stability.

(*a*) $p_d(s)=0$ has no roots in the right-half plane ;

(*b*) $e(s)=0$ has no roots in the right-half plane ; and

(*c*) $z_{\hat{r}}(s)=0$ has no roots in the right-half plane.

Suppose now that a Nyquist D-contour, as shown in Fig. 2, is drawn on each of the m copies of $\mathfrak{C}$ before they are pieced together to form the Riemann surface $\mathfrak{R}_{\hat{r}}$ on which $\hat{r}(s)$ is defined. When the surface is formed the set of Nyquist D-contours combine to form a set of closed Jordan contours enclosing right-half plane regions on $\mathfrak{R}_{\hat{r}}$. The extended Principle of the Argument (MacFarlane and Postlethwaite 1977 c) can then be applied to each right-half plane region on $\mathfrak{R}_{\hat{r}}$. Therefore for a particular right-half plane region, not necessarily simply connected but with a boundary made up from Nyquist D-contours, we have that the difference between the number of zeros and poles of the algebraic function $\hat{r}(s)$ in the region, is equal to the number of clockwise encirclements of the origin in $\mathfrak{C}$ by the image of the boundary curves, under $\hat{r}(s)$, for that particular region. If we therefore consider all the right-half plane regions and apply the extended Argument Principle to each, we have that

$$N_{\hat{r}} = Z_{\hat{r}} - P_{\hat{r}} \tag{8.16}$$

(i) $N_{\hat{r}}$ is the net sum of clockwise encirclements of the origin in $\mathfrak{C}$ by the set of image curves, under $\hat{r}(s)$, of the set of curves formed on $\mathfrak{R}_{\hat{r}}$ by piecing together the appropriate set of Nyquist D-contours when forming $\mathfrak{R}_{\hat{r}}$;

(ii) $Z_{\hat{r}}$ is the number of right-half plane zeros of $\hat{r}(s)$; and

(iii) $P_{\hat{r}}$ is the number of right-half plane poles of $\hat{r}(s)$.

If we now consider the inverse characteristic gain function $\hat{g}(s)$ in the same way, we find that

$$N_{\hat{g}} = Z_{\hat{g}} - P_{\hat{g}} \tag{8.17}$$

where :

(i) $N_{\hat{g}}$ is the net sum of clockwise encirclements of the origin in $\mathfrak{C}$ by the inverse characteristic gain loci ;

(ii) $Z_{\hat{g}}$ is the number of right-half plane zeros of $\hat{g}(s)$; and

(iii) $P_{\hat{g}}$ is the number of right-half plane poles of $\hat{g}(s)$.

We can therefore combine eqns. (8.16) and (8.17) to give

$$N_{\hat{r}} - N_{\hat{g}} = Z_{\hat{r}} - P_{\hat{r}} - Z_{\hat{g}} + P_{\hat{g}} \tag{8.18}$$

which, using eqn. (8.11), becomes

$$N_{\hat{r}} - N_{\hat{g}} = Z_{\hat{r}} - Z_{\hat{g}} \tag{8.19}$$

One of the conditions (condition (*c*)) for closed-loop stability is $Z_{\hat{r}}=0$, which is therefore equivalent to

$$N_{\hat{r}}-N_{\hat{g}}=-Z_{\hat{g}} \tag{8.20}$$

or

$$N_{\hat{r}}-N_{\hat{g}}=-P_{g} \tag{8.21}$$

using eqn. (3.6), where P_g is the number of right-half plane poles of $g(s)$. This leads us to consider combining conditions (*b*) and (*c*) for closed-loop stability into the single equivalent condition

$$N_{\hat{r}}-N_{\hat{g}}=-P_{G} \tag{8.22}$$

where P_G is the number of right-half plane poles of $G(s)$.

8.1. *Necessity and sufficiency of condition* (8.22) *for closed-loop stability when* $p_d(s)=0$ *has no roots in the right-half plane*

From eqn. (5.1) we have that

$$P_G=P_g+e \tag{8.23}$$

where e is the number of right-half plane zeros of $e(s)$, and combining this with eqn. (8.19), and using eqn. (3.6), we have that

$$N_{\hat{r}}-N_{\hat{g}}=Z_{\hat{r}}+e-P_G \tag{8.24}$$

To establish the necessity of condition (8.22) suppose that

$$N_{\hat{r}}-N_{\hat{g}}\neq -P_G$$

Then from eqn. (8.24) this implies that

$$Z_{\hat{r}}\neq 0 \quad \text{and/or} \quad e\neq 0$$

Hence we conclude, in view of the known conditions for stability, that condition (8.22) is necessary for closed-loop stability.

For sufficiency, suppose that

$$N_{\hat{r}}-N_{\hat{g}}=-P_G$$

then from eqn. (8.24) we must have that

$$Z_{\hat{r}}=e=0$$

and hence that the system is closed-loop stable.

We have therefore shown that when $p_d(s)$ has no zeros in the right-half plane

$$N_{\hat{r}}-N_{\hat{g}}=-P_G$$

is a necessary and sufficient condition for closed-loop stability.

From eqn. (8.7) we see that the image under $\hat{r}(s)$ of the set of Nyquist D-contours is related to the inverse characteristic gain loci by a unit shift. Consequently $N_{\hat{r}}$ can be determined by counting the number of clockwise encirclements of the inverse characteristic gain loci about the critical point. This completes the proof of the inverse Nyquist stability criterion (statement (1)).

Note that if the system has a scalar feedback gain k introduced, as shown in Fig. 4, then the following relationships hold

$$R(s)^{-1} = kI_m + G(s)^{-1} \tag{8.25}$$

and

$$\hat{r}(s) = k + \hat{g}(s) \tag{8.26}$$

so that in the stability criterion the critical point is taken as $(-k + j0)$.

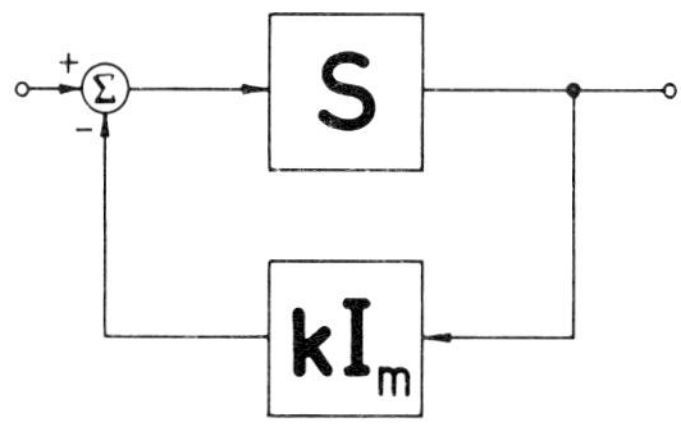

Figure 4. Multivariable feedback configuration with scalar gain k.

Statement (2) of the inverse Nyquist stability criterion is proved as follows. From eqn. (8.16), and the pole-zero relationships of § 5, we are led to consider combining conditions (*b*) and (*c*) for closed-loop stability into the single equivalent condition

$$N_{\hat{r}} = -Z_G \tag{8.27}$$

where Z_G is the number of right-half plane zeros of $G(s)$.

8.2. *Necessity and sufficiency of condition* (8.27) *for closed-loop stability when* $p_d(s) = 0$ *has no roots in the right-half plane*

From eqns. (3.5), (5.2) and (8.11) we have that

$$P_{\hat{r}} = P_{\hat{g}} = Z_g = Z_G - e \tag{8.28}$$

where Z_g is the number of right-half plane zeros of $g(s)$. If we then combine this with eqn. (8.16) we find that

$$N_{\hat{r}} = Z_{\hat{r}} + e - Z_G \tag{8.29}$$

To establish the necessity of condition (8.27) suppose that

$$N_{\hat{r}} \neq -Z_G$$

Then from eqn. (8.29) this implies that

$$Z_{\hat{r}} \neq 0 \quad \text{and/or} \quad e \neq 0$$

and hence we conclude that condition (8.27) is necessary for closed-loop stability

For sufficiency, suppose that

$$N_{\hat{r}} = -Z_G$$

then from eqn. (8.29) we must have that

$$Z_{\hat{r}} = e = 0$$

and hence that the system is closed-loop stable.

We have therefore shown that when $p_d(s)$ has no zeros in the right-half plane

$$N_{\hat{r}} = -Z_G$$

is a necessary and sufficient condition for closed-loop stability, where $N_{\hat{r}}$ is equivalent to the number of clockwise encirclements of the inverse characteristic gain loci about the critical point. This completes the proof of the generalized inverse Nyquist stability criterion (statement (2)).

9. An extension of the criterion to a more complicated feedback configuration

Let us now consider the feedback configuration shown in Fig. 5, where S_1 and S_2 are linear multivariable dynamical systems described by the state-space-model defining triples $\{A_1, B_1, C_1\}$ and $\{A_2, B_2, C_2\}$ with proper transfer function matrices $G_1(s)$ and $G_2(s)$, respectively.

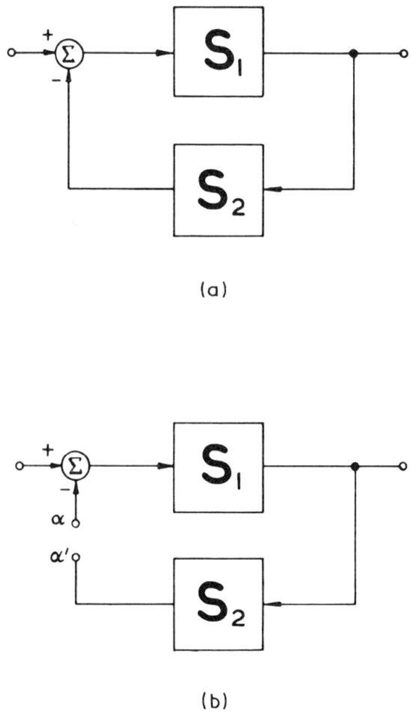

Figure 5. A more complicated feedback configuration. (*a*) Closed-loop; (*b*) open-loop.

If the configuration has no open-loop uncontrollable and/or unobservable modes whose characteristic frequencies lie in the right-half plane, then the generalized inverse Nyquist stability criterion can be stated in the following two ways.

(1) The configuration will be closed-loop stable if and only if the net sum of anticlockwise encirclements of the critical point $(-1+j0)$ by the set of inverse characteristic gain loci of the return-ratio matrix $L(s)[=G_2(s)G_1(s)]$, minus the net sum of anticlockwise encirclements of the origin by the set of inverse characteristic gain loci of $L(s)$, is equal to the total number of right-half plane poles of $G_1(s)$ and $G_2(s)$.

(2) The configuration will be closed-loop stable if and only if the net sum of anticlockwise encirclements of the critical point $(-1+j0)$ by the set of inverse characteristic gain loci of $L(s)$ is equal to the total number of right-half plane zeros of $G_1(s)$ and $G_2(s)$.

The poles and zeros of $G_1(s)$ and $G_2(s)$ are considered, rather than the poles and zeros of $L(s)$, because in forming the product $G_2(s)G_1(s)$ the possibility arises of pole-zero cancellations between the cascade systems ; see Desoer and Chan (1975). The proof of the generalized inverse Nyquist stability criterion, for this more complicated configuration, will therefore follow similar lines to those used in the proof of the generalized Nyquist stability criterion, given by MacFarlane and Postlethwaite (1977 a), and is therefore omitted here.

10. Example

As an illustrative example consider the multivariable feedback configuration of Fig. 4 with

$$G(s)=\frac{1}{1{\cdot}25(s+1)(s+2)}\begin{bmatrix}(s-1) & s \\ -6 & (s-1)\end{bmatrix}$$

The pole and zero polynomials for $G(s)$ are

$$p_G(s)=1{\cdot}25^2(s+1)(s+2)$$

and

$$z_G(s)=1$$

respectively, and therefore we have that

$$P_G=0 \quad \text{and} \quad Z_G=0$$

The inverse characteristic gain loci for $G(s)$ are shown in Fig. 6.

Let $N(\hat{\mathfrak{R}}, x)$ denote the net sum of clockwise encirclements of the point x by the inverse characteristic gain loci $\hat{\mathfrak{R}}$. Then if we take the critical point as $(-k, 0)$, and apply the inverse Nyquist stability criterion (statement (1)), the closed-loop system will be stable if and only if

$$N(\hat{\mathfrak{R}}, -k)-N(\hat{\mathfrak{R}}, 0)=-P_G=0$$

Alternatively, if we apply statement (2) of the stability criterion, the closed-loop system will be stable if and only if

$$N(\hat{\mathfrak{R}}, -k)=-Z_G=0$$

From Fig. 6 we therefore find that the system is closed-loop stable for

$$-1{\cdot}875 < k \leqslant 0$$
$$0 \leqslant k < 1{\cdot}25$$

and

$$2{\cdot}5 < k < \infty$$

Note that the first range of stability corresponds to positive feedback.

To illustrate the use of the inverse characteristic gain loci for handling feedback dynamics we will look at the stabilizing effect of derivative feedback. Consider the feedback configuration shown in Fig. 7, where

$$G(s) = \frac{1}{1{\cdot}25(s+1)(s+2)}\begin{bmatrix} (s-1) & s \\ -6 & (s-2) \end{bmatrix}$$

and

$$H(s) = 0{\cdot}25sI_2$$

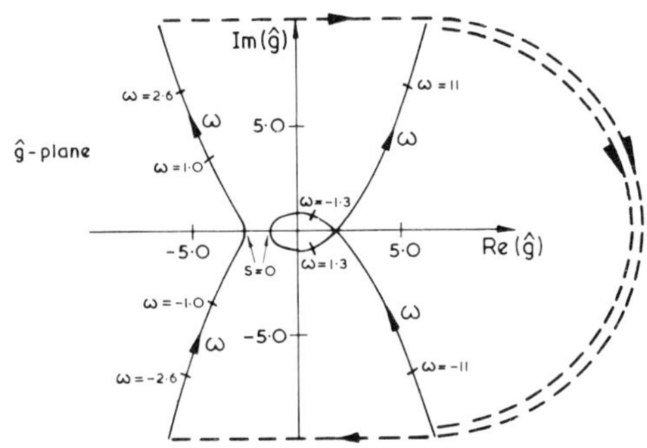

Figure 6. Inverse characteristic gain loci for $G(s)$.

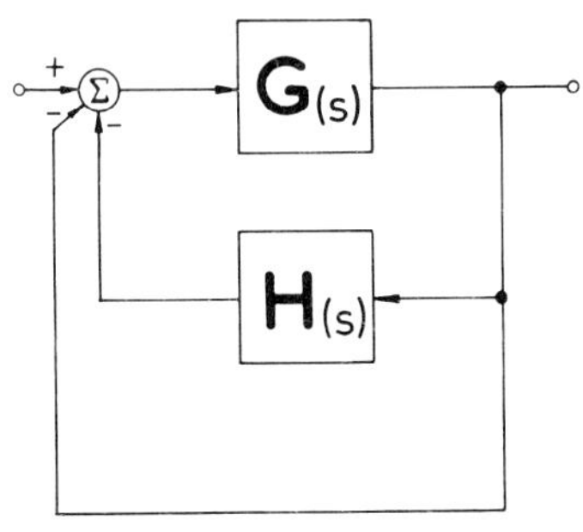

Figure 7. Multivariable feedback configuration with dynamic feedback.

Note that this configuration is equivalent to that of Fig. 5 with

$$G_1(s) = G(s) \quad \text{and} \quad G_2(s) = (1 + 0{\cdot}25s)I_2$$

The pole and zero polynomials for $G_1(s)$ and $G_2(s)$ are

$$p_{G_1}(s) = 1{\cdot}25^2(s+1)(s+2)$$
$$p_{G_2}(s) = 1$$
$$z_{G_1}(s) = 1$$
$$z_{G_2}(s) = (1 + 0{\cdot}25s)^2$$

and the inverse characteristic gain loci for the return-ratio matrix $L(s) = G_2(s)G_1(s)$ are shown in Fig. 8. If we assume a scalar feedback gain k, in each of the feedback loops of Fig. 7, then we can take the critical point as $(-k, 0)$, and apply the inverse Nyquist stability criterion of § 9 to give the following conditions for stability. The closed loop system is stable for

$$-\infty < k \leqslant 0 \qquad \text{(positive feedback)}$$
$$0 \leqslant k < 1.25$$

and

$$2{\cdot}5 < k < \infty$$

Note that the closed-loop system is now stable for all negative values of k.

The transfer function matrix $G(s)$ used in this section was also used by

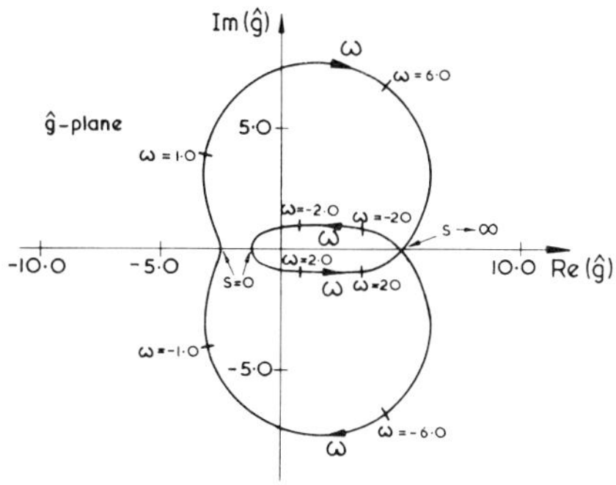

Figure 8. Inverse characteristic gain loci for $L(s)$.

MacFarlane and Postlethwaite (1977 a). In that paper the Riemann surface for $g(s)$ was constructed, and the characteristic frequency loci (root loci) were shown to be the 180° phase contours of $g(s)$. Obviously the Riemann surface for $\hat{g}(s)$ is topologically the same as for $g(s)$, the phase contours and magnitude contours (in decibels) on one surface being the negative of those on the other surface. Note that the 180° phase contours of $\hat{g}(s)$ will also define the characteristic frequency loci which will start at the zeros of $\hat{g}(s)$ and terminate at its poles.

The inverse Nyquist stability criterion has therefore been shown to be a potentially useful tool in the analysis and design of linear multivariable feedback systems. It is also interesting to note that the criterion has recently been used by Mees and Rapp (1977) to establish stability criteria for multiple-loop *non-linear* feedback systems.

ACKNOWLEDGMENTS

I would like to thank Professor A. G. J. MacFarlane, who supervised this work and made many helpful comments and criticisms.

REFERENCES

BLISS, G. A., 1966 (reprint of 1933 original), *Algebraic Functions* (New York : Dover).

DESOER, C. A., and CHAN, W. S., 1975, *J. Franklin Inst.*, **300,** 335.

MACFARLANE, A. G. J., 1970 a, *Proc. Inst. elect. Engrs*, **117,** 2037 ; 1970 b, *Dynamical System Models* (London : Harrap).

MACFARLANE, A. G. J., and POSTLETHWAITE, I., 1977 a, *Int. J. Control*, **25,** 81 ; 1977 b, *Ibid.*, **26,** 265 ; 1977 c, *Ibid.* (in the press).

MEES, A. I., and RAPP, P. E., 1977, *Proc. IFAC Fourth Multivariable Technological Systems Symposium*, Fredericton, Canada (in the press).

POSTLETHWAITE, I., 1977, *Int. J. Control*, **25,** 677.

ROSENBROCK, H. H., 1970, *State Space and Multivariable Theory* (London : Nelson). 1974, *Computer-Aided Control System Design* (London : Academic Press).

SPRINGER, G., 1957, *Introduction to Riemann Surfaces* (Reading, Massachusetts : Addison-Wesley).

Extended Principle of the Argument

A. G. J. MACFARLANE and I. POSTLETHWAITE

An extension of the Principle of the Argument is developed in order to justify its application (MacFarlane and Postlethwaite 1977) to an algebraic function defined on an appropriate Riemann surface.

1. Introduction

In a recent paper (MacFarlane and Postlethwaite 1977) a proof of a generalized Nyquist stability criterion was presented which was based on the use of the Principle of the Argument applied to an algebraic function defined on an appropriate Riemann surface. The required extension of the Principle of the Argument does not seem to be readily available in the literature. Therefore, to justify its use in the previous paper, an appropriately extended Principle of the Argument will be developed here.

The extension required is non-trivial, with two main problems to be overcome. The first of these arises from the fact that, in general, the Riemann surface of an algebraic function will be multiply connected. For an example of this source of difficulty consider a characteristic gain function $g(s)$ associated with a 2×2 rational matrix transfer function $G(s)$. Suppose that $g(s)$ has four branch points, and that each branch point is associated with a cycle of two sheets (Bliss 1966); then the genus number (Bliss 1966, Springer 1957) of the associated Riemann surface is one. Further, suppose that these branch points are disposed in the frequency plane (s-plane) in the way shown in Fig. 1. Then, taking two copies of the complex number sphere (Riemann number sphere), making cuts between the branch points, and forming the topological equivalent of the Riemann surface in the usual way (Springer 1957) one obtains a torus, as illustrated in Figs. 2 and 3. The region Ω, shown shaded on this torus and having a boundary $\partial\Omega$, corresponds to the interiors of the pair of Nyquist

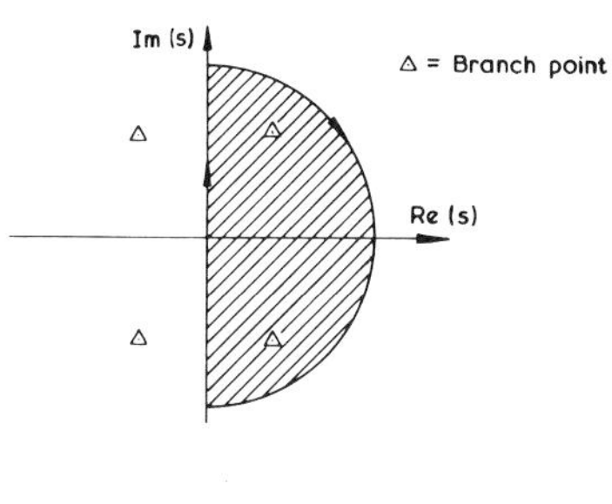

Figure 1. Frequency plane.

D-contours (shown in Fig. 2) for the original complex number spheres out of which the torus was constructed. To cope with such a situation we must ensure that the extended version of the Argument Principle holds for a region of a Riemann surface which is non-simply connected and whose boundary $\partial\Omega$ consists of several distinct closed Jordan contours on the surface. This basic extension of the principle is achieved via a suitable generalization of the Cauchy residue theorem and is covered in § 2.

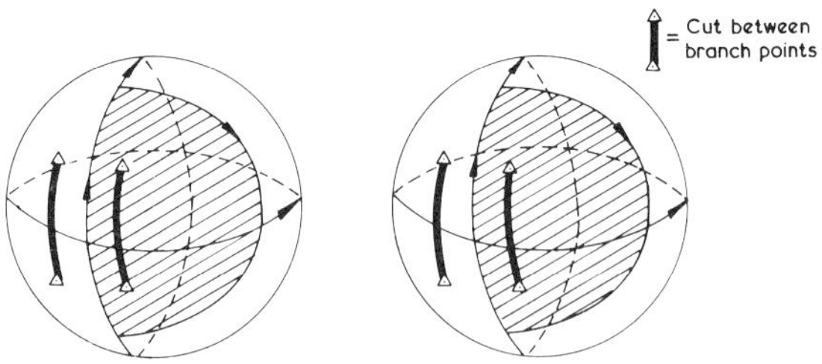

Figure 2. Two copies of the complex number sphere.

The second obstacle to be overcome in a derivation of a suitably extended Argument Principle is more directly associated with the branch points of an algebraic function ; in this case with their effect on the calculation of residues via contour integrals taken round the boundary $\partial\Omega$ of a region Ω having branch points in its interior. This problem is overcome by a change of variable and is treated in § 3.

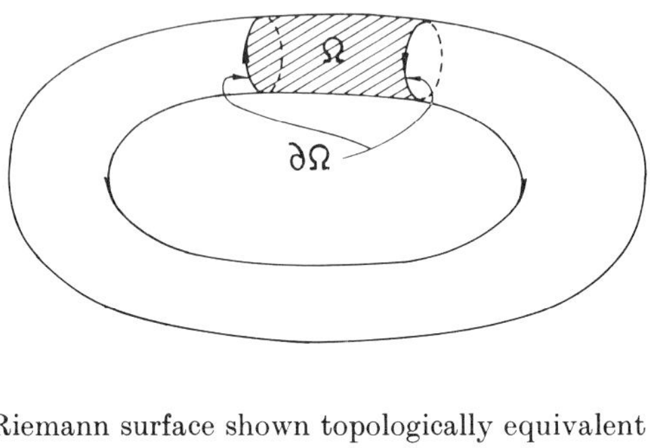

Figure 3. Riemann surface shown topologically equivalent to a torus.

In § 4 the results of §§ 2 and 3 are combined to give the required extended Principle of the Argument. Finally, in § 5, a few remarks are made concerning its application to the proof of the generalized Nyquist stability criterion given by MacFarlane and Postlethwaite (1977).

2. Generalized form of Cauchy's residue theorem

The basic requirement is a generalization of Cauchy's residue theorem to complex functions defined on the Riemann surface of some known algebraic function, where the region of integration may be non-simply connected and have a boundary consisting of one or more closed Jordan contours. Such a generalization may be found in Bliss (1966) ; the main theorem is given below with slight rephrasing and change of notation for the context of this paper. Two observations may be useful in helping the reader unfamiliar with Riemann surface theory to understand this theorem.

(i) Each non-singular place (or point) on the Riemann surface of an algebraic function $f(s)$ is uniquely defined by a pair of values (s, f) which satisfy a known algebraic equation

$$\Delta(s, f)=0$$

Thus at each point on the Riemann surface the value of a complex function $\Psi(s, f)$ of the *pair* of complex variables s and f is defined.

(ii) The Riemann surface for an algebraic function is an orientable surface. Thus, one can unambiguously define a positive sense in which a boundary $\partial\Omega$ can be traversed so as to ' enclose ' some region Ω. The simplest way in which to visualize this procedure is to imagine someone walking round the boundary $\partial\Omega$ on a two-dimensional manifold in which Ω lies (embedded in a familiar space of three dimensions). The positive sense of traversal may then be taken as that in which someone walking *forward* round any portion of $\partial\Omega$ will always have Ω on his *right-hand* side. In Fig. 3 the boundary $\partial\Omega$ is shown with a positive orientation.

2.1. *Generalized residue theorem*

Let $\mathscr{R}$ be the Riemann surface of an algebraic function $f(s)$ and Ω a portion of $\mathscr{R}$ not necessarily simply connected but having a boundary $\partial\Omega$ consisting of one or more closed Jordan contours not passing through any singular places on $\mathscr{R}$. Then, if a function $\Psi(s, f)$ of the places on $\mathscr{R}$ is analytic on Ω and $\partial\Omega$ except possibly for a finite number of singular points in Ω, we have

$$\frac{1}{2\pi j}\int_{\partial\Omega}\Psi(s, f)ds=\Sigma \text{ residues of } \Psi(s, f) \text{ in } \Omega \tag{2.1}$$

where the boundary $\partial\Omega$ is traversed in the positive sense with respect to Ω.

To derive the extended Argument Principle we will first apply the generalized residue theorem to the function $f'(s)/f(s)$, where $f'(s)$ denotes the derivative of $f(s)$ with respect to s, to give

$$\frac{1}{2\pi j}\int_{\partial\Omega}\frac{f'(s)}{f(s)}ds=\Sigma \text{ residues of } \frac{f'(s)}{f(s)} \text{ in } \Omega \tag{2.2}$$

Note that $f(s)$ must have no poles or branch points (either of which are referred to as singular places) on the boundary $\partial\Omega$. For the purposes of deriving an Argument Principle we will now also exclude any zeros of $f(s)$ from $\partial\Omega$. The next step is to calculate the residues of $f'(s)/f(s)$ in Ω.

3. Calculation of residues

The poles, zeros and branch points of $f(s)$ are singularities of $f'(s)/f(s)$, a fact which will become clearer as this section progresses. We will consider all the branch points of $f(s)$ but it turns out that only those which are also poles or zeros of $f(s)$ are relevant.

In the neighbourhood of a pole or a zero (*not* including those which are associated with a branch point) $f(s)$ can be represented by a series which defines a single-valued function. For a pole p_i of order t_{p_i} the algebraic function can be represented by

$$f(s) = \sum_{n=-t_{p_i}}^{\infty} a_n(s-p_i)^n, \quad a_{-t_{p_i}} \neq 0 \tag{3.1}$$

i.e.

$$f(s) = \frac{\phi(s)}{(s-p_i)^{t_{p_i}}} \tag{3.2}$$

where $\phi(p_i) \neq \infty$ and $\phi(s)$ is analytic in the neighbourhood of p_i. This gives

$$\frac{f'(s)}{f(s)} = \frac{-t_{p_i}}{(s-p_i)} + \frac{\phi'(s)}{\phi(s)} \tag{3.3}$$

and since $\phi'(s)/\phi(s)$ is analytic in the neighbourhood of p_i the function $f'(s)/f(s)$ has a simple pole of residue $-t_{p_i}$ at $s = p_i$.

For a zero z_i of order t_{z_i} the algebraic function can be represented by the series

$$f(s) = \sum_{n=t_{z_i}}^{\infty} c_n(s-z_i)^n, \quad c_{t_{z_i}} \neq 0 \tag{3.4}$$

i.e.

$$f(s) = (s-z_i)^{t_{z_i}}\theta(s) \tag{3.5}$$

where $\theta(z_i) \neq 0$ and $\theta(s)$ is analytic in the neighbourhood of z_i. This gives

$$\frac{f'(s)}{f(s)} = \frac{t_{z_i}}{(s-z_i)} + \frac{\theta'(s)}{\theta(s)} \tag{3.6}$$

and we see that $f'(s)/f(s)$ has a simple pole of residue t_{z_i} at $s = z_i$.

To determine the residues at a branch point it is necessary to make a change of variable. (See Bliss 1966, p. 81.) The procedure is as follows.

In the neighbourhood of a finite branch point b_i the algebraic function can be represented by a series of the form (Knopp 1947)

$$f(s) = \sum_{n=-\infty}^{\infty} d_n[{}^r\surd(s-b_i)]^n \tag{3.7}$$

where r is the number of sheets which form a cycle at the branch point. If the branch point is a pole of order $t_{p_i}{}^b$ and consists of a cycle of r sheets we have the series expansion

$$f(s) = \sum_{n=-t_{p_i}{}^b}^{\infty} d_n[{}^r\surd(s-b_i)]^n, \quad d_{-t_{p_i}{}^b} \neq 0 \tag{3.8}$$

Similarly if the branch point is a zero of order $t_{z_i}{}^b$ we have that

$$f(s) = \sum_{n=t_{z_i}{}^b}^{\infty} d_n[{}^r\sqrt{(s-b_i)}]^n, \quad d_{t_{z_i}{}^b} \neq 0 \tag{3.9}$$

otherwise

$$f(s) = \sum_{n=0}^{\infty} d_n[{}^r\sqrt{(s-b_i)}]^n, \quad d_0 \neq 0 \tag{3.10}$$

These expansions are multivalued and therefore not suitable for determining residues. If, however, we make the substitution

$$s = b_i + x^r \tag{3.11}$$

we find that the series expansions (3.8), (3.9) and (3.10) become

$$f(x) = \sum_{n=-t_{p_i}{}^b}^{\infty} d_n x^n, \quad d_{-t_{p_i}{}^b} \neq 0 \tag{3.12}$$

$$f(x) = \sum_{n=t_{z_i}{}^b}^{\infty} d_n x^n, \quad d_{t_{z_i}{}^b} \neq 0 \tag{3.13}$$

and

$$f(x) = \sum_{n=0}^{\infty} d_n x^n, \quad d_0 \neq 0 \tag{3.14}$$

respectively, where it is to be understood that

$$f(x) \equiv f(b_i + x^r) \tag{3.15}$$

The substitution has therefore mapped the algebraic function onto the x-plane where it can be represented by a single-valued series expansion.

By definition (Bliss 1966) the residue at b_i of $f'(s)/f(s)$ analytic in a neighbourhood of b_i except possibly at b_i itself is

$$(\text{residue})_{b_i} \triangleq \frac{1}{2\pi j} \int_C \frac{f'(s)}{f(s)}\, ds \tag{3.16}$$

where C is a closed positively oriented Jordan contour on $\mathscr{R}$ bounding a neighbourhood of b_i in which $f'(s)/f(s)$ is analytic except possibly at b_i. If we substitute for s from (3.11) we find that

$$\begin{aligned}(\text{residue})_{b_i} &\triangleq \frac{1}{2\pi j} \int_C \frac{f'(s)}{f(s)}\, ds \\ &= \frac{1}{2\pi j} \int_{C_x} \frac{f'(x)}{f(x)} \frac{dx}{ds}\, ds \\ &= \frac{1}{2\pi j} \int_{C_x} \frac{f'(x)}{f(x)}\, dx \end{aligned} \tag{3.17}$$

where C_x is a closed positively oriented Jordan contour on the x-plane bounding a neighbourhood of the origin. The residue of $f'(s)/f(s)$ at a branch point b_i is therefore equal to the residue of $f'(x)/f(x)$ at the origin. The function $f'(x)/f(x)$ is single valued and therefore its residue is easily obtained from its series expansion. Following the same procedure as before we note that $f'(x)/f(x)$ in the neighbourhood of the origin has :

(i) a simple pole with residue $-t_{p_i}{}^b$; if the branch point is a pole of order $t_{p_i}{}^b$; or

(ii) a simple pole with residue $t_{z_i}{}^b$; if the branch point is a zero of order $t_{z_i}{}^b$;

otherwise $f'(x)/f(x)$ is analytic.

4. Extended Principle of the Argument

Combining the results of § 3 with that of eqn. (2.2) we have that

$$\frac{1}{2\pi j}\int_{\partial\Omega}\frac{f'(s)}{f(s)}\,ds=[\sum_i t_{z_i}+\sum_i t_{z_i}{}^b]-[\sum_i t_{p_i}+\sum_i t_{p_i}{}^b]$$

$$={}^{\#}Z-{}^{\#}P \tag{4.1}$$

where ${}^{\#}Z$ and ${}^{\#}P$ are, respectively, the total number of poles and zeros of $f(s)$ in Ω enclosed by the boundary $\partial\Omega$ traversed in the positive sense with respect to Ω, and where Ω lies in the Riemann surface of the algebraic function $f(s)$. Note that multiple poles and zeros must be counted as many times as their orders indicate.

The left-hand side of eqn. (4.1) is equivalent to the net sum of the clockwise encirclements of a curve set Γ about the origin of the f-plane where Γ is the image of $\partial\Omega$ under $f(s)$. Denoting this net sum of clockwise encirclements by $N(\Gamma, 0)$, we finally obtain the required extended Principle of the Argument in the form

$$N(\Gamma, 0)={}^{\#}Z-{}^{\#}P \tag{4.2}$$

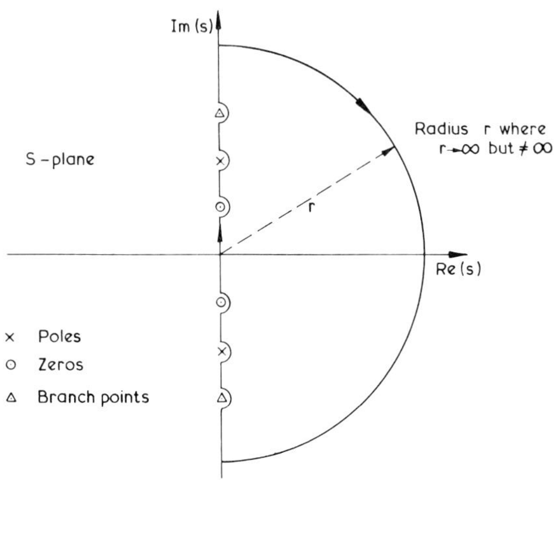

Figure 4. Standard Nyquist D-contour.

5. Remarks on the generalized Nyquist stability criterion (MacFarlane and Postlethwaite 1977)

In the proof of the generalized Nyquist stability criterion given by MacFarlane and Postlethwaite (1977), the closed Jordan contours on the Riemann surface were made up from standard Nyquist D-contours, one on each sheet forming the surface. The D-contour was depicted as having no poles on its boundary but it should also have excluded zeros and branch points from the boundary, as shown here in Fig. 4.

Also it was assumed that on the Riemann surface the right-half plane regions would be simply connected having a single closed Jordan contour boundary made up from one or more D-contours. This is not always true since it is possible to obtain right-half plane regions with multiple boundaries as illustrated in Fig. 3. However, because the multiple boundaries will always be made up from D-contours the results of applying the extended Principle of the Argument will still be the same.

References

Bliss, G. A., 1966, *Algebraic Functions*, reprint of 1933 original (New York : Dover).
Knopp, K., 1947, *Theory of Functions*, Part 2 (New York : Dover).
MacFarlane, A. G. J., and Postlethwaite, I., 1977, *Int. J. Control*, **25**, 81.
Springer, G., 1957, *Introduction to Riemann Surfaces* (Reading, Massachusetts : Addison-Wesley).

Characteristic frequency functions and characteristic gain functions

A. G. J. MACFARLANE and I. POSTLETHWAITE

In the transform analysis of linear systems the two classical approaches use complex functions to study gain as a function of frequency (the Nyquist–Bode approach), and to study frequency as a function of gain (the Evans root locus approach). The central feature of any transform approach to the analysis of linear multivariable systems is the use of transfer-function matrices, and it is the primary purpose of this paper to explain how the ideas of studying gain as a function of frequency, and frequency as a function of gain, can be extended to the multivariable case by associating with transfer-function matrices characteristic gain functions and characteristic frequency functions.

The characteristic gain function and characteristic frequency function are algebraic functions, each of which is defined on an appropriate Riemann surface. Considering the Riemann surface domain for the characteristic frequency function, a stability criterion for linear multivariable feedback systems is established which does not require a net sum of encirclements of the critical point.

1. Introduction

Complex numbers are used for two distinct purposes in the transform-domain analysis of linear systems : to represent frequencies and to represent gains. The two basic classical approaches to the analysis and design of linear single-loop feedback systems use complex functions to study gain as a function of frequency (the Nyquist–Bode approach), and frequency as a function of gain (the Evans root locus approach). The central feature of any transform approach to the analysis of linear multivariable systems is the use of transfer-function matrices, and it has been shown in a previous paper (MacFarlane and Postlethwaite 1977) how such a matrix-valued function of a complex variable can be associated with algebraic function theory (Bliss 1966). If $L(s)$ is a return-ratio matrix for a linear multivariable feedback system (MacFarlane 1970), then it can be associated with a set of characteristic algebraic functions by means of the characteristic equation

$$\Delta(l, s) = \det [lI_m - L(s)] = 0 \qquad (1.1)$$

where I_m is a unit matrix of order m, the order of the square matrix $L(s)$. For simplicity of exposition in what follows, and because this is in any case the usual situation for transfer-function matrices arising from practical situations, it is assumed that $\Delta(l, s)$ is irreducible over the field of rational functions in the complex variable s.

If eqn. (1.1) is considered in terms of $\Delta(l, s)$ being regarded as a polynomial in l with coefficients which are rational functions of s, then it defines an algebraic function $l(s)$. This algebraic function will be called the *characteristic gain function* of the matrix $L(s)$. Alternatively, however, we may equally well consider eqn. (1.1) in terms of $\Delta(l, s)$ being regarded as a polynomial in s with coefficients which are rational functions of l. In this case it can be used to define another algebraic function $s(l)$ which will be called the *characteristic frequency function* of the matrix $L(s)$. Thus, associated with an irreducible characteristic equation like (1.1) are two characteristic algebraic functions: a characteristic gain function and a characteristic frequency function, each of which has as its domain an appropriate Riemann surface (Springer 1957). The Riemann surface associated with the characteristic gain function will be called the s-surface or frequency surface (since the domain of the gain function is comprised of frequencies). The Riemann surface associated with the characteristic frequency function of $L(s)$ will be called the l-surface or gain surface (since the domain of the frequency function is comprised of gains). A method of constructing the frequency surface for a given rational transfer-function has been given by MacFarlane and Postlethwaite (1977) and is briefly reviewed in § 3, and a method of constructing the corresponding gain surface is given in § 5.

The frequency surface (domain of $l(s)$), which should be intuitively thought of as built out of characteristic frequency values, contains the multivariable root loci for $L(s)$. As shown by MacFarlane and Postlethwaite (1977) these are, with the exception of some possible single-point loci, the 180°-phase contours of the characteristic gain function on the appropriate frequency surface. This is illustrated by an example in § 3 where a frequency surface is shown with a collection of constant-phase and constant-amplitude contours of the algebraic function $l(s)$. As a convenient and consistent terminology, we may call the multivariable root loci *characteristic frequency loci* and the characteristic loci used in previous treatments of the generalized Nyquist stability criterion may accordingly be called *characteristic gain loci*. It is shown in § 4 how these characteristic gain loci can be constructed from the calibrated frequency surface. If the gain surface is shown as a collection of constant-phase and constant-amplitude contours of $s(l)$, then the $\pm 90°$ phase contours will define the characteristic gain loci. This is illustrated by example in § 5, and in § 6 it is shown how to obtain the characteristic frequency loci from a calibrated gain surface.

A stability criterion for linear multivariable feedback systems is given in § 7 in a form which does not require a net sum of encirclements of the critical point. This is derived using arguments similar to those used by Saeks (1975) and DeCarlo and Saeks (1977). Finally, in § 8, a discussion is given of the relationship between the constant-real-part and constant-imaginary-part contours of $s(l)$ on a gain surface and the degree of relative stability of a closed-loop system.

2. Characteristic frequencies and characteristic gains

The archetypal feedback situation is when a set of outputs from one part of a system are connected back to a corresponding number of inputs to another part of the system. When such feedback connections are made the system characteristic frequencies are altered. A prime objective of all frequency-response approaches to the linear feedback problem is to find ways of relating closed-loop characteristic frequencies to open-loop gains.

Suppose, as shown in Fig. 1, we have a linear dynamical system S and select m outputs to be connected back to m selected inputs. Let the transfer-function matrix relating the selected set of inputs and outputs be $G(s)$, and suppose that this is a proper matrix of rational functions in s with a state space realization corresponding to the triple of matrices $\{A, B, C\}$. As shown in Fig. 1, let the outputs of S be connected back to the inputs through a gain matrix $g^{-1}I_m$, where g is a complex scalar gain and I_m is a unit matrix of order m. The basic idea behind what follows is that we can look at this situation from two points of view : we can examine the gain g as a function

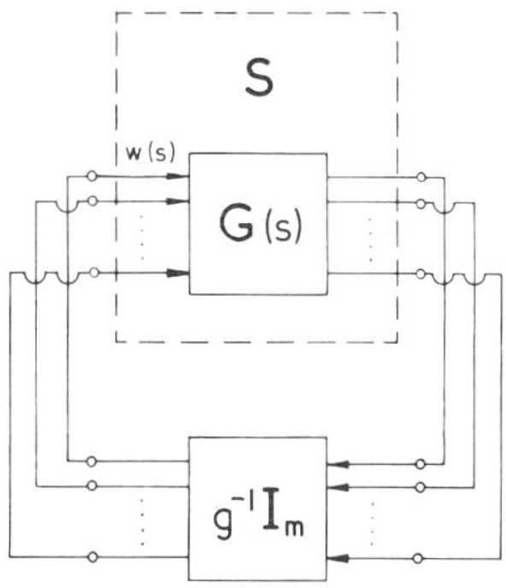

Figure 1. Arrangement for illustration of relationship between characteristic gains and characteristic frequencies.

of frequency s, or we can study the system closed-loop frequencies as a function of the gain g. As shown in Fig. 1, let $w(s)$ be the vector of transform inputs to $G(s)$. Then we must have that

$$G(s)w(s) = g(s)w(s) \tag{2.1}$$

and so, for a fixed value of frequency, $\bar{s}$ say, the corresponding values of gain $\{g_i(\bar{s}) : i = 1, 2, \ldots, m\}$ must be eigenvalues of the complex matrix $G(\bar{s})$. As the frequency variable s takes all possible values, we can calculate the corresponding values of a characteristic gain function $g(s)$; this will obviously be an m-valued function of the complex variable s. This characteristic gain function will be defined via the equation

$$\det [gI_m - G(s)] = 0 \tag{2.2}$$

We can also take a fixed value of gain, $\bar{g}$ say, and determine the corresponding set of closed-loop characteristic frequencies $\{s_i(\bar{g}) : i = 1, 2, \ldots, n\}$, where n is the order of the realization of $G(s)$. Thus we can investigate the properties of a characteristic frequency function $s(g)$. This may be defined, as discussed in § 5, by an equation of the form

$$\det [sI_n - S(g)] \tag{2.3}$$

where

$$S(g) = [A + g^{-1}BC] \tag{2.4}$$

The characteristic frequency function $s(g)$ is obviously an n-valued function of the complex variable g.

In the remainder of this paper we will mostly use a return-ratio matrix $L(s)$ for the purposes of discussion, rather than a plant transfer-function matrix $G(s)$, since return-ratio matrices are more directly related to a statement of stability results.

3. Characteristic frequency loci (root loci) on the *s*-surface

The feedback configuration considered is shown in Fig. 2 (*a*). S_1 has m inputs and S_2 has m outputs and they are both taken to be linear dynamical

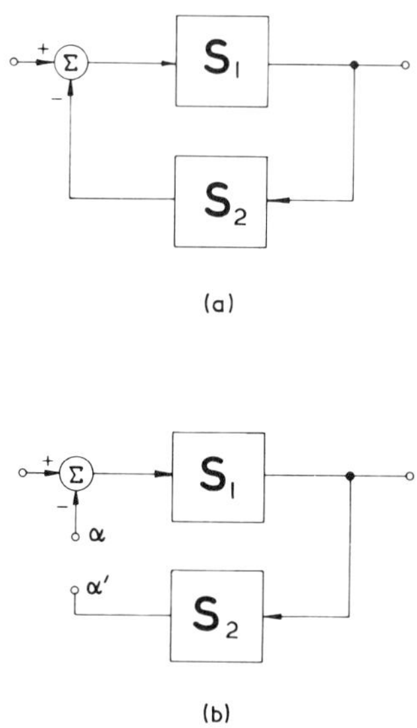

Figure 2. Multivariable feedback configuration. (*a*) Closed loop; (*b*) open loop.

systems having proper transfer-function matrices $G_1(s)$ and $G_2(s)$ with state-space realizations specified by the matrix triples $\{A_1\ B_1, C_1\}$ and $\{A_2, B_2, C_2\}$ respectively. The state-space description of the overall open-loop system is then given by the triple $\{A, B, C\}$, where

$$\left.\begin{aligned} A &= \begin{bmatrix} A_1 & 0 \\ B_2C_1 & A_2 \end{bmatrix} \\ B &= \begin{bmatrix} B_1 \\ 0 \end{bmatrix} \\ C &= [0\ \ C_2] \end{aligned}\right\} \tag{3.1}$$

and

Suppose all the feedback loops are broken at the point shown in Fig. 2 (*b*). Then the corresponding return-difference matrix (MacFarlane 1970) for this break point is

$$F(s)=I_m+L(s) \tag{3.2}$$

where

$$L(s)=G_2(s)G_1(s) \tag{3.3}$$

is the return-ratio matrix for this break point. The characteristic gain functions for $L(s)$ are obtained from the equation

$$\Delta(l,\, s)=\det\,[lI_m-L(s)]=0 \tag{3.4}$$

The stability of the linear multivariable feedback system depends largely on the characteristic function $l(s)$, and a study of a plot of the constant-phase and constant-magnitude contours of this function on the s-surface is very instructive. A method for constructing the s-surface has been given by MacFarlane and Postlethwaite (1977), and is briefly restated here.

The method is based on finding the eigenvalues of the return-ratio matrix $L(s)$ for a grid of values covering the s-plane and then sorting these eigenvalues into a continuous form along certain lines of the grid. If the return-ratio matrix is of order $m\times m$, then m arrays are used to store the sorted eigenvalues. When the calculations and sorting are complete the m arrays represent the m sheets required to build the Riemann surface. The necessary cuts on a sheet are identified by a discontinuity between two lines of eigenvalues, and the 'stitching' together of the sheets becomes obvious by matching eigenvalues on one side of a cut on one sheet to those on another.

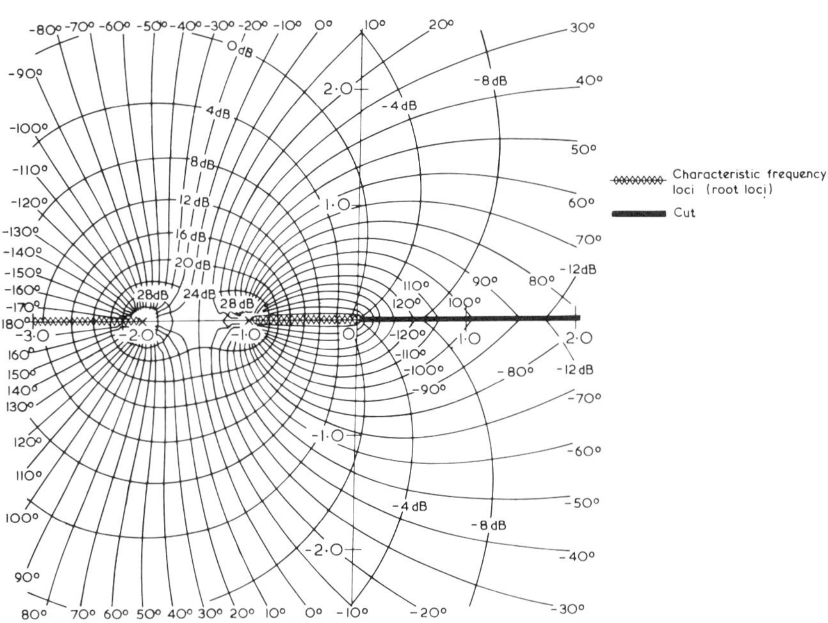

Figure 3. Sheet 1 of the s-surface for the return-ratio matrix of § 3.

As an illustrative example (MacFarlane and Postlethwaite 1977) the s-surface for

$$L(s) = \frac{1}{1\cdot 25(s+1)(s+2)} \begin{bmatrix} s-1 & s \\ -6 & s-2 \end{bmatrix} \tag{3.5}$$

characterized by constant-phase and constant-magnitude contours of $l(s)$ is shown in Figs. 3 and 4.

It has been shown (MacFarlane and Postlethwaite 1977) that for the multivariable feedback configuration of Fig. 5 the root loci are defined by solutions of the following equations :

$$p_x(s) = 0 \tag{3.6}$$

$$p_d(s) = 0 \tag{3.7}$$

$$e(s) = 0 \tag{3.8}$$

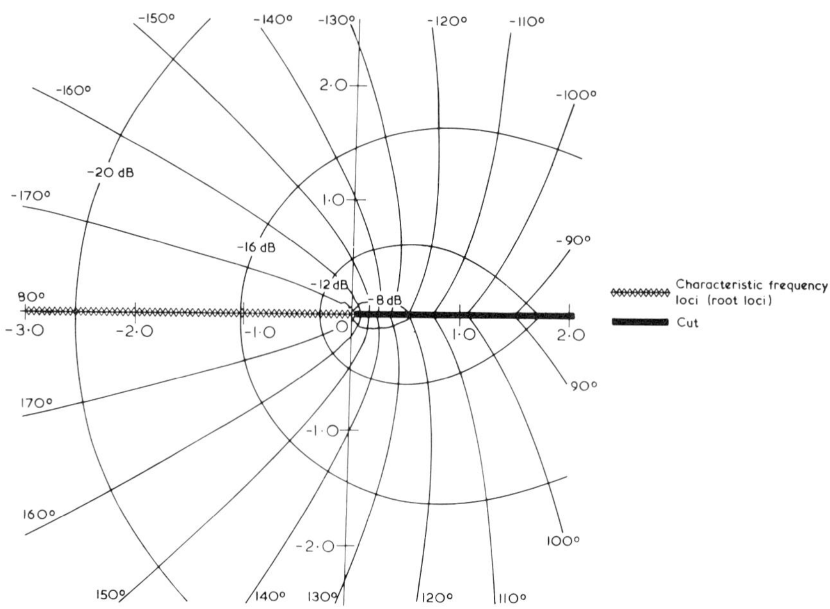

Figure 4. Sheet 2 of the s-surface for the return-ratio matrix of § 3.

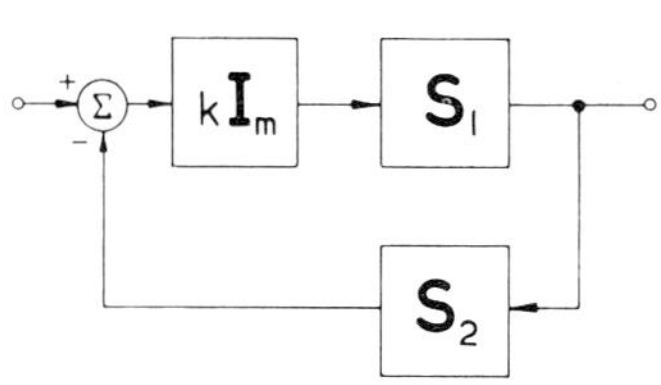

Figure 5. Multivariable feedback configuration with scalar gain k.

and

$$l(s) = -\frac{1}{k} \tag{3.9}$$

for all values of the scalar gain k. Here we have that

(i) $p_x(s)$ has as its zeros those poles of $G_1(s)$ and $G_2(s)$ which are lost when $L(s)$ is formed,

(ii) $p_d(s)$ has as its zeros the decoupling zeros of the overall system associated with the set of eigenvalues of A (A being the overall system matrix) which correspond to modes of the system which are uncontrollable and/or unobservable, and

(iii) $e(s)$ has as its zeros those poles of $L(s)$ which are not poles of the characteristic function $l(s)$.

Equation (3.9) is a direct generalization of the defining equation for a single-loop root locus, while for $e(s) \neq 1$, and/or $p_d(s) \neq 1$, and/or $p_x(s) \neq 1$ there will also be degenerate loci each consisting of a single point. It is clear therefore that with the exception of possible solutions from eqns. (3.6), (3.7) and (3.8) the multivariable root loci are simply the 180°-phase loci of the characteristic gain function $l(s)$ on its associated Riemann surface. This is illustrated in Figs. 3 and 4.

4. Determination of the characteristic gain loci from the *s*-surface

In practice the characteristic gain loci are generated as the loci in the complex plane traced out by the eigenvalues of the matrix $L(s)$ as s traverses the imaginary axis, $s = j\omega$. Therefore from an s-surface plot the characteristic gain loci are determined by the values of the constant-phase and constant-magnitude contours as they cross the $s = j\omega$ axes. Consequently if the contour patterns in the vicinity of the imaginary axes are relatively dense an accurate plot of the characteristic gain loci can be obtained.

5. Characteristic gain loci on the *l*-surface

In constructing the s-surface it is straightforward to substitute for s in the characteristic equation

$$\Delta(l, s) \triangleq \det\,[lI_m - L(s)] = 0 \tag{5.1}$$

and find l since this is a standard eigenvalue problem. To construct the Riemann surface for $s(l)$ using the same procedure as for the s-surface it is desirable to be able to find s easily for a given l; eqn. (5.1) does not allow this. A new relationship between s and l is therefore required. This is obtained using the state-space description of the overall system as follows.

From eqn. (3.9), that is

$$l(s) = -\frac{1}{k} \tag{5.2}$$

we have a relationship which gives the dependence of the closed-loop poles on the scalar feedback gain k. The reason for k being chosen to be scalar is so that the solutions of s trace out the standard root locus plot. In general, k can be considered as complex, and in considering eqn. (5.2) we will take this to be the case.

From the state-space point of view the closed-loop poles are given by

$$\det [sI_n - A + kBC] = 0 \tag{5.3}$$

where A, B and C are state-space matrices for the overall system. At a closed-loop pole eqn. (5.2) holds and therefore we can substitute for k in eqn. (5.3) to give

$$\det \left[sI_n - A - \frac{1}{l} BC \right] = 0 \tag{5.4}$$

This is the required relationship, since for a given l the required values of s can now be calculated as the eigenvalues of the matrix $[A + (1/l)BC]$.

Note that eqn. (5.4) also gives solutions, when they exist, for eqns. (3.6), (3.7) and (3.8).

The procedure for calculating the l-surface can now follow the same lines as for the s-surface. For a grid of values covering the l-plane the eigenvalues of the matrix $[A + (1/l)BC]$ are found, sorted along specified lines of the grid, and stored in n arrays, where n is the number of states for the overall system, If $p_x(s) \neq 1$, and/or $e(s) \neq 1$, and/or $p_d(s) \neq 1$, then some arrays will exist having the same value of s stored in every position. The surface made up from the arrays (sheets) with varying values of s is the required l-surface.

The l-surface for the example used in § 3 is shown characterized by constant phase and magnitude contours of $s(l)$ in Figs. 6 and 7. Now the characteristic

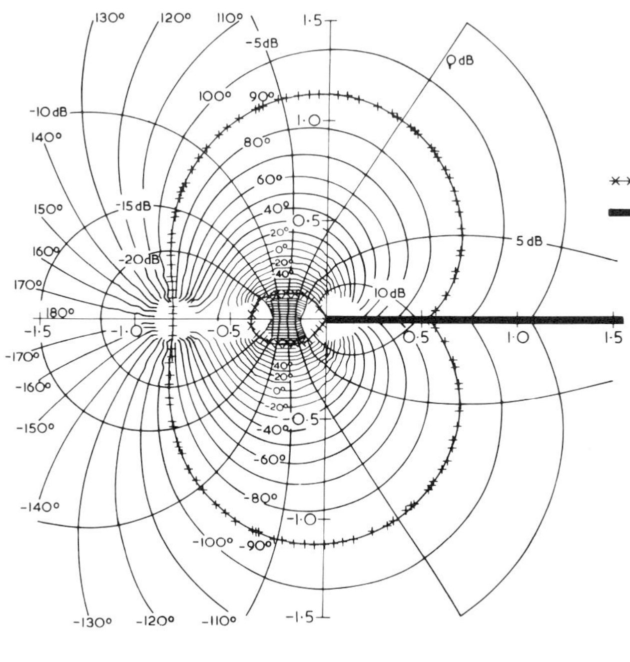

Figure 6. Sheet 1 of the l-surface for the return-ratio matrix of §3.

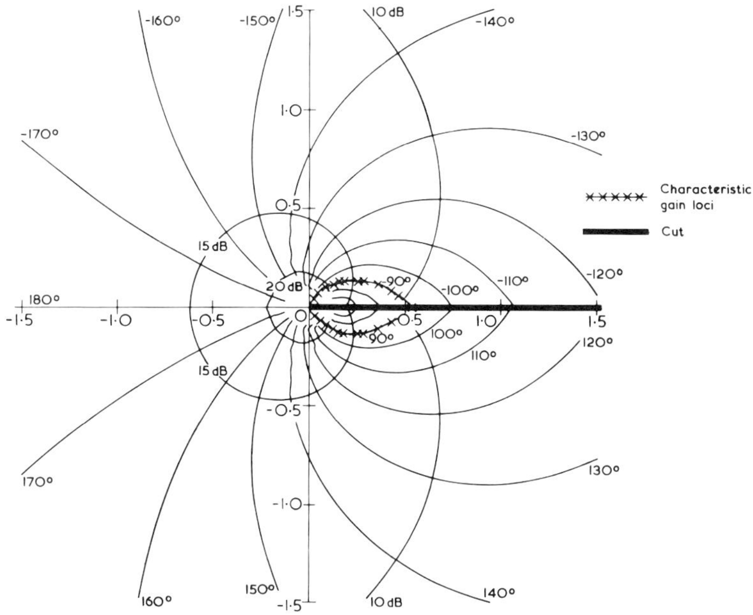

Figure 7. Sheet 2 of the l-surface for the return-ratio matrix of §3.

gain loci of $L(s)$ are the images of the imaginary axis ($s = j\omega$) and therefore on the l-surface the $\pm 90°$ phase contours trace out the characteristic gain loci ; this is illustrated in Figs. 6 and 7.

6. Determination of the characteristic frequency loci (root loci) from the *l*-surface plot

Apart from possible single-point loci, the root loci are simply the values of s at which the characteristic gain function has a phase of 180°. Therefore from the l-surface plot the root loci are determined by the values of the constant-phase and constant-magnitude contours as they cross the negative real axis on each sheet. Consequently if the contour patterns along the negative real axes are relatively dense, then an accurate plot of the root loci can be obtained.

7. Stability inference using a calibrated *l*-surface

Let the system of Fig. 2 (*a*) have no open-loop uncontrollable and/or unobservable modes whose characteristic frequencies lie in the right half-plane. Then by the generalized Nyquist stability criterion (MacFarlane and Postlethwaite 1977) this configuration will be closed-loop stable if and only if the net sum of anti-clockwise encirclements of the critical point $(-1+j0)$ by the set of characteristic gain loci of the return-ratio matrix $L(s)(=G_2(s)G_1(s))$ is equal to the total number of right half-plane poles of $G_1(s)$ and $G_2(s)$.

Work has been done by Saeks (1975), and DeCarlo and Saeks (1977), that suggests that having to count the number of encirclements of the critical point

$(-1+j0)$ is unnecessary. They argue that on a Riemann surface the image of the inside of the standard Nyquist D-contour will be one or more simply connected regions. If on removing the critical points $(-1+j0)$ from the surface these simply-connected regions can each be continuously deformed to a point, then this implies that the characteristic function $l(s)$ is never equal to -1 in the right half-plane. (Note that from eqn. (3.9) $l(s)=-1$ in the right half-plane corresponds to an unstable closed-loop pole.) Using the l-surface construction these arguments are now combined with knowledge of the polynomials $p_x(s)$ and $e(s)$ to give a multivariable feedback stability criterion which does not involve the number of encirclements of the critical point by the characteristic gain loci.

MacFarlane and Postlethwaite (1977) have shown that the following four conditions are necessary and sufficient for closed-loop stability :

(*a*) $e(s)=0$ has no roots in the right half-plane ;
(*b*) $f(s)$ has no zeros in the right half-plane (where $f(s)$ is the algebraic function corresponding to the return-difference matrix $F(s)$), i.e. $l(s) \neq -1$ in the right half-plane ;
(*c*) $p_x(s)=0$ has no roots in the right half-plane ;
(*d*) $p_d(s)=0$ has no roots in the right half-plane.

Each sheet of an l-surface plot characterized by constant-phase and constant-magnitude contours is divided by branches of the characteristic gain loci into regions corresponding to the image of the left-half s-plane and that of the right-half s-plane. Therefore, given such a calibrated surface, one can see at a glance whether $l(s)$ is equal to -1 in the right half-plane without the need for counting encirclements. An equivalent statement of the generalized Nyquist stability criterion is therefore as follows.

Let the system have no open-loop uncontrollable and/or unobservable modes whose characteristic frequencies lie in the right half-plane. Then the system is closed-loop stable if and only if on the l-surface there are no '-1' points in regions corresponding to the right half-plane, and $p_x(s)$ and $e(s)$ have no zeros in the right half-plane.

Note that if the method proposed in § 5 for constructing the l-surface is used, then the zeros of $p_x(s)$, $e(s)$ and $p_d(s)$ are given explicitly as the values of s invariant with l, and therefore covering a whole array.

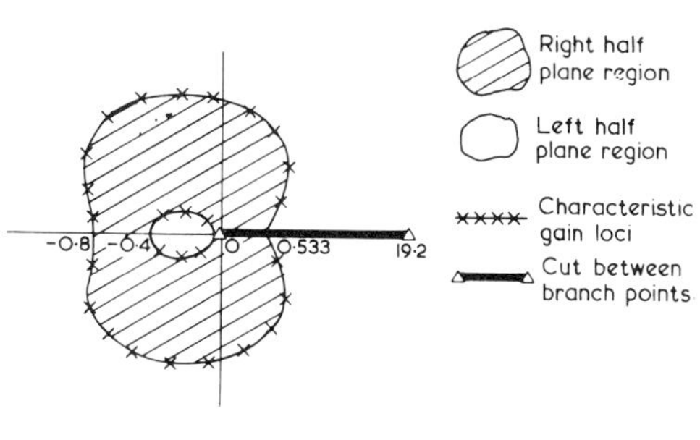

Figure 8. Sketch of Fig. 6 emphasizing right half and left half-plane regions.

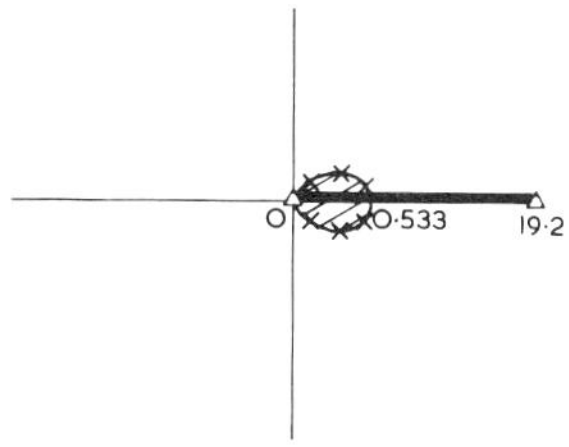

Figure 9. Sketch of Fig. 7 emphasizing right half and left half-plane regions (symbols as in Fig. 8).

To illustrate the stability criterion the example from § 3 is again considered. The l-surface is shown in Figs. 6 and 7, and in Figs. 8 and 9 sketches are made of the sheets to emphasize the right half-plane and left half-plane regions. Introducing a scalar gain k into the forward path, as shown in Fig. 5, gives critical points at $-1/k+j0$. As we increase k positively from zero, the critical points move from $-\infty$ towards the origin on each sheet. On sheet 1 the critical point is in a right half-plane region for $1{\cdot}25<k<2{\cdot}5$ while on sheet 2 for positive k the critical point never moves into a right half-plane region. Therefore, because we also have $p_x(s)=1$ and $e(s)=1$, the closed-loop system is stable for $0 \leqslant k < 1{\cdot}25$ and $2{\cdot}5<k<\infty$.

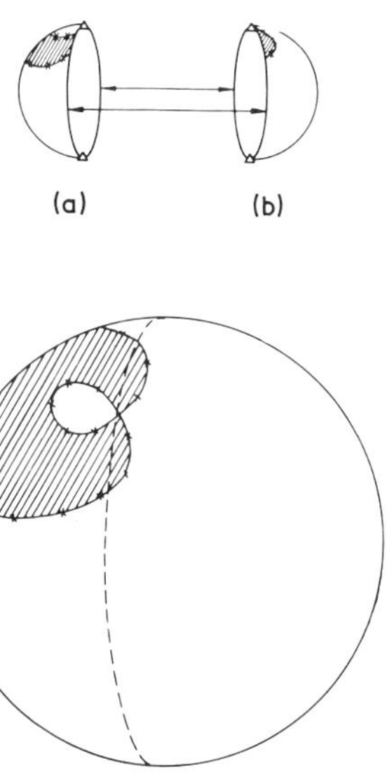

Figure 10. (*a*) Sheet 1 cut along line joining branch points and deformed into a hemisphere; (*b*) sheet 2 cut along line joining branch points and deformed into a hemisphere; (*c*) the complete l-surface topologically represented as a sphere by suitably joining up the hemispheres constructed from sheet 1 and sheet 2.

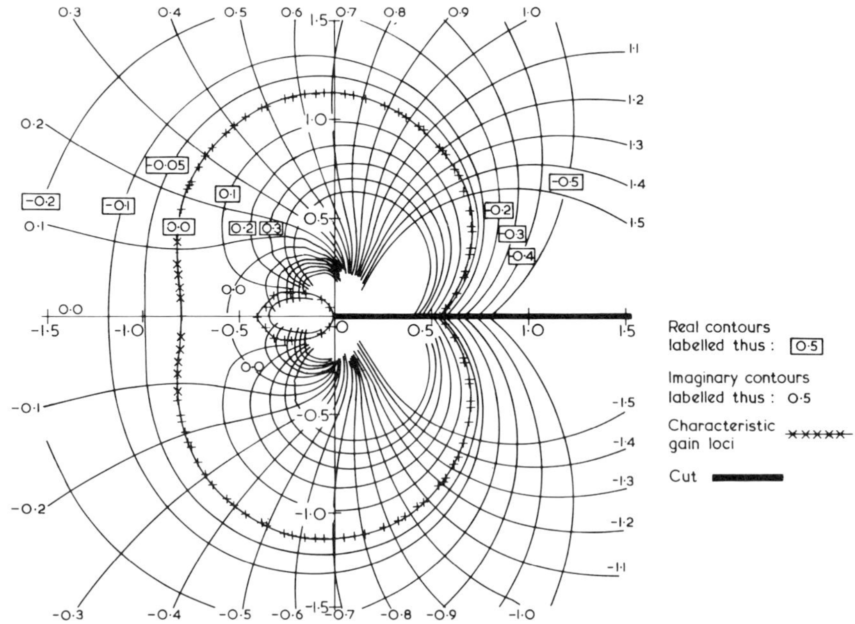

Figure 11 (*a*). Sheet 1 of the *l*-surface characterized by constant real and imaginary contours.

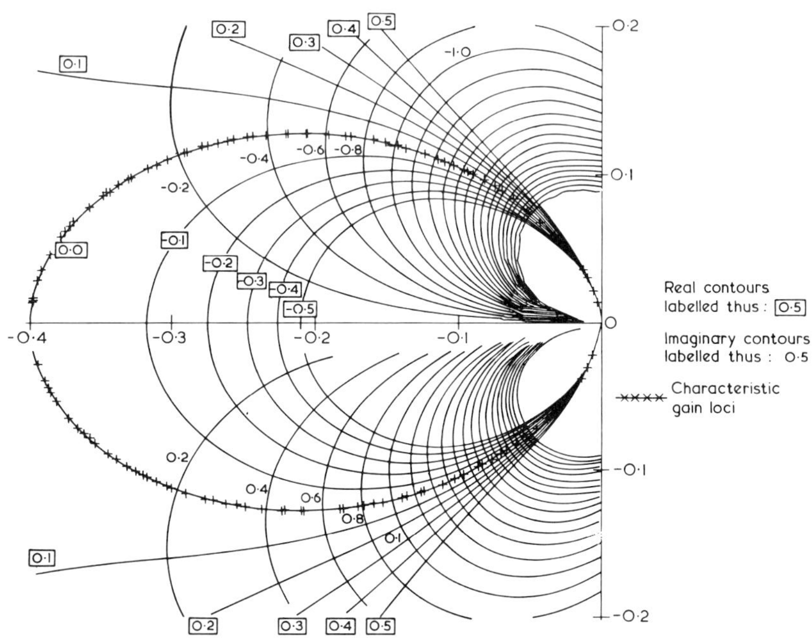

Figure 11 (*b*). Blown-up region close to the origin of Fig. 11 (*a*).

If k is increased negatively from zero the critical points move from ∞ towards the origin on each sheet. On both sheets the critical points are in right half-plane regions for $-\infty < k \leqslant -1{\cdot}875$. Therefore for negative k, which corresponds to positive feedback, the closed-loop system is stable for $-1{\cdot}875 < k \leqslant 0$.

Topologically, Riemann surfaces can be considered as spheres with handles (Springer 1957). The l-surface for this example is topologically equivalent to a simple sphere as shown in Fig. 10. It can be seen that the image of the inside of the Nyquist D-contour, i.e. the right half-plane region, is a simply-connected region on the surfaces as conjectured by DeCarlo and Saeks (1977).

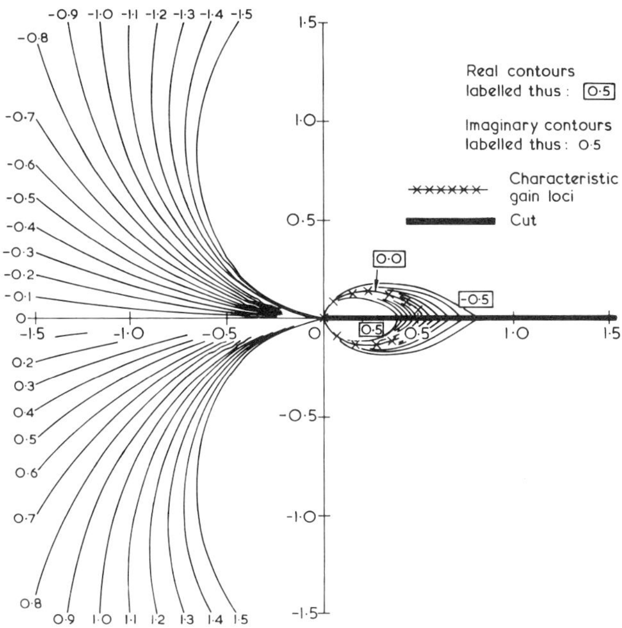

Figure 12. Sheet 2 of the l-surface characterized by constant real and imaginary contours.

If the calibrated gain surface is projected onto the complex gain plane **C**, we have the normal presentation of the characteristic gain loci, plus the superposition of contours representing both right half-plane and left half-plane regions. Stability can now be predicted by considering the right half-plane regions in relation to a single critical point $(-1+j0)$. However, this presentation will in general be difficult to comprehend because of the overlapping of contours. Therefore, although the counting of encirclements, in the generalized Nyquist stability criterion (MacFarlane and Postlethwaite 1977), is not fundamental to system stability, it does afford the simplest method for predicting closed-loop stability in the complex gain plane **C**.

8. Relative stability from an l-surface plot characterized by constant real and constant imaginary contours

From an l-surface plot of constant real and constant imaginary contours of $s(l)$ it is possible to accurately determine the closed-loop poles. This is illustrated by an example using the return-ratio matrix of § 3 to find the dominant closed-loop poles.

Let these dominant closed-loop poles be given by

$$s_c = \alpha \pm j\beta$$

Then α is the smallest (in magnitude) negative real contour that passes through any one of the critical -1 points, and β is the corresponding imaginary contour. The l-surface plot characterized by constant real and imaginary contours is shown in Figs. 11 (*a*), 11 (*b*) and 12 from which

$$\alpha \simeq -0{\cdot}05$$

and

$$\beta = 0$$

so that

$$s_c \simeq -0{\cdot}05$$

By hand calculation the dominant closed-loop pole is

$$s_c = -0{\cdot}0528$$

The l-surface representation therefore seems a useful tool for the engineer in practice, in that it enables him to assess stability at a glance and also to give a measure of the relative stability of the system by picking out the dominant closed-loop pole.

References

Bliss, G. A., 1966 (reprint of 1933 original), *Algebraic Functions* (New York : Dover Publications).
DeCarlo, R., and Saeks, R., 1977, *Int. J. Control*, **26,** 279.
MacFarlane, A. G. J., 1970, *Proc. Inst. elect. Engrs*, **117,** 2037.
MacFarlane, A. G. J., and Postlethwaite, I., 1977, *Int. J. Control*, **25,** 81.
Saeks, R., 1975, *Trans. I.E.E.E. Circuits Syst.*, **22,** 780.
Springer, G., 1957, *Introduction to Riemann Surfaces* (Reading, Massachusetts : Addison-Wesley).

Association between branch points and stationary points on the gain and frequency surfaces

IAN POSTLETHWAITE

In this paper it is shown that a branch point on a gain surface corresponds to a branch point or stationary point on the corresponding frequency surface and vice versa. The topological equivalence between the two surfaces is also established.

1. Introduction

In recent research on complex variable methods for the analysis and design of multivariable feedback systems (MacFarlane and Postlethwaite 1977, MacFarlane *et al.* 1977, Postlethwaite 1978), a main theme has been to associate with a system two sets of algebraic functions : characteristic gain functions and characteristic frequency functions which are defined on Riemann surface domains known as frequency surfaces and gain surfaces respectively. The Riemann surface of an algebraic function is topologically equivalent to a sphere with ' handles ', and the number of handles is known as the *genus number* of the surface (Springer 1957). The genus number may prove to be an important characteristic of a system and it is felt that it may be related to ' decoupling ', that is, the transformation of a multivariable system to a system which is effectively a set of single-input, single-output systems. A single-input, single-output system has a corresponding trivial (one-sheeted) frequency surface of genus zero, so that to decouple a multivariable (m-input, m-output) system would be equivalent to reducing an m-sheeted frequency surface, with a genus number greater than or equal to zero, to a set of m trivial frequency surfaces each of zero genus. Consequently the establishment of relationships between branch point singularities (whose nature and location determine such topological properties as the genus number) and decoupling could be a fruitful line of research. As a first step in this direction an interesting relationship is developed in this paper which shows that a branch point on a gain surface corresponds to a branch point or stationary point on the corresponding frequency surface and vice versa. The topological equivalence between the two surfaces is also established.

2. Association between branch points and stationary points on the gain and frequency surfaces

Let us suppose that from an open-loop gain matrix $G(s)$ we have obtained (via $\det [gI - G(s)] = 0$) an algebraic equation in terms of the complex gain variable g and the complex frequency variable s, given by

$$F \triangleq f(g, s) = 0 \tag{2.1}$$

Let us also consider the following equations involving the partial derivatives of F:

$$\left(\frac{\partial F}{\partial g}\right)_s = 0 \tag{2.2}$$

and

$$\left(\frac{\partial F}{\partial s}\right)_g = 0 \tag{2.3}$$

The pairs of values $[g_i, s_i]$ simultaneously satisfying eqns. (2.1) and (2.2) are the branch points on the frequency surface (denoted by $\Re_s$) and the corresponding multiple roots of g; the pairs of values $[g_i, s_i]$ simultaneously satisfying eqns. (2.1) and (2.3) are the branch points on the gain surface (denoted by $\Re_g$) and the corresponding multiple roots of s.

If we consider g as a function of s, the total derivative of F with respect to s is

$$\frac{dF}{ds} = \left(\frac{\partial F}{\partial s}\right)_g + \left(\frac{\partial F}{\partial g}\right)_s \frac{dg}{ds} \tag{2.4}$$

Alternatively we can consider s as a function of g, and obtain

$$\frac{dF}{dg} = \left(\frac{\partial F}{\partial g}\right)_s + \left(\frac{\partial F}{\partial s}\right)_g \frac{ds}{dg} \tag{2.5}$$

But from eqn. (2.1), F is identically zero and therefore its total derivatives must also be zero, and from eqns. (2.4) and (2.5) this implies that

$$\left(\frac{\partial F}{\partial s}\right)_g = -\left(\frac{\partial F}{\partial g}\right)_s \frac{dg}{ds} \tag{2.6}$$

and

$$\left(\frac{\partial F}{\partial g}\right)_s = -\left(\frac{\partial F}{\partial s}\right)_g \frac{ds}{dg} \tag{2.7}$$

If a pair $[g_i, s_i]$ corresponds to a branch point on the gain surface then $[g_i, s_i]$ satisfies eqn. (2.3) which implies from eqn. (2.6) that either

$$\left(\frac{\partial F}{\partial g}\right)_s = 0 \quad \text{or} \quad \frac{dg}{ds} = 0$$

at this pair. Therefore on the frequency surface the place $[g_i, s_i]$ corresponds to either a branch point or a stationary point.

Similarly if a pair $[g_j, s_j]$ corresponds to a branch point on the frequency surface then $[g_j, s_j]$ satisfies eqn. (2.2) which implies from eqn. (2.7) that either

$$\left(\frac{\partial F}{\partial s}\right)_g = 0 \quad \text{or} \quad \frac{ds}{dg} = 0$$

at this pair. Therefore on the gain surface the place $[g_j, s_j]$ corresponds to either a branch point or a stationary point.

3. Topological equivalence between $\Re_s$ and $\Re_g$

Associated with an algebraic equation

$$f(g, s) = 0 \tag{3.1}$$

is a single abstract Riemann surface, $\mathfrak{R}$, whose places are the pairs of values $[g_i, s_i]$ which are the totality of solutions of (3.1). The gain and frequency surfaces, $\mathfrak{R}_g$ and $\mathfrak{R}_s$, are concrete representations of $\mathfrak{R}$ obtained by mapping from $\mathfrak{R}$ to n complex g-planes or m complex s-planes respectively, where m and n are the maximum exponents of g and s in (3.1). Therefore, since the genus number is a topological property associated with $\mathfrak{R}$, it seems intuitively obvious that $\mathfrak{R}_g$ and $\mathfrak{R}_s$ have this same genus, and hence are topologically equivalent.

The frequency surface $\mathfrak{R}_s$ is the domain of definition for the algebraic function $g(s)$ defined via the equation

$$f(g, s) = a_0(s)g^m + a_1(s)g^{m-1} + \dots + a_m(s) = 0 \tag{3.2}$$

The gain surface $\mathfrak{R}_g$ is the domain of definition for the algebraic function $s(g)$ defined via the equation

$$\phi(s, g) = b_0(g)s^n + b_1(g)s^{n-1} + \dots + b_n(s) = 0 \tag{3.3}$$

But $\phi(s, g)$ is simply $f(g, s)$ rewritten as a polynomial in s with coefficients which are polynomials in g, and therefore one again expects $\mathfrak{R}_s$ and $\mathfrak{R}_g$ to be topologically equivalent. The equivalence is in fact established by the trivial application of a powerful theorem in algebraic function theory (Bliss 1966) which effectively says that if two algebraic curves or equations

$$f(g, s) = 0 \quad \text{and} \quad \phi(\xi, \eta) = 0$$

are equivalent under a birational transformation then the genus number of $\mathfrak{R}_s$ is equal to the genus number of $\mathfrak{R}_\eta$. The equations are equivalent under a birational transformation if there is a one to one correspondence between places on $\mathfrak{R}_s$ and $\mathfrak{R}_\eta$ defined by equations of the form

$$\xi = \xi(g, s), \quad \eta = \eta(g, s)$$

with inverses of the form

$$g = g(\xi, \eta), \quad s = s(\xi, \eta)$$

where the four functions on the right are rational functions in their arguments.

4. Example

As an illustrative example consider the open-loop gain matrix

$$G(s) = \frac{1}{1{\cdot}25(s+1)(s+2)} \begin{bmatrix} s-1 & s \\ -6 & s-2 \end{bmatrix}$$

from which the following algebraic equation, in the complex gain variable g and the complex frequency variable s, is obtained

$$f(g, s) = 1{\cdot}25^2(s+1)(s+2)g^2 - 1{\cdot}25(2s-3)g + 1 = 0$$

This equation defines two algebraic functions $g(s)$ and $s(g)$. The Riemann surface domain for the characteristic gain function $g(s)$, that is the frequency surface $\mathfrak{R}_s$, is shown in figs. 1 and 2 ; the gain surface $\mathfrak{R}_g$ corresponding to the characteristic frequency function $s(g)$ is shown in figs. 3 and 4.

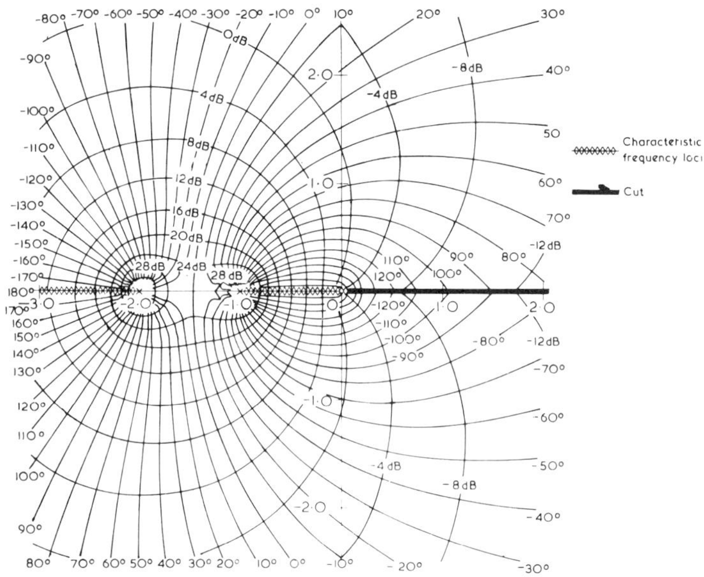

Figure 1. Sheet one of the frequency surface.

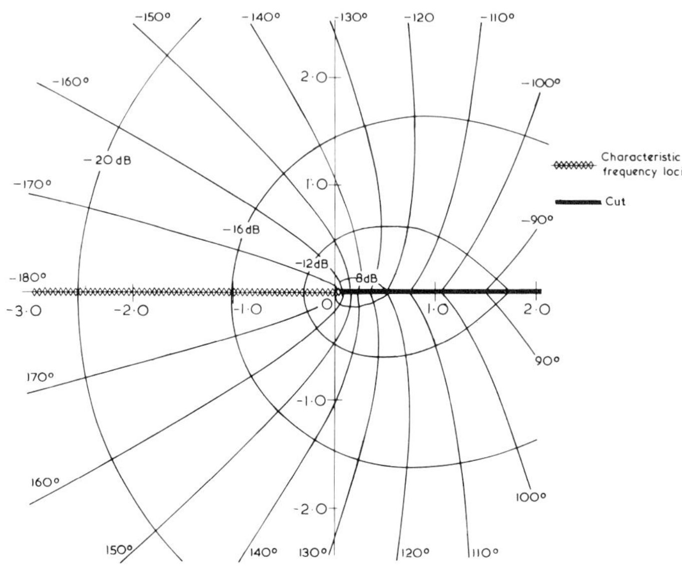

Figure 2. Sheet two of the frequency surface.

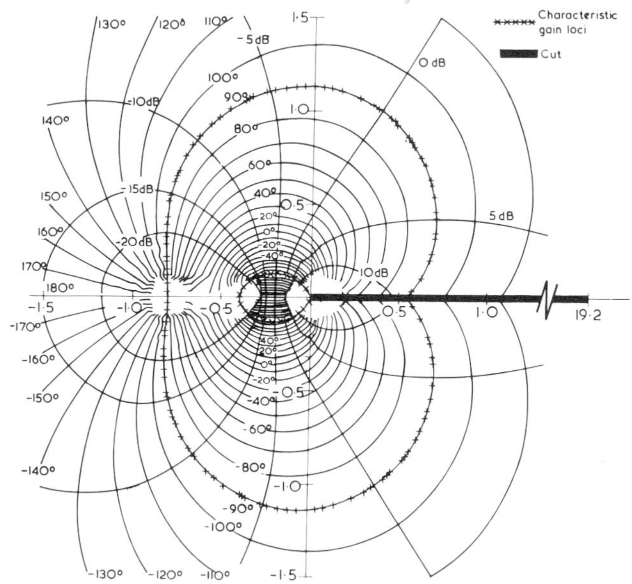

Figure 3. Sheet one of the gain surface.

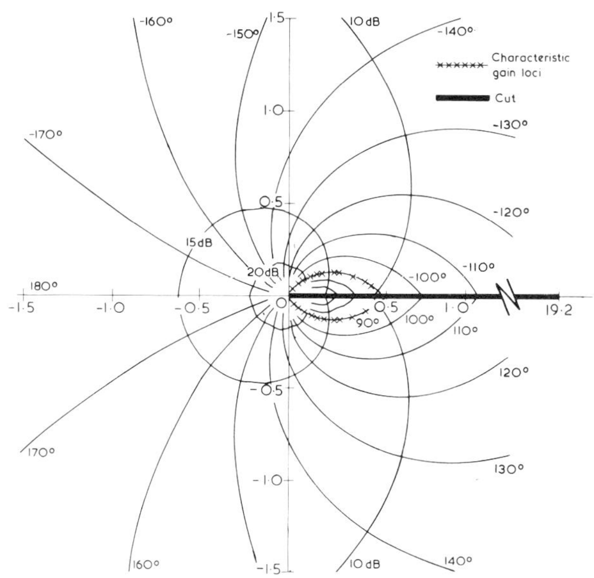

Figure 4. Sheet two of the gain surface.

On the frequency surface there is a branch point at $s \simeq 0{\cdot}042$ $(g \simeq -0{\cdot}549)$ and also at infinity $(g=0)$. (Note that there are always an even number of branch points.) The place $[-0{\cdot}549, 0{\cdot}042]$ corresponds to a stationary point on sheet one of the gain surface ; see Fig. 3. The place $[0, \infty]$ corresponds to a branch point on the gain surface ; see Figs. 3 and 4.

On the gain surface there are two finite branch points at $g=0$ $(s=\infty)$ and and $g=19{\cdot}2$ $(s \simeq -1{\cdot}458)$. The place $[0, \infty]$ corresponds to a branch point on the frequency surface ; see Figs. 1 and 2. The place $[19{\cdot}2, -1{\cdot}458]$ corresponds to a stationary point on sheet one of the frequency surface ; see Fig. 1.

It is clear that both surfaces have a genus of zero since in both instances the sheets can be deformed into hemispheres so that the cut runs along the equator and then the hemispheres can be suitably joined into a sphere (Springer 1957). The genus can also be found numerically using the following expression

$$\text{genus number} = \tfrac{1}{2} \sum (r-1) - z + 1$$

where r is the number of sheets in a cycle at a particular branch point, the sum being taken over all branch points, and z is the number of sheets forming the surface. Therefore the genus number for the frequency surface is

$$\tfrac{1}{2}(1+1) - 2 + 1 = 0$$

and similarly for the gain surface.

Acknowledgment

This work was supported by the S.R.C. and Trinity Hall, Cambridge.

References

Bliss, G. A., 1966 (reprint of 1933 original), *Algebraic Functions* (New York : Dover), p. 137.

MacFarlane, A. G. J., Kouvaritakis, B., and Edmunds, J. M., 1977, International Forum on Alternatives for Multivariable Control, Chicago, U.S.A.

MacFarlane, A. G. J., and Postlethwaite, I., 1977, *Int. J. Control*, **26,** 265.

Postlethwaite, I., 1978, Ph.D. Thesis, University of Cambridge, England.

Springer, G., 1957, *Introduction to Riemann Surfaces* (Reading, Massachusetts : Addison–Wesley).

Characteristic gains, characteristic frequencies and stability

J. M. EDMUNDS

The purpose of this paper is to indicate how characteristic gain loci (or generalized Nyquist diagrams) can be used to count the number of characteristic frequencies (or closed-loop poles) which occur in a given area of the frequency plane for any given feedback gain. The relationship between the gain loci and the characteristic frequency loci is shown to hold for a wide range of linear systems, both sampled and continuous, including those with time delays. The relationship between the frequency loci and the inverse gain loci (or inverse Nyquist diagrams) is also deduced. These results are then used to show how for many systems the frequency loci can be used to determine the only possible ranges of feedback gain to obtain a stable closed-loop system. It is also indicated why in many real systems the closed-loop system should be stable at all but the edges of these possible ranges of feedback gain. The approach used avoids the complex integration which is usually used to justify the Nyquist diagrams (Bode 1945). One example demonstrating the superiority of state-space models over rational polynomial Laplace transfer-function models in the determination of the stability of multivariable systems is given. Some examples of gain loci, inverse gain loci, frequency loci and inverse frequency loci for systems with and without time delays are included.

1. Characteristic gains and characteristic frequencies

The names characteristic frequency loci (instead of root loci) and characteristic gain loci (generalized Nyquist diagrams) have been used as suggested by MacFarlane (MacFarlane and Postlethwaite 1977). These names have often been abbreviated in this paper to frequency loci and gain loci.

The characteristic frequencies of a closed-loop system, of the type shown in Fig. 1, for a given feedback gain $-I/g$ are the frequencies for which the closed-loop system has infinite gain. These characteristic frequencies are the closed-loop poles. The characteristic gains of a system for a given frequency are the eigenvalues of the open-loop system frequency response matrix at that frequency. Hence if one of the characteristic gains is g a feedback of $-I/g$ will ensure that the system has a closed-loop pole at the given frequency. The

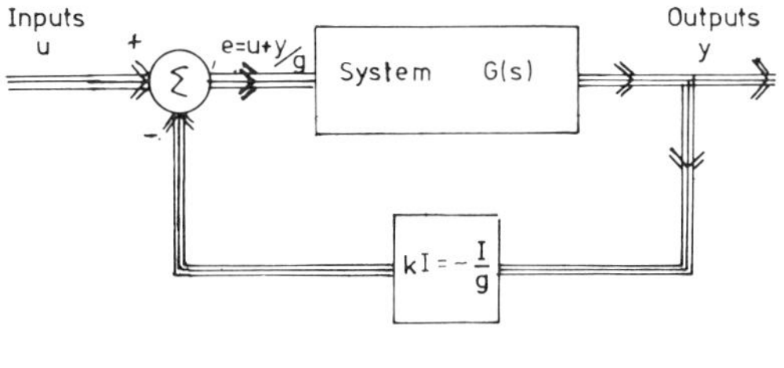

Figure 1. General form of multivariable feedback system.

normal single-input Nyquist diagrams are the plots of the characteristic gains for values of frequency around the Nyquist D-contour, and the root loci are the plots of the characteristic frequencies for gains on the negative real axis. The characteristic gains will be denoted by $g(s)$ since they are functions of frequency, and the characteristic frequencies will be denoted by $s(g)$ since they are functions of gain. These two functions are just different ways of expressing the relationships between the characteristic gains and frequencies, and so both contain the same information about the system. For most linear systems these functions are continuous and differentiable; the one main problem is that they are not usually single-valued. The number of characteristic gains for a given frequency is equal to the number of inputs, and the number of characteristic frequencies is equal to the number of open-loop poles.

The continuity of the functions $g(s)$ and $s(g)$ means that a line in the gain plane gets mapped to a set of lines in the frequency plane, and a line in the

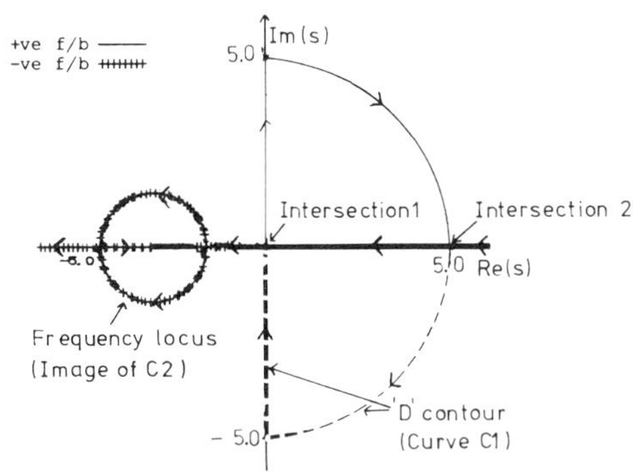

(a)

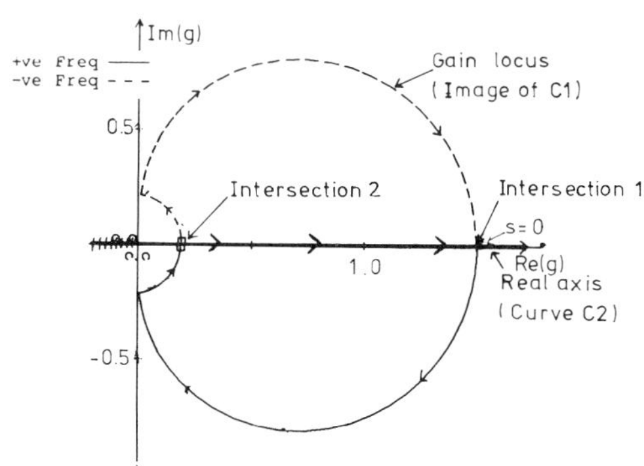

(b)

Figure 2. (a) Root locus and Nyquist D-contour. The system is $(s+3)/(s+1)(s+2)$. (b) Corresponding Nyquist diagram for the system $(s+3)/(s+1)(s+2)$.

frequency plane gets mapped to a set of lines in the gain plane. Intersecting lines in the frequency plane are mapped to intersecting lines in the gain plane. In order to determine the closed-loop stability or instability of the system, we need to know the number of closed-loop poles in the right-half plane, so the curves to consider on the frequency plane are the root loci and the Nyquist D-contour. Calling the Nyquist D-contour curve C1 and the real axis of the gain plane curve C2, the gains for which closed-loop poles cross into the right-half plane are given by the points at which one of the images of C2 intersect curve C1 (Fig. 2 (*a*)). In the gain plane the corresponding points are where one of the images of the curve C1 intersect the curve C2 (Fig. 2 (*b*)). Thus the gains at which closed-loop poles enter or leave the right-half plane can be found by considering the points at which one of the characteristic gain loci crosses the real axis of the gain plane. The positive part of the real axis corresponds to positive feedback. Since the function $g(s)$ is differentiable at almost all points, the gain loci also indicate if the closed-loop pole is entering or leaving the right-half plane for increasing feedback gain. Points on the right of the curve C1 are mapped to points on the right of the images of C1, so if a pole is entering the right-half plane for increasing feedback, points to the right of the gain locus must correspond to larger feedback gains than to the left. Figure 3 shows a more complicated single-input example with three poles going to infinity for large feedback gains. Hence the gain loci can be used to count the number of closed-loop right-half-plane poles for different feedback gains.

For sampled systems the origin-centred unit radius circle on the z-plane is used instead of the Nyquist D-contour. The ' unstable ' section of the z-plane consists of all the points with modulus greater than or equal to one.

The number of closed-loop system poles in the unstable section of the frequency plane for a given feedback gain k (Fig. 1) is equal to :

The number of unstable roots for any given feedback gain k'

+ the number of crossings from the left to the right of the gain loci while going along the real axis of the gain plane from $-1/k'$ to $-1/k$

− the number of crossings from the right to the left of the gain loci while going along the real axis of the gain plane from $-1/k'$ to $-1/k$.

Here the right- and left-hand sides of the gain loci are defined by the directions associated with the loci by going round the boundary of the unstable section of the frequency plane in the direction for which the unstable section is on the right of the boundary. It will be noticed that if, while proceeding from left to right along the axis of the gain plane, a gain locus crosses from below, then an extra pole is becoming unstable. Similarly if one crosses from above, a pole is leaving the unstable section of the frequency plane. When going from right to left, gain loci crossing from above correspond to extra unstable poles and from below correspond to poles leaving the unstable section. For an illustration of the relationship between the axis crossings and the number of closed-loop right-half-plane poles consider the Figs. 3 (*a*) and 3 (*b*).

The layout of the remainder of this paper is as follows. In § 2 the continuity of the root loci is established. In § 3 it is shown how the gain

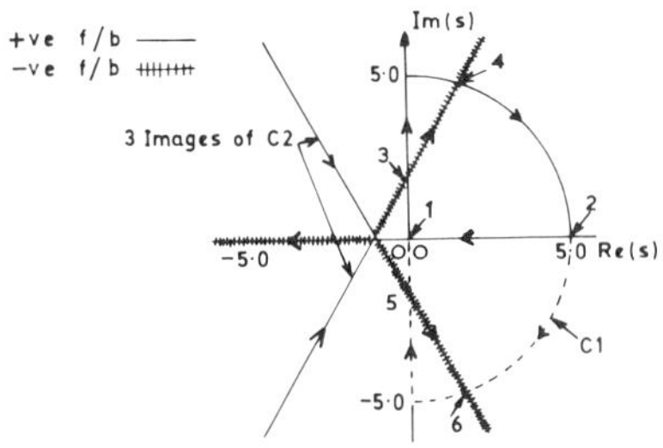

(*a*)

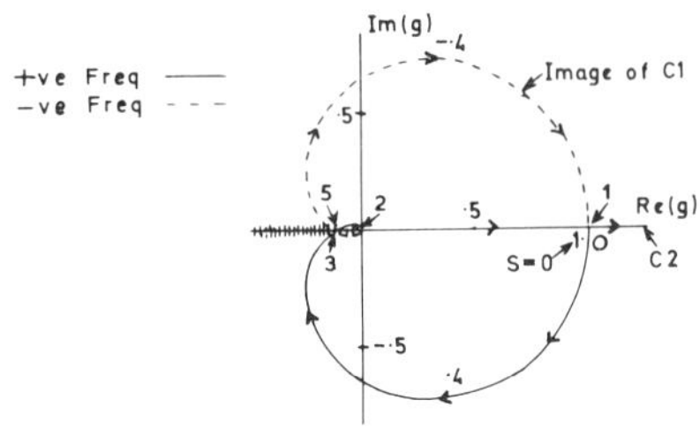

(*b*)

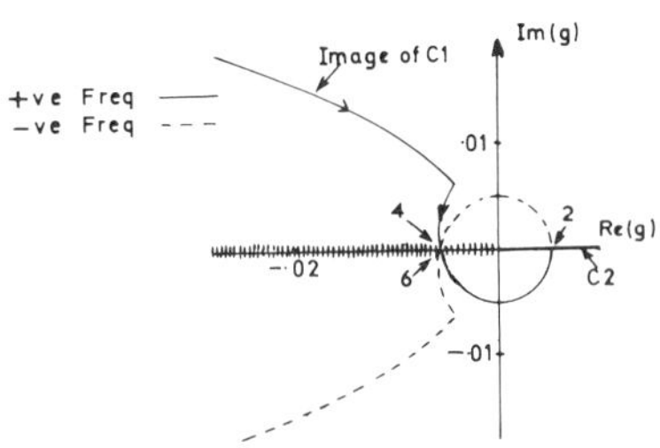

(*c*)

Figure 3. (*a*) Root locus and Nyquist D-contour. The system is $1/(s+1)(s+1)(s+1)$. (*b*) Corresponding Nyquist diagram for the system $1/(s+1)(s+1)(s+1)$. (*c*) Corresponding Nyquist diagram magnified.

loci and the inverse gain loci can be used to count the number of right-half-plane poles for varying feedback gains. In § 4 it is shown that the gain loci indicate the stability or instability of the closed-loop system. In § 5 more details of the contour C1 to be used to define the gain loci are given. Section 6 gives some examples of systems for which the root loci are continuous, and therefore for which the gain loci indicate stability ; this includes systems with time delays. In § 7 it is shown that the criterion can be used for systems which are made up of several sub-systems each of which has one of the forms suggested in § 6. Section 8 gives one way of calculating the frequency loci, and § 9 gives examples of gain and frequency loci. In § 10 it is indicated that it may be best to use state-space models instead of transfer-function models when dealing with multivariable systems, since extra states can be easily introduced when manipulating transfer functions.

2. Characteristic frequency loci

Suppose a system can be described by a transfer-function matrix $G(s)$ (Fig. 1) where each of the elements of $G(s)$ is dependent only on the frequency. Then for a given negative feedback gain $-I/g$ the closed-loop system poles, which are affected by feedback, are at the frequencies for which one of the eigenvalues of $G(s)$ is equal to g. Usually the closed-loop system is stable if none of these movable closed-loop poles, and none of the poles which are totally unaffected by this feedback, are in the right-half plane. When each of the elements of the frequency response is a differentiable function of frequency, it will be found that the closed-loop system poles migrate along continuous loci as the feedback gain is increased. These paths are the characteristic frequency loci, and consist of all the points in the frequency plane for which one of the eigenvalues of $G(s)$ is real.

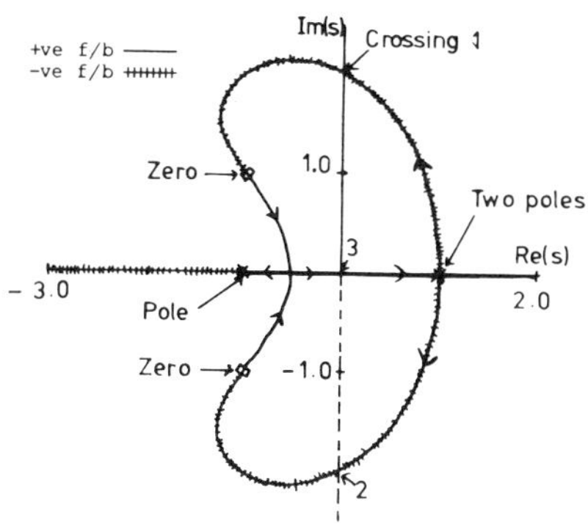

Figure 4. Characteristic frequency locus for Example 1.

Example 1

Figure 4 is an example of a frequency locus for a simple single-input system with poles at -1, 1 and 1, and zeros at $1+j$ and $1-j$. The frequency loci start at the poles for small negative feedback gains, and migrate to the zeros for large negative gains. They then continue from the zeros for large positive feedback gains and return to the poles for small positive feedback gains.

In this paper it shall be assumed that the frequency plane can be divided into two distinct sections, one of which we shall call unstable, and the other we shall call stable. This allows the same discussion to apply to continuous and sampled systems. For continuous systems, these areas are the right- and left-half planes respectively. For sampled-data systems, the areas are the outside and inside of the origin-centred unit circle in the z-plane. For details of the boundaries, and what to do with singularities and other difficult points on these boundaries see § 5. Sufficient conditions shall be applied to the frequency response within the unstable section of the frequency plane to ensure that the frequency loci within the unstable section are continuous and do not split to form extra loci. A particular part of the unstable section of the plane can be ignored if it is known that there cannot be any characteristic gains, in the range of gains being considered, within the area. This usually permits the point at infinity to be omitted from the unstable part of the plane, thus allowing the usual Nyquist contour to be used as the boundary. Once the continuity of the frequency loci has been established, the only way that closed-loop poles can enter or leave the unstable section of the frequency plane is by crossing the boundary of this unstable section. The system shown in Fig. 1 will then have no unstable closed-loop poles for a particular negative feedback gain $-I/g$ if the number of frequency loci leaving the unstable section for feedback values less than $-I/g$ is equal to the number of open-loop unstable poles plus the number of roots entering the unstable section for gains less than $-I/g$.

2.1. *Continuity of the characteristic frequency loci*

The proof (for an alternative proof see Appendix A) of the continuity of the frequency loci for systems with all of the elements of $G(s)$ regular except at a finite set of simple poles has been split into four parts :

(1) It is first noted that at almost all points the eigenvalues of $G(s)$ are differentiable, and hence continuous. At any such point the frequency loci must be continuous.

(2) Secondly it is seen that, given any area of the frequency plane on the boundary of which the eigenvalues are continuous and differentiable functions, the number of frequency loci entering is the same as the number leaving. This ensures that the frequency loci do not split to form extra branches. It also ensures that these loci form continuous lines which do not have ends, assuming that positive and negative feedback gains are considered in the construction of the loci.

(3) Next it is noted that the eigenvalues are continuous functions at all points except the poles of the elements of $G(s)$, and hence the eigenvalues are continuous functions along the frequency loci. This prevents sudden jumps in the feedback gain occurring while proceeding along the frequency loci.

(4) Near the poles of $G(s)$ the eigenvalues can be approximated by an algebraic function, and so the finite eigenvalues must be continuous at such points. Therefore the only possible discontinuities in the frequency loci must occur at the poles of the individual elements of the transfer function matrix $G(s)$ and must correspond to zero feedback gain.

(1) *Differentiability of the characteristic gains*

Suppose that each of the elements of $G(s)$ is regular except at a finite number of isolated singularities (i.e. each element of $G(s)$ is continuous and differentiable except at a finite number of poles). The eigenvalues of $G(s)$ are the values of g for which :

$$\det [gI - G(s)] = 0 \tag{1}$$

Therefore the eigenvalues are the roots of some equation of the form :

$$P_0(s) + P_1(s)g + \ldots + P_{n-1}(s)g^{n-1} + g^n = 0 \tag{2}$$

Here each of the coefficients $P_i(s)$ is a sum of products of the elements of $G(s)$, and hence is regular except at a set of isolated points (i.e. since each of the elements of G is continuous and differentiable, each of the functions $P_i(s)$ is continuous and differentiable). If s is not one of these isolated points, then the eigenvalues for points $(s+ds)$ with ds small, are approximately the roots of the polynomial :

$$P_0(s) + P_1(s)g + \ldots + g^n + ds[P'_0(s) + P'_1(s)g + \ldots + P'_{n-1}(s)g^{n-1}] \tag{3}$$

Here each of the coefficients $P'_i(s)$ are the differentials of the coefficients $P_i(s)$. We want to show that the individual eigenvalues can be considered as continuous differentiable functions of s except at isolated points in the unstable section of the frequency plane, and that they are continuous at all finite values.

Suppose then that g is an eigenvalue of $G(s)$, i.e. a root of eqn. (2). Then substituting $(g+dg)$ for g in polynomial (3) gives a polynomial :

$$\begin{aligned} &P_0(s) + P_1(s)(g+dg) + \ldots + (g+dg)^n \\ &\qquad + ds[P'_0(s) + P'_1(s)(g+dg) + \ldots + P'_{n-1}(s)(g+dg)^{n-1}] \quad (4)\\ &(\text{for small } dg) = P_0(s) + P_1(s)g + \ldots + P_{n-1}(s)g^{n-1} + g^n \\ &\qquad + dg[P_1(s) + 2P_2(s)g + \ldots + (n-1)P_{n-1}(s)g^{n-2} + ng^{n-1}] \\ &\qquad + ds[P'_0(s) + \ldots + P'_{n-1}(s)g^{n-1}] \\ &\quad = a\,dg + b\,ds \quad (5) \end{aligned}$$

Therefore, assuming that neither a or b is zero or infinite an eigenvalue of $G(s+ds)$ is :

$$g - \frac{b\,ds}{a} \tag{6}$$

Hence the eigenvalues are continuous and differentiable functions of s except at a set of isolated points in the unstable section of the frequency plane. So at almost all points within the unstable section of the frequency plane, the frequency loci are continuous and do not split since, if one of the frequency loci goes through a point s where the eigenvalue is a differentiable function of frequency, there is a unique direction in the frequency plane to ensure that the change in the eigenvalue is real. (That is ds has to be chosen so that $b\,ds/a$ is real.)

(2) *Frequency loci entering and leaving areas of the frequency plane*

Consider an area in the frequency plane on the boundary of which all the eigenvalues of $G(s)$ are continuous differentiable functions, with non-zero gradient. The eigenvalues around this boundary can be thought of as several continuous single-valued functions of distance from a given point on the boundary. There may be less of these functions than there are eigenvalues of $G(s)$, since there may be branch points joining some of the sheets of eigenvalues together. However each of these functions will be continuous and differentiable and periodic, with a period which is some multiple of the length of the boundary. The imaginary part of any one of these functions is a continuous differentiable function of the distance round the boundary. Therefore the imaginary part is zero a countable number of times, within the repeating length of the function, and crosses from being positive to being negative the same number of times as it crosses from being negative to positive. But frequency loci leaving the area correspond to the imaginary part changing from negative to positive, if we go round the boundary anti-clockwise ; and frequency loci entering the area correspond to the imaginary part changing from positive to negative. Thus the number of frequency loci entering is the same as the number leaving.

Proof that frequency loci leaving the area correspond to the imaginary part of one of the eigenvalues changing from negative to positive when going round the boundary anti-clockwise.

A frequency locus leaving the area at some point s' implies that, if the real eigenvalue of $G(s)$ is g, there is a line through s' along which one of the eigenvalues of $G(s)$ is real and has values greater than g for values of s outside the area. However the eigenvalues are differentiable functions at s', so angles are preserved, and the picture of the frequency locus near to the point s' must be as shown in Fig. 5. Therefore the imaginary part of the eigenvalue must be positive for frequencies on the boundary reached by going anti-clockwise from s' if these frequencies are sufficiently close to s'. Similarly for frequency loci entering the area, the imaginary part must change from positive to negative.

Therefore we conclude that the frequency loci cannot split to form extra branches at any point in the frequency plane, and also that the frequency loci cannot end since if a branch of the locus ended it would be possible to draw a circle around the end point such that more frequency loci entered the circle than left it.

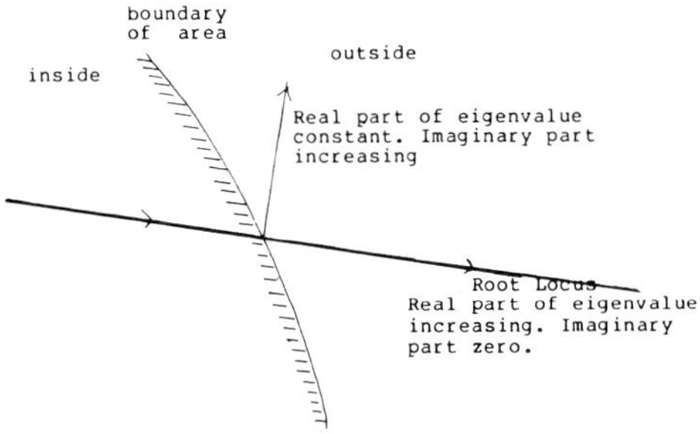

Figure 5. Magnified view of a branch of the root locus leaving an area.

(3) *Continuity of the eigenvalues*

Suppose that all of the elements of $G(s)$ are finite for some frequency s'. Then each of the coefficients of polynomial (2) is finite, and hence all the eigenvalues of $G(s')$ are finite. Also all the elements of $G(s)$ are continuous and differentiable and so, at points near to s', the eigenvalues of $G(s)$ are the roots of the polynomial (3), with the term involving ds small. Therefore the roots of this new polynomial are close to the roots for $s=s'$; hence the eigenvalues are continuous functions of frequency.

(4) *Near to poles of elements of* $G(s)$

Suppose one or more of the elements of $G(s)$ has a pole at $s=s'$. Then each element of $G(s)$ can be approximated close to s' by a Laurent series:

$$G_{i,j}(s)=a_0(s-s')^{-m}+a_1(s-s')^{-(m-1)}+\ldots+a_m \tag{7}$$

Each of the coefficient functions P_i in eqn. (2) can be approximated by a similar series in $(s-s')$.

Let:

$$z=1/(s-s') \tag{8}$$

The eigenvalues are now approximated by an algebraic function in z and g, and so near to s' each of the eigenvalues can be approximated by a series (Knopp 1947):

$$\sum_{n=-m}^{\infty} C_n[p\sqrt{(1/z)}]^n \tag{9}$$

Thus the finite eigenvalues of $G(s)$ near to $s=s'$ are continuous, and the only discontinuities in the frequency loci occur for infinite eigenvalues of $G(s)$, which correspond to no feedback.

Since the frequency loci are continuous and there are no isolated points where one of the eigenvalues is real and finite the number of roots in the unstable section of the plane for a given feedback gain k must be equal to:

The number of unstable roots for any given feedback gain k'

+ the number of roots which enter the unstable section while changing the feedback gain from k' to k

− the number of roots which leave the unstable section while changing the feedback gain from k' to k.

Roots can only enter or leave the unstable section of the frequency plane by crossing the boundary of this section.

3. Gain loci and inverse gain loci

In this section it is shown that, if the frequency loci are continuous, the gain loci and inverse gain loci can be used to indicate the number of closed-loop poles in the unstable part of the frequency plane.

3.1. *Characteristic gain loci*

We shall define the gain loci to be the eigenvalues of $G(s)$ for s going round the boundary of the unstable section of the frequency plane in such a direction that the unstable section is on the right-hand side of the boundary. These are a set of complex functions of the position on the boundary, and can be plotted as a set of curves on a 'gain plane', i.e. a plane on which the two axes represent the real and imaginary parts of the eigenvalues of the system transfer function matrix or the characteristic gains. (For a single-input, single-output continuous-time system the gain loci are just the familiar Nyquist diagrams.) The points where frequency loci enter or leave the unstable area of the frequency plane correspond to places where the gain loci cross the real axis in the gain plane. These gain loci can be used to count the number of roots leaving or entering the unstable section of the frequency plane. A root entering the unstable region of the frequency plane with changing feedback gain corresponds to crossing from the left- to right-hand side of the gain locus while moving along the real axis of the gain plane. This can be seen by the same argument as was used in § 2 to determine whether a branch of a frequency locus was entering or leaving a circle. The gain locus in Fig. 6 corresponds to the frequency locus in Fig. 4.

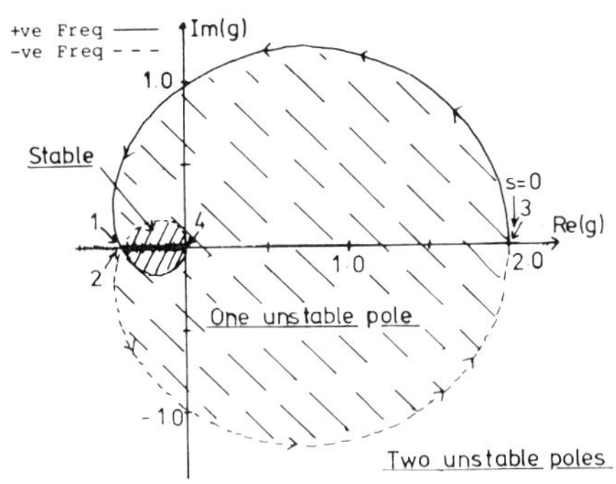

Figure 6. Characteristic gain loci for Example 1, displayed on the gain plane.

The number of closed-loop system poles in the unstable section of the frequency plane for a given feedback gain k is equal to :

The number of unstable roots for any given feedback gain k'

+ the number of crossings from the left to the right of the gain loci while going along the real axis of the gain plane from $-1/k'$ to $-1/k$

− the number of crossings from the right to the left of the gain loci while going along the real axis of the gain plane from $-1/k'$ to $-1/k$.

Here the right- and left-hand sides of the gain loci are defined by the directions associated with the loci by going round the boundary of the unstable section of the frequency plane in the direction for which the unstable section is on the right-hand side of the boundary. It will be noticed that if, while proceeding from left to right along the axis of the gain plane, a gain locus crosses from below, then an extra pole is becoming unstable. Similarly if one crosses from above, a pole is leaving the unstable section of the frequency plane. When going along the real axis in the gain plane from right to left, gain loci crossing from above correspond to extra unstable poles and from below correspond to poles leaving the unstable section. It will also be noticed that if $G(\bar{s})=\overline{G(s)}$ and the unstable section of the frequency plane is symmetrical about the real axis (both of these conditions occur with Laplace and z-transforms), then gain loci corresponding to mapping of the boundary of the unstable section in the upper half of the frequency plane can be used to determine the number of unstable poles. For example with Laplace transfer functions the plots of the characteristic gains along the upper half of the imaginary axis can be used to determine stability. With these half loci the number of crossings while proceeding along the real axis is found and then doubled (Fig. 7). If one of these half gain loci starts on the real axis then going through this point should be counted as half of a crossing.

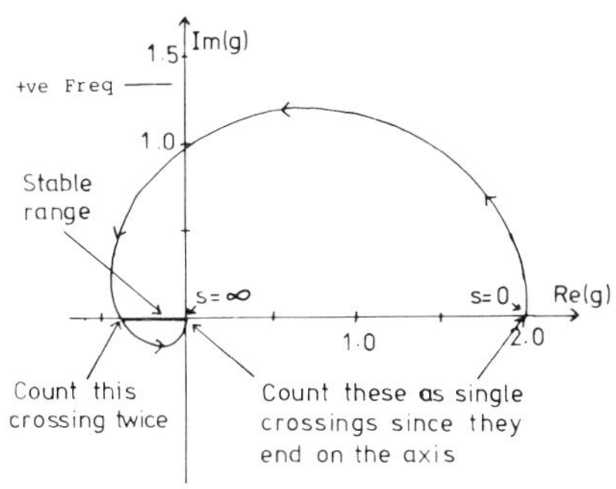

Figure 7. Half characteristic gain loci for Example 1.

Stability at a given feedback gain k requires that there must be no closed-loop unstable roots. (For further details of stability in terms of closed-loop pole positions see § 4.)

There are several special cases of interest :

(1) When the number of open-loop unstable poles is known ; in this case k' can be taken as zero and so axis crossings between $-$infinity and $-1/k$ have to be considered. This corresponds to the usual Nyquist criterion.

(2) When the number of zeros in the unstable part of the frequency plane is known ; in this case crossings of the axis between zero and $-1/k$ can be considered. This corresponds to the modified Nyquist criterion for the cases with a known number of right-half-plane zeros.

(3) When neither the number of poles nor the number of zeros in the unstable part of the frequency plane is known, the gain loci can be used to indicate the only possible ranges of feedback gain to give stability. This is because they can be used to indicate which ranges of feedback gain have the smallest number of unstable poles.

Note that the number of crossings can be counted either way along the real axis if a proper count of crossings at infinity is made.

3.2. *Inverse characteristic gain loci*

These shall be defined as the inverses of the gain loci, and shall be plotted on the inverse gain plane. They are formed by plotting the eigenvalues of the inverse of $G(s)$ while going round the boundary of the unstable section of the frequency plane in the positive direction defined previously. These are the same as the inverses of the eigenvalues of $G(s)$; therefore the places where these inverse loci cross the real axis also correspond to the feedback gains at which closed-loop poles cross the boundary of the unstable section of the frequency plane. Inversion is a differentiable operation, so the right-hand side of the inverse gain loci corresponds to the right-hand side of the gain loci and the right-hand side of the boundary of the unstable section of the frequency plane. Hence the inverse gain loci can be used to count the number of unstable closed-loop poles in a very similar manner to that used in connection with the direct gain loci, the only change being that the counting is now done on the section of the real axis between $-k'$ and $-k$ instead of between $-1/k'$ and $-1/k$ (Fig. 8).

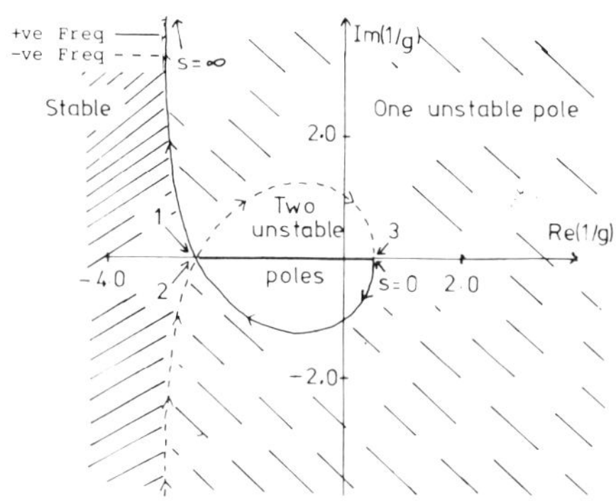

Figure 8. Inverse characteristic gain loci for Example 1.

The number of closed-loop system poles in the unstable section of the frequency plane for a given feedback gain k is equal to:

The number of unstable roots for any given feedback gain k'

+ the number of crossings from the left to the right of the inverse gain loci while going along the real axis of the inverse gain plane from $-k'$ to $-k$

− the number of crossings from the right to the left of the inverse gain loci while going along the real axis of the inverse gain plane from $-k'$ to $-k$.

Here the right- and left-hand sides of the inverse loci are defined by the directions associated with the loci by going round the boundary of the unstable section of the frequency plane in the direction for which the unstable section is on the right of the boundary. Figure 8 shows the inverse gain loci corresponding to Fig. 4.

There are several special cases of interest :

(1) When the number of open-loop unstable poles is known ; here k' can be taken as zero and so axis crossings between zero and $-k$ have to be considered. This corresponds to the usual inverse Nyquist criterion.

(2) When the number of zeros in the unstable part of the frequency plane is known ; here crossings of the axis between $-$infinity and $-k$ have to be considered. This corresponds to the modified inverse Nyquist criterion for the cases with a known number of right-half-plane zeros.

(3) When neither the number of poles nor the number of zeros in the unstable part of the frequency plane is known, the inverse gain loci can be used to indicate the only possible ranges of feedback gain to give stability. This is because they can be used to indicate which ranges of feedback gain have the smallest number of unstable poles.

4. Stability in terms of characteristic frequencies

4.1. *Continuous systems*

It has been shown that it is possible for certain unusual systems to be unstable even if the frequency response is finite in the closed right-half plane (Cook 1973). Hence the gain loci defined by going round the edge of the closed right-half plane do not completely indicate stability or instability for all linear systems. However if the unstable section of the frequency plane is taken as the closed right-half plane then the following observations can be made.

4.1.1. *Instability of linear systems*

If the system has a characteristic frequency in the open right-half plane the system will be unstable, since there is a finite decaying input which produces an infinite output. Similarly if the system has an unobservable or uncontrollable unstable state the system can still be considered unstable since eventually non-linearities in any real system will reveal this state. This unobservable or uncontrollable state could be considered to give an isolated immovable characteristic frequency. Therefore the characteristic frequencies can be used to indicate that certain ranges of feedback gain will give rise to an unstable closed-loop system.

4.1.2. *Finite state systems*

If a system can be described by a finite-state linear model, i.e. a set of finite-state space matrices, and all the closed-loop system poles, including any unobservable or uncontrollable ones, have real parts less than zero, the system is exponentially stable (definition 4.1.5. in Desoer and Vidyasagar 1975). Hence the closed-loop system impulse response decays exponentially. The impulse response of any of the states will also be exponentially stable.

4.1.3. *Proper systems*

Desoer and Vidyasagar have proved (theorem 4.4.14 of Desoer and Vidyasagar 1975; N.B. their theorem and proof is more general than stated here) that if a system is Laplace transformable, is proper, and the closed-loop system impulse response is uniquely defined, then the following holds.

The closed-loop system is L_p stable if and only if

(1) there exists $N(\cdot)$ and $D(\cdot)$ in A^{n*n} such that

$$G(s) = N(s)D^{-1}(s) \tag{10}$$

with $N(s)$ and $D(s)$ obeying certain non-cancellation conditions. A matrix function belongs to A^{n*n} if all its elements are in a convolution algebra A.

(2)

$$\inf_{\mathrm{Re}\, s > 0} \{\det [I + F(s)G(s)]\} > 0 \tag{11}$$

This type of stability is roughly associated with the fact that the absolute value of the closed-loop system impulse response has a finite integral between zero and infinity, and the closed-loop system is proper. The first condition for this type of stability is a very slight restriction on the type of system model that can be used, and the second condition is that there are no characteristic frequencies in the closed right-half plane, or almost in the closed right-half plane for the given feedback. The characteristic gain loci can be used to test this second condition.

Suppose that the gain loci imply that the closed-loop system has no closed right-half-plane poles for the particular feedback gain $-1/k$, and that a small circle can be drawn around the $-1/k$ point in the gain plane such that it does not cut any of the gain loci (Fig. 9). Then if the radius of this circle is r and s is in the closed right-half plane, then any eigenvalue of $G(s)$ must be at least a distance r from $-1/k$. Therefore factorizing polynomial (2) and substituting $-1/k$ for g gives that:

$$\text{Modulus } [\det |I + FG(s)|] > |kr|^n \tag{12}$$

and so the second condition is satisfied.

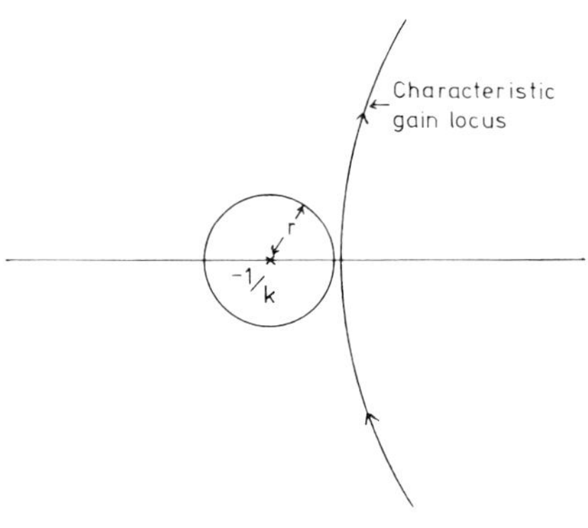

Figure 9. For stability it must be possible to draw a small circle round the critical point without touching the characteristic gain loci.

4.2. *Sampled systems*

For sampled-data systems the unstable region of the frequency plane is all the plane except for the inside of the unit disk ; i.e. a system is unstable if it has a characteristic frequency with magnitude greater than or equal to one, and a linear system is stable if all of its characteristic frequencies have magnitude less than one. The complications considered in the continuous case are not being considered here since only rational polynomial transfer function matrices are going to be considered for the sampled case.

5. Areas of the frequency plane to define characteristic gain loci

5.1. *For stability*

For determining stability, the right-half plane and the outside of the unit disk are considered for the continuous and sampled systems respectively, as described in § 4. Since the right-half plane has not got a finite length boundary the usual Nyquist contour will be used with the semi-circle of a sufficiently large radius to ensure that in the rest of the frequency plane the characteristic gains cannot be in the range that we are considering (Fig. 10).

5.2. *Detours to avoid poles, zeros, branch points etc.*

It was specified in § 2.1 that there should be no poles, zeros, branch points or points with zero gradient on the boundary of the unstable section of the frequency plane, and that if there were any such points small detours should be made to avoid them. These small detours should be into the part of the frequency plane usually counted as stable, in order to ensure that no actual unstable closed-loop characteristic frequencies are allowed by accident. However as these detours become small the characteristic gain loci and inverse gain loci should approach those with no detours, so it should be possible to use the loci defined without the detours to determine the stability of the system.

The only times when this limiting argument does not hold are when there is an open-loop pole on the boundary with the direct gain loci being used, and when there is a zero on the boundary with the inverse gain loci being used. In these cases the decision on how many poles are leaving or entering the unstable section should be made by considering the inverse gain loci and gain loci respectively. If difficulty is found in determining the number of axis crossings in some small region of the gain plane, it should be remembered that between any two frequencies a given characteristic locus can only have a net number of crossings equal to one or zero.

5.3. *To ensure a particular degree of system damping*

The gain loci basically give a way of checking the number of closed-loop system poles in a given area of the frequency plane. Hence for continuous-time systems if the area used to define these loci is the right-half plane, they define the stability of the system. However if, for some reason, other areas of the frequency plane are of interest, and the loci are being constructed from a model of the system rather than by doing frequency response tests, then these areas can be used to define similar loci which indicate the numbers of closed-loop poles enclosed. An example of the definition of such an area

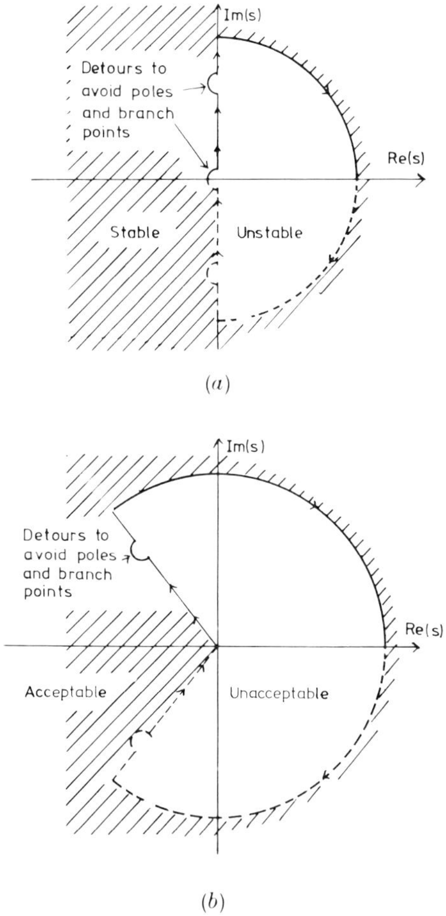

Figure 10. (*a*) Nyquist D-contour on the frequency plane (*s*-plane). (*b*) Contour modified to ensure a particular degree of damping.

would be if it was decided that it was undesirable to have states with less than a given amount of damping ; the boundary for defining the loci would then be two lines through the origin together with an arc of a large circle enclosing most of the right-half plane (Fig. 10). For a damping ratio not less than R the line used should be that with a phase ' theta ' such that cosine (theta) $= R$ (Takahashi, Rabins and Auslander 1970).

6. Some system models

In this section some examples of the type of systems which satisfy the conditions for the characteristic gain loci to indicate stability are given. According to Desoer (Desoer and Vidyasagar 1975), arguments due to L.

Schwartz have shown that most linear time invariant systems fall into the category covered by his proof of stability, and hence the characteristic gain loci can be used with most linear systems arising from practical investigations.

6.1. *Rational polynomial transfer function matrices*

Suppose each element of the Laplace transfer-function matrix is a rational polynomial in s; then each of the elements of $G(s)$ is regular except at the roots of its denominator polynomial. There is only a finite number of such poles, and hence only a finite number of unstable poles. Therefore the characteristic gain locus criterion of § 3 can be used to determine the closed-loop system stability if the number of open-loop poles is known. Similarly gain loci can be used with rational z-transfer-function matrices.

6.2. *State-space matrices*

Suppose that a system is described by:

$$\left.\begin{aligned} \dot{\mathbf{x}} = A\mathbf{x} + B\mathbf{u} \quad \text{or} \quad (sI - A)\mathbf{x} = B\mathbf{u} \\ \text{and} \qquad \mathbf{y} = C\mathbf{x} + D\mathbf{u} \end{aligned}\right\} \tag{13}$$

where

$\mathbf{x}$ is a vector of states;
$\mathbf{u}$ is a vector of inputs;
$\mathbf{y}$ is a vector of outputs;
A, B, C and D are real matrices; and
s is the complex frequency variable.

Then

$$\mathbf{y} = C(sI - A)^{-1}B\mathbf{u} + D\mathbf{u}$$

The open-loop singularities are the values of s for which the determinant of $(sI - A)$ is zero, or the eigenvalues of the matrix A, and so can be counted. The transfer-function matrix is equal to:

$$G(s) = C(sI - A)^{-1}B + D \tag{14}$$

Therefore each of the elements of $G(s)$ can be represented by a rational polynomial in s, and hence is regular except at the singularities, so the gain loci can be used to determine the number of closed-loop unstable poles for any given feedback gain. For the case of sampled state-space matrices, or transition matrices the above discussion holds with s replaced by z, a time advance operator.

It will be noted that the stability which is implied when using this type of model is exponential stability of the output and of the states.

6.3. *Laplace transfer functions with time delays*

A time delay of T units of time can be represented in the continuous frequency domain by an operator of the form:

$$\exp(-Ts) \tag{15}$$

Suppose that a system can be described by a matrix of Laplace transfer functions each of which is a rational polynomial in s together with a time delay associated with it ; i.e. each element has the form :

$$G_{i,j}(s)=\exp(-T_{i,j}s)\,\frac{a_{0,i,j}+a_{1,i,j}s+\ldots+a_{m,i,j}s^{m_{i,j}}}{b_{0,i,j}+b_{1,i,j}s+\ldots+b_{n,i,j}s^{n_{i,j}}} \tag{16}$$

Here the coefficients $a_{i,j}$ and $b_{i,j}$ and the variable $T_{i,j}$ are different for the different elements $G_{i,j}(s)$. The rational polynomial in s is regular except at a finite number of poles, and so can only contribute a finite number of singularities in the right-half plane. The time delay contributes no open-loop unstable poles since the magnitude of this term is always less than unity in the right-half plane. Each element of $G(s)$ is differentiable at all but a finite number of points in the right-half plane since a product of differentiable functions is differentiable. Therefore if the number of open-loop unstable poles can be counted, the gain loci can be used to determine the stability of the closed-loop system. Note that exponential functions are periodic along the imaginary axis, so the gain loci will usually encircle the origin an infinite number of times. For proper systems ($n_j > m_j$ for all i, j in eqn. (16)) this will not matter too much since the rational polynomial will ensure that the gain loci will approach the origin for large values of s, and they will only cross the real axis a finite number of times for real values less than any given $-1/k$. However if the system is improper ($n=m$), the gain loci can cross the real axis an infinite number of times for some non-zero value of feedback gain, and so care must be taken when interpreting the gain loci.

6.4. *State-space models with time delays*

Suppose that a system is described by :

$$\dot{\mathbf{x}}=A\mathbf{x}+B\mathbf{u} \tag{17}$$

and

$$\mathbf{y}=C\mathbf{x}+D\mathbf{u} \tag{18}$$

where

- $\mathbf{x}$ is a vector of states ;
- $\mathbf{u}$ is a vector of inputs ;
- $\mathbf{y}$ is a vector of outputs ;
- A is a real matrix ;
- B is a matrix of operators with each element having the following effect on the corresponding input

$$B_{i,j}\mathbf{u}_j(t)=b_0\mathbf{u}_j(t)+b_1\mathbf{u}_j(t-T_1)+\ldots+b_n\mathbf{u}_j(t-T_n)$$

  Here the T_i are positive time delays, i.e. each element of B is a sum of a finite number of time delay operators ;

- C is a matrix of operators similar in form to B ; and
- D is a matrix of operators similar in form to B.

The transfer function takes the form

$$G(s)=C(s)(sI-A)^{-1}B(s)+D(s) \tag{19}$$

Here each of the elements of $B(s)$, $C(s)$ and $D(s)$ is a regular function of s for all places in the right-half plane. So the elements of $G(s)$ are regular in the right-half plane except for any values of s which are eigenvalues of the matrix A. Hence the number of open-loop unstable poles can be counted, and the gain loci can then be used to determine the number of closed-loop unstable poles.

6.5. *Some distributed systems*

Many sorts of distributed systems have transfer-functions in the form of Laplace transforms which can be described as sums of exponentials divided by other sums of exponentials. If each of the elements of the transfer-function matrix is of this form then each element will be differentiable for almost all frequency values, and so the gain loci can be used to count closed-loop poles passing between the stable and unstable sections of the frequency plane. In certain cases it may be possible to determine the number of right-half-plane poles for some value of feedback gain, and so determine the stability of the closed-loop system for other feedback gains.

7. Combining sub-systems

An actual system may be made up of several sub-systems where each sub-system has one of the forms suggested in § 6. In this section the rules governing the use of gain loci for such combined systems are considered.

7.1. *Combining in parallel*

Suppose that a system is made from two sub-systems in parallel, and each of these sub-systems can be described by transfer-function matrices with elements which are regular at all but a distinct set of points in the unstable section of the frequency plane (Fig. 11). Since transfer functions of systems in parallel just add, the system's transfer-function matrix is just the sum of the transfer-function matrices of the individual elements, i.e.

$$G(s) = G_1(s) + G_2(s) \tag{20}$$

However the sum of two regular functions is regular, and so gain loci can be used to determine the number of characteristic frequencies in the unstable section of the frequency plane. The number of open-loop unstable modes that the system has is the sum of the number of open-loop unstable modes of the two sub-systems.

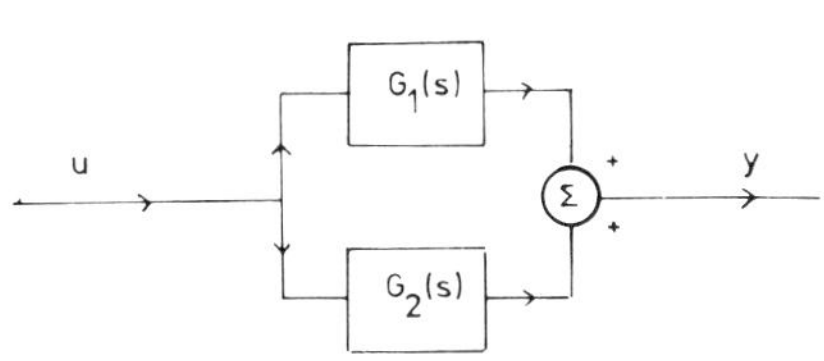

Figure 11. Combining sub-systems in parallel. $G(s) = G_1(s) + G_2(s)$.

7.2. *Combining in series*

Suppose that a system is composed of two sub-systems in series and that each of the sub-systems can be represented by Laplace transfer-function matrices, the elements of which are regular in the unstable section of the frequency plane at all but a distinct set of points (Fig. 12). The Laplace transfer-function matrix of the system is the product of the transfer-function matrices of the sub-systems. Therefore the elements of $G(s)$ are formed by adding several products of the elements of the transfer-function matrices of

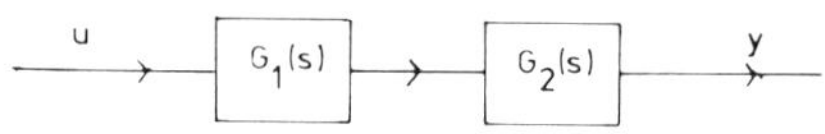

Figure 12. Combining sub-systems in series. $G(s)=G_1(s)G_2(s)$.

the sub-systems. Hence the elements of $G(s)$ are differentiable at all but a distinct set of points in the unstable section of the frequency plane, since the product of two differentiable functions is differentiable. Therefore the gain loci can be used to determine the number of roots moving between the stable and unstable sections of the frequency plane as the feedback gain is changed. The number of open-loop unstable poles is equal to the sum of the open-loop unstable poles of the individual elements.

7.3. *Combining in a feedback configuration*

Suppose that a system is composed of two sub-systems, one of which forms a forward path between the input and output, and the other of which acts as as internal feedback loop for the system (Fig. 13). The transfer function of the combination of these sub-systems is :

$$G(s)=\frac{\mathbf{y}(s)}{\mathbf{u}(s)}=[I+G_1(s)G_2(s)]^{-1}G_1(s) \tag{21}$$

At the points where $[I+G_1(s)G_2(s)]$ is not singular, each of the elements of this matrix is differentiable, and so each of the elements of the inverse is differentiable. Therefore each of the elements of $G(s)$ is differentiable, and so the gain loci can be used to indicate the gains for which closed-loop poles pass between the stable and unstable sections of the frequency plane. The

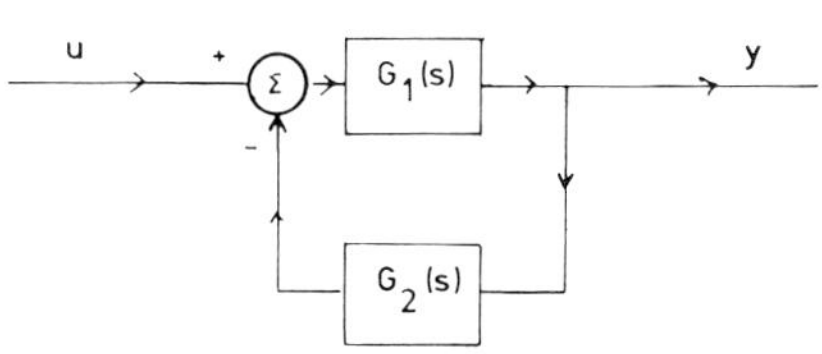

Figure 13. Combining sub-systems in a feedback configuration. $G(s)=(I+G_1G_2)^{-1}G_1$.

number of open-loop unstable poles for $G(s)$ can be determined by counting the number of unstable open-loop poles of the sub-systems and then using the gain loci for the product of the sub-systems.

8. A method of constructing characteristic frequency loci

It was noted in § 2 that the eigenvalues as a function of frequency form continuous functions, so the characteristic frequency loci can be constructed by the following steps :

(1) Decide the area of interest in the frequency plane. For the present discussion this is assumed to be a rectangular area.

(2) Choose a rectangular grid of points covering this area, and evaluate the characteristic gains associated with each of these points (Fig. 14).

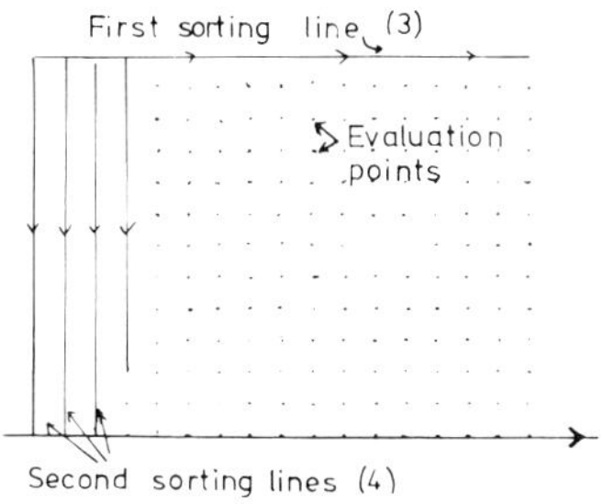

Figure 14. Evaluating $G(s)$ at a grid of points on the s-plane, and sorting the results to make the surfaces as continuous as possible.

(3) Starting at one of the corners of the grid go across the grid rearranging the characteristic gains to make the functions as near continuous as possible. In the limit as the grid size is decreased these functions become continuous.

(4) Starting at each of the points on the line which has been sorted out for continuity at step 3, sort out the characteristic gains to be as continuous as possible down the grid.

(5) Record the sorting which would make the characteristic gains across the grid continuous, but do not rearrange them. With single-input systems this step is not needed, but for a multi-input system it is necessary, since discontinuities can occur in this direction. These discontinuities occur when a branch point is passed ; this is a point where two or more of the characteristic gains are equal. The discontinuities then correspond to cuts in the frequency plane. Since problems in sorting also occur due to the finite grid size, particularly when passing system poles, it is an advantage to check for cases where two cuts, joining the same two sheets, are produced at adjacent lines of grid points, since in this situation the cuts are probably just due to a sorting error at some point, and can both be removed.

(6) Go through the grid of points rectangle by rectangle, sheet by sheet, ensuring that the four corner values chosen make the sheet as continuous as possible in both directions, and draw any part of the -180 degree phase contours which are found to go through the rectangle. This can be done by splitting the square into two triangles, checking the sides of each triangle for crossings by this phase contour, and then interpolating to obtain better estimates of any such crossing places.

9. Examples of the various forms of loci

In this section examples of gain loci, inverse gain loci, frequency loci and, for completeness, inverse frequency loci are given for systems with and without time delays. The inverse frequency loci have not been mentioned before in this paper, but are just the inverses of the frequency loci or root loci.

Example 1

This example is the one used in §§ 2 and 3 to give simple illustrations of the gain and frequency loci. The transfer function is :

$$G(s)=\frac{(s+1+j)(s+1-j)}{(s+1)(s-1)(s-1)} \tag{22}$$

The frequency loci are shown in Fig. 4, the inverse frequency loci in Fig. 15, the gain loci in Fig. 6 and the inverse gain loci in Fig. 8. The points where roots are leaving the right-half plane have been numbered in order to help identify which axis crossings correspond.

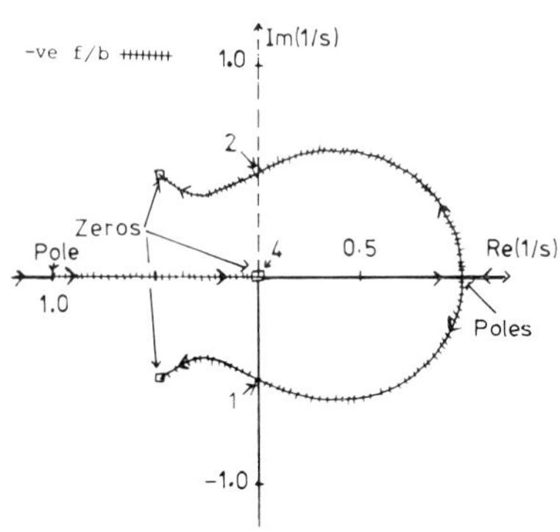

Figure 15. Inverse frequency loci for Example 1.

Example 2. *Single-input system with a time delay*

This is an open-loop unstable system with a single time delay. The Laplace transfer function is :

$$G(s)=\exp(-0{\cdot}1s)\,\frac{(s+1+j)(s+1-j)}{(s+1)(s-1)(s-1)} \tag{23}$$

Figure 16 (*a*) shows the frequency locus for small values of s. It can be seen that a pair of poles leave the right-half plane for a negative feedback gain of about 2·8, and at frequencies of + and $-2{\cdot}3$. Figure 16 (*b*) shows the frequency loci for a larger part of the frequency plane. It can be seen that the time delay has caused extra branches of the frequency loci. These occur at approximately intervals of $10{\cdot}0\pi$ along the imaginary axis of the frequency plane. Figures 16 (*c*) and 16 (*d*) show the corresponding inverse frequency loci. Figures 16 (*e*) and 16 (*f*) show the Nyquist diagram or gain locus. Figures 16 (*g*) and 16 (*h*) show the inverse gain locus.

Example 3. *Stable single-input system with an echo*

In this example an echo with dynamics has been added to a simple system. It will be seen on the characteristic frequency loci that if the real part of the

Figure 16

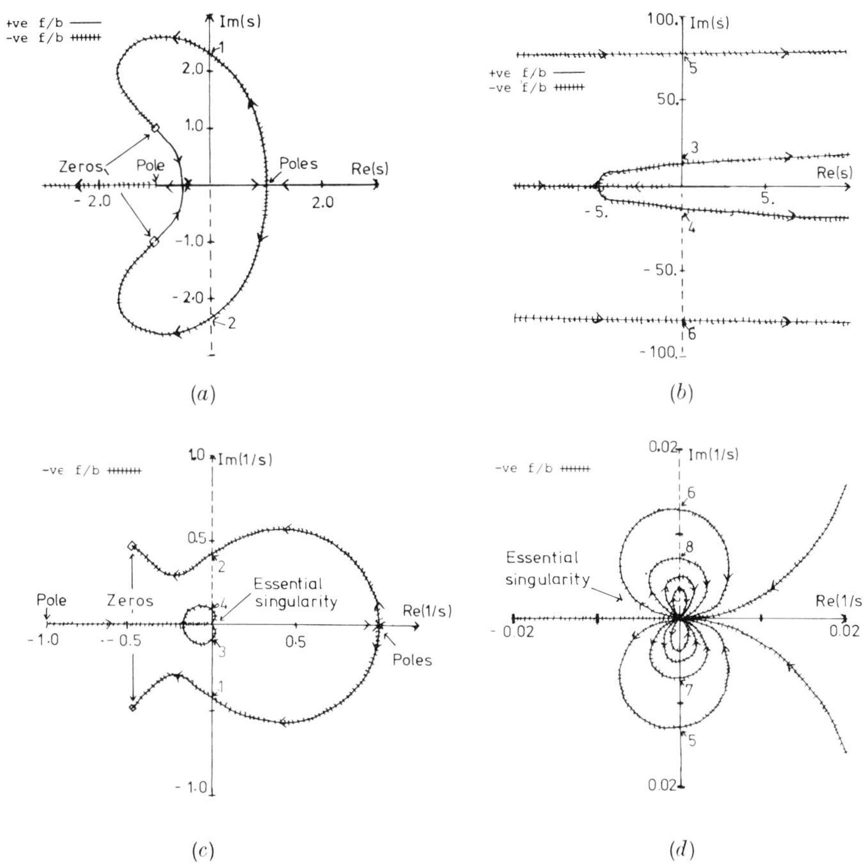

(*figure continued over*)

Fig. 16 (*continued*)

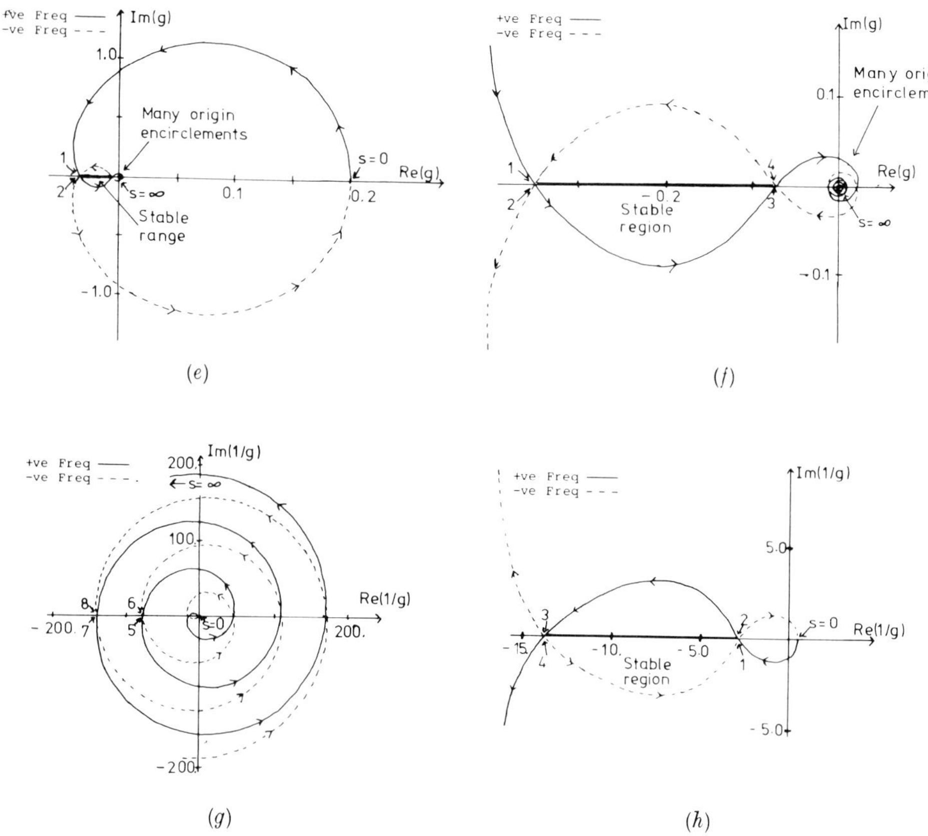

Figure 16. (*a*) Characteristic frequency loci for Example 2, which has a small time delay. (*b*) Characteristic frequency loci for Example 2, which has a small time delay. (*c*) Inverse characteristic frequency loci for Example 2, which has a small time delay. (*d*) Inverse characteristic frequency loci for Example 2, which has a small time delay. (*e*) Characteristic gain loci for Example 2, which has a small time delay. (*f*) Characteristic gain loci for Example 2, which has a small time delay. (*g*) Inverse characteristic gain loci for Example 2, which has a small time delay. (*h*) Inverse characteristic gain loci for Example 2, which has a small time delay.

frequency is large and negative, the time-delay term dominates the frequency response, but if it is large and positive, or if the imaginary part is large, the term without the time delay dominates. The Laplace transfer function is :

$$G(s)=\frac{1}{(s+1)}+\exp(-1{\cdot}5s)\,\frac{1}{(s+1)(s+1)(s+1)} \tag{24}$$

Figures 17 (*a*) and 17 (*b*) show the characteristic frequency loci. Note that now l rge number of loci start off from minus infinity and head for the right-half plane, but now the zeros which were at plus infinity in Example 2 have been moved, mainly into the left-half plane, so most of the root loci never reach the right-half plane. Figures 17 (*c*)–(*g*) show the corresponding inverse frequency loci, gain loci and inverse gain loci.

Example 4. *Unstable two-input system*

This is a constructed example of a system with two inputs, and two open loop unstable poles. The Laplace transfer-function matrix is :

$$G(s)=\begin{bmatrix} \dfrac{2}{(s-1)(s+2)} & \dfrac{-2}{(s-3)(s+2)} \\[2ex] \dfrac{0\cdot5}{(s-1)(s+0\cdot5)} & \dfrac{3}{(s-3)} \end{bmatrix} \tag{25}$$

Figure 17

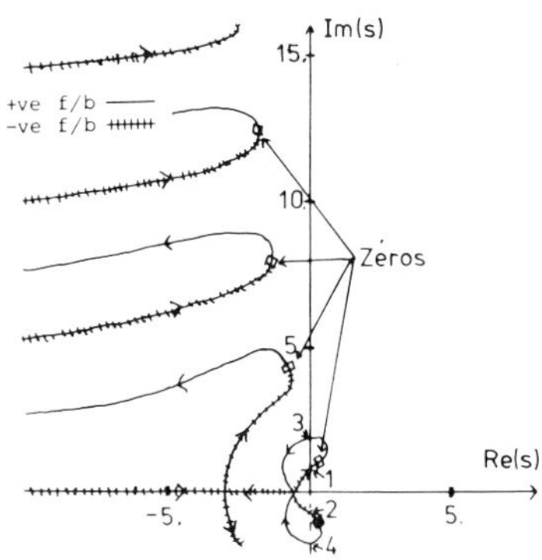

(*a*)

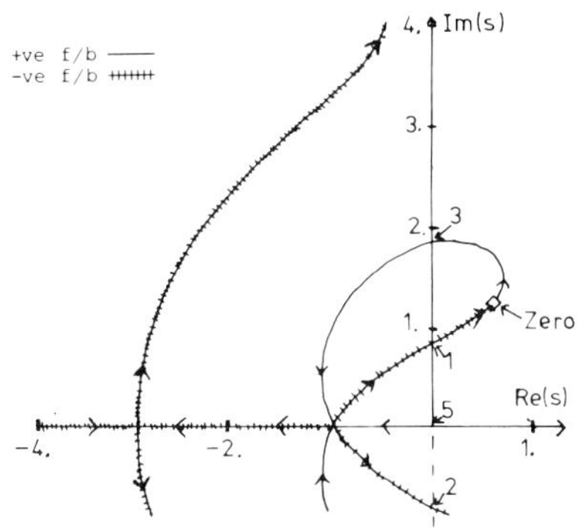

(*b*)

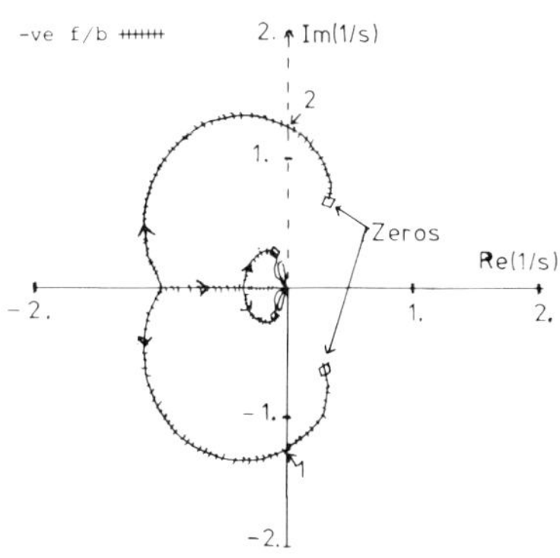

(*c*)

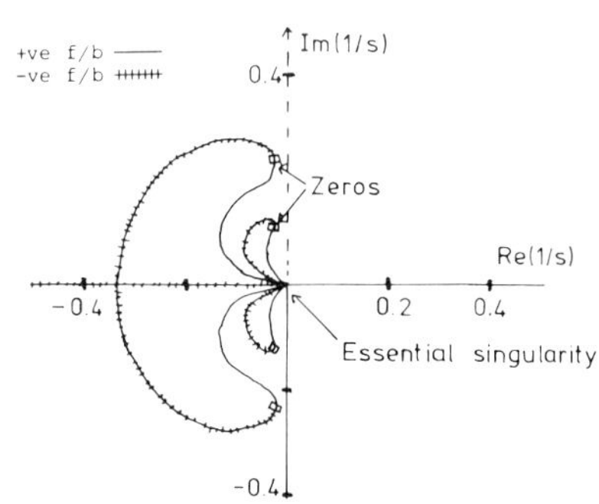

(*d*)

(*figure continued over*)

Fig. 17 (*continued*)

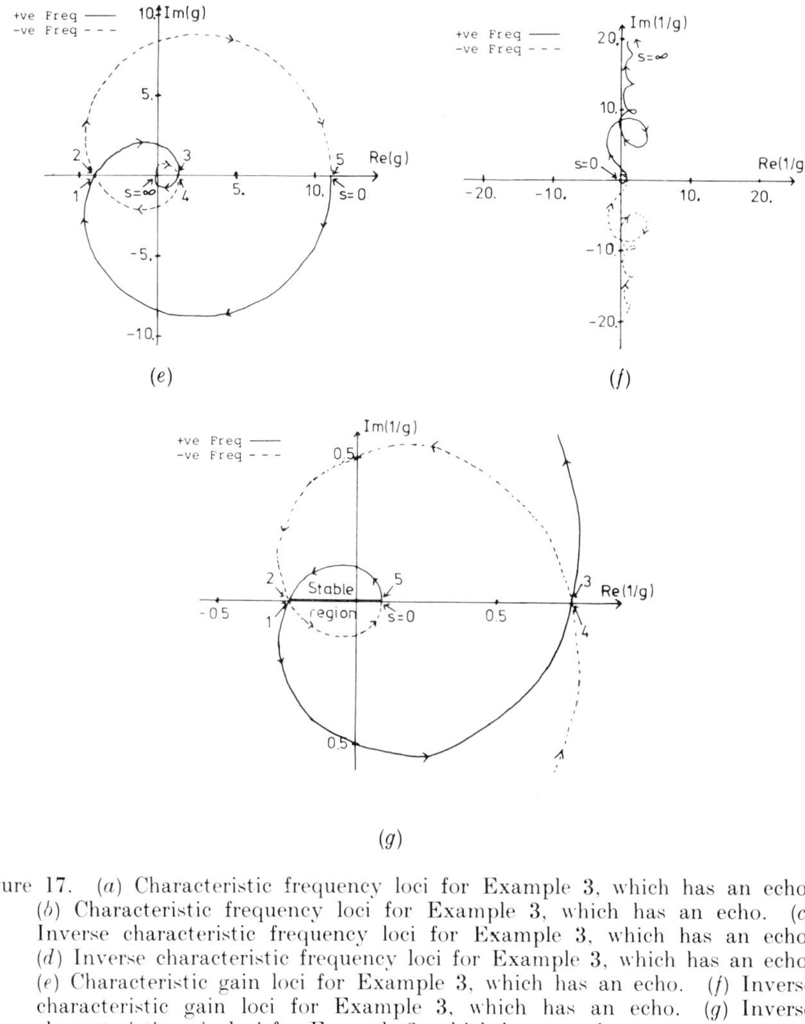

(*e*) (*f*) (*g*)

Figure 17. (*a*) Characteristic frequency loci for Example 3, which has an echo. (*b*) Characteristic frequency loci for Example 3, which has an echo. (*c*) Inverse characteristic frequency loci for Example 3, which has an echo. (*d*) Inverse characteristic frequency loci for Example 3, which has an echo. (*e*) Characteristic gain loci for Example 3, which has an echo. (*f*) Inverse characteristic gain loci for Example 3, which has an echo. (*g*) Inverse characteristic gain loci for Example 3, which has an echo.

Figures 18 (*a*) and 18 (*b*) show the frequency loci. Notice that one of the frequency loci starts on one sheet and then goes through a branch point at about 0·73 onto the other sheet. In doing this it folds back on itself. It can also be seen from Fig. 18 (*d*) that the gain loci do not have to start on the real axis as in the single input case.

Example 5. Unstable two-input example with time delays

This example has been constructed to show some of the possible effects of having several different time delays in a multivariable system. This system

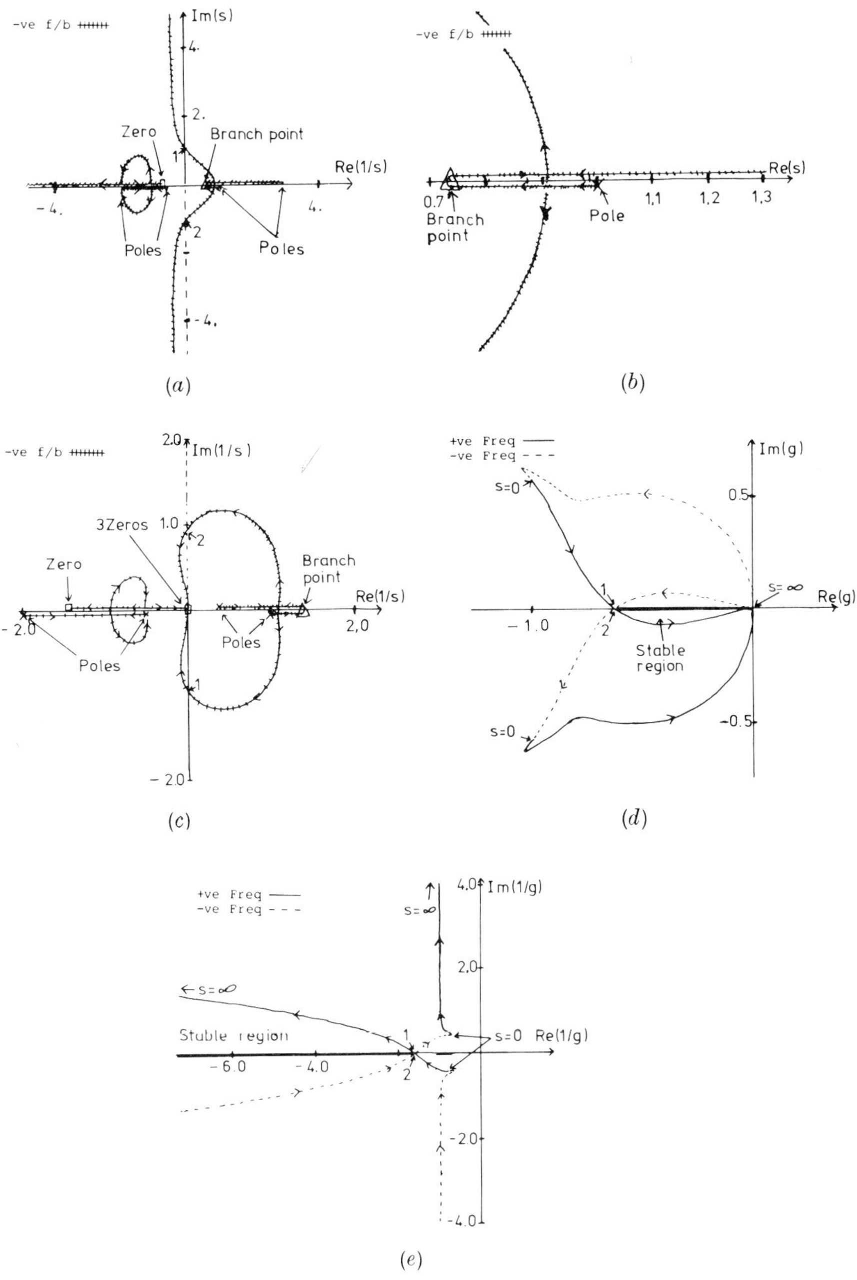

Figure 18. (*a*) Characteristic frequency loci for Example 4, on the *s*-plane. (*b*) Characteristic frequency loci for Example 4, showing a branch point. (*c*) Inverse characteristic frequency loci for Example 4. (*d*) Characteristic gain loci for Example 4; note that these loci can start off the real axis for multivariable systems. (*e*) Inverse characteristic gain loci for Example 4.

Figure 19

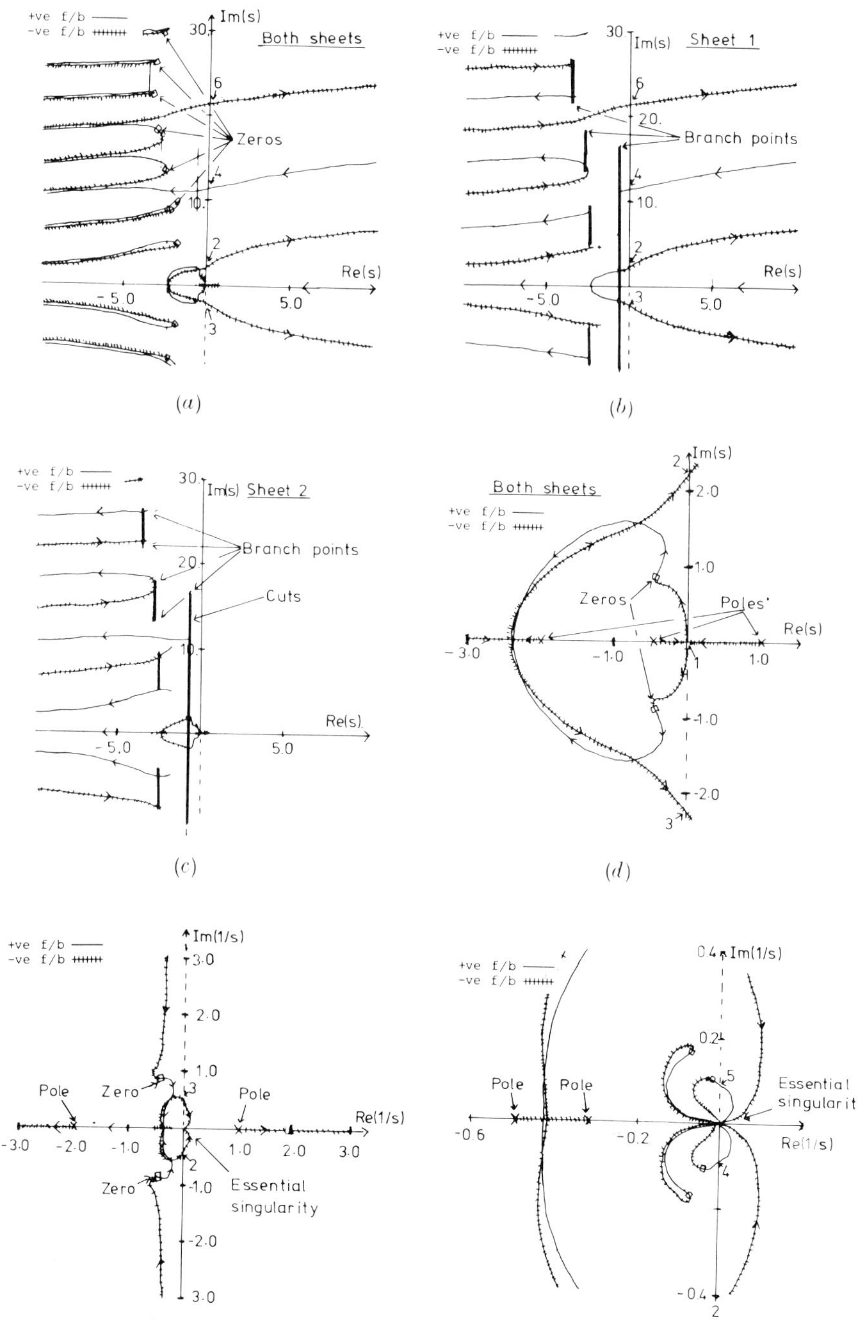

(*figure continued over*)

Fig. 19 (*continued*)

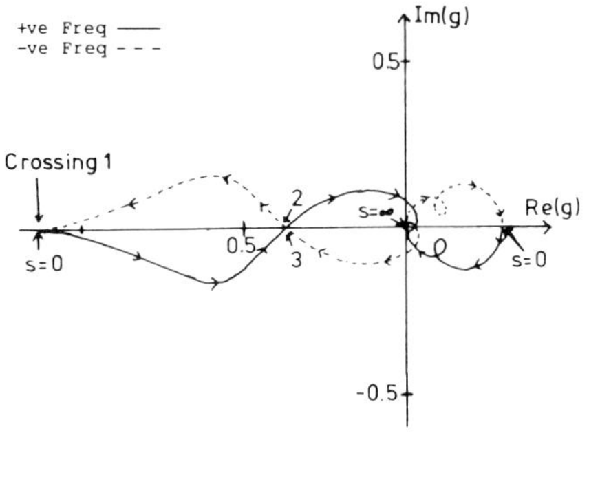

(*g*)

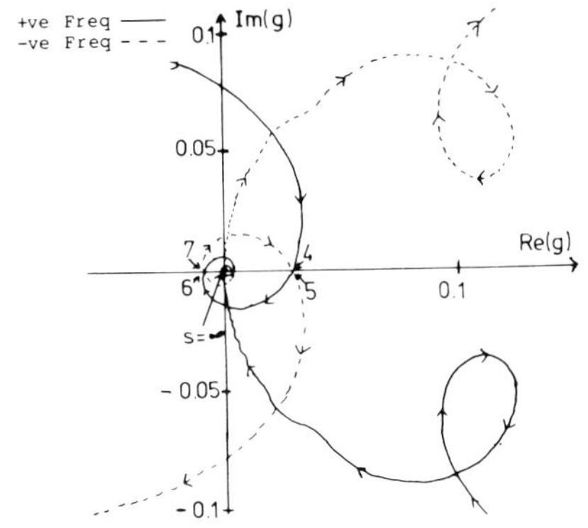

(*h*)

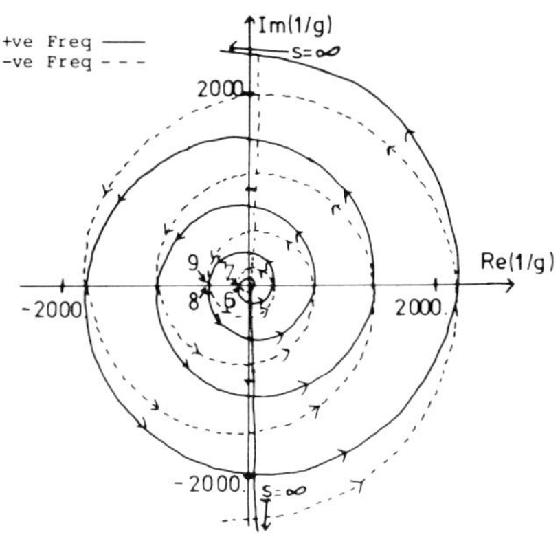

(*i*)

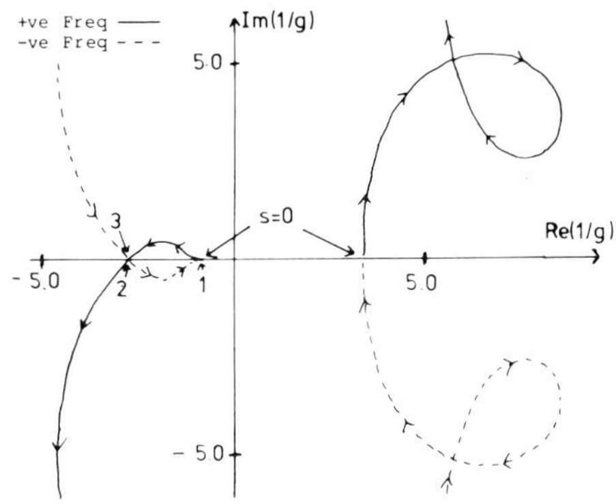

(*j*)

Figure 19. (*a*) Characteristic frequency loci on both sheets of the Riemann surface for Example 5, which has three different time delays. (*b*) Characteristic frequency loci on one sheet of the Riemann surface for Example 5, which has three different time delays. (*c*) Characteristic frequency loci on the other sheet of the Riemann surface for Example 5, which has three different time delays. (*d*) Characteristic frequency loci on both sheets of the Riemann surface for Example 5, which has three different time delays. (*e*) Inverse characteristic frequency loci for Example 5, which has three different time delays. (*f*) Inverse characteristic frequency loci for Example 5, which has three different time delays. (*g*) Characteristic gain loci for Example 5, which has three different time delays. (*h*) Characteristic gain loci for Example 5, which has three different time delays. (*i*) Inverse characteristic gain loci for Example 5, which has three different time delays. (*j*) Inverse characteristic gain loci for Example 5, which has three different time delays.

has infinite numbers of zeros and branch points. It has also been arranged that the frequency response has a point with zero gradient close to the imaginary axis of the frequency plane, at about 0·003. The Laplace transfer function used is :

$$G(s)=\begin{bmatrix} \dfrac{4\exp(-0{\cdot}3s)}{(s+4)(s-1)} & \dfrac{-0{\cdot}5\exp(-1{\cdot}1s)}{(s+3)(s+2)} \\[2ex] \dfrac{\exp(-0{\cdot}7s)}{(s+0{\cdot}5)(s-1)} & \dfrac{0{\cdot}5}{(s+3)} \end{bmatrix} \tag{26}$$

It can be seen from Fig. 19 that an infinite set of root loci start at minus infinity on both sheets of the Riemann surface and head towards the right-half plane. However since there are smaller time delays also many of these loci go through zeros and return to minus infinity. It has been arranged so that different time delays dominate the frequency response for the two sheets in the right-half plane. This results in an infinite number of right-half-plane loci on one of the sheets, and only a single branch on the other. Thus one of the gain loci makes an infinite number of origin encirclements, the other making none.

Example 6. *Poles on the imaginary axis*

This example is to illustrate the sort of gain loci which occur when there are poles on the boundary of the unstable section of the frequency plane. The transfer function used is :

$$G(s)=\frac{(s+1)}{(s+j)(s-j)} \tag{27}$$

It will be noticed that the characteristic gain loci now pass through infinity. However the inverse gain loci indicate that these two poles go into the left-half plane for small values of negative feedback (Fig. 20).

Example 7. *Sampled data example*

This example illustrates the use of characteristic gain and frequency loci for sampled systems. The system described by eqn. (25) in Example 4 is used with two different sample times indicating how the loci gradually change with increasing sample periods. The following state-space realization of the transfer function was used :

$$\dot{\mathbf{x}}=A\mathbf{x}+B\mathbf{u}$$
$$\mathbf{y}=C\mathbf{x}+D\mathbf{u}$$

with

$$\left.\begin{aligned} A&=\begin{bmatrix} 1{\cdot}0 & 0{\cdot}0 & 0{\cdot}0 & 0{\cdot}0 \\ 2{\cdot}0 & -2{\cdot}0 & -2{\cdot}0 & 0{\cdot}0 \\ 0{\cdot}0 & 0{\cdot}0 & 3{\cdot}0 & 0{\cdot}0 \\ 0{\cdot}5 & 0{\cdot}0 & 0{\cdot}0 & -0{\cdot}5 \end{bmatrix}, \quad B=\begin{bmatrix} 1{\cdot}0 & 0{\cdot}0 \\ 0{\cdot}0 & 0{\cdot}0 \\ 0{\cdot}0 & 1{\cdot}0 \\ 0{\cdot}0 & 0{\cdot}0 \end{bmatrix} \\ C&=\begin{bmatrix} 0{\cdot}0 & 1{\cdot}0 & 0{\cdot}0 & 0{\cdot}0 \\ 0{\cdot}0 & 0{\cdot}0 & 3{\cdot}0 & 1{\cdot}0 \end{bmatrix}, \quad D=\begin{bmatrix} 0{\cdot}0 & 0{\cdot}0 \\ 0{\cdot}0 & 0{\cdot}0 \end{bmatrix} \end{aligned}\right\} \tag{28}$$

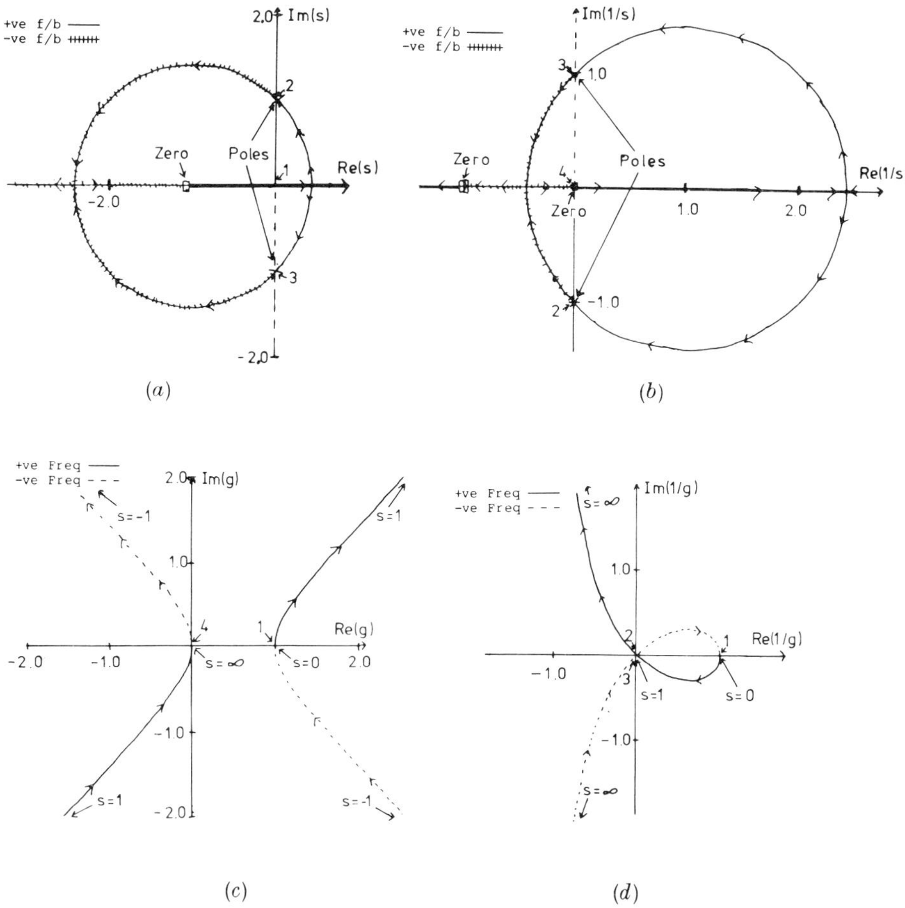

Figure 20. (*a*) Characteristic frequency loci for Example 6, which has open-loop poles on the imaginary axis of the frequency plane. (*b*) Inverse characteristic frequency loci for Example 6, which has open-loop poles on the imaginary axis of the frequency plane. (*c*) Nyquist diagram for Example 6, which has open-loop poles on the imaginary axis of the frequency plane. (*d*) Inverse Nyquist diagram for Example 6, which has open-loop poles on the imaginary axis of the frequency plane.

The sampled system can be described by a set of transition matrices such that :

$$\left.\begin{aligned} z\mathbf{x} &= A'\mathbf{x} + B'\mathbf{u} \\ \mathbf{y} &= C\mathbf{x} + D\mathbf{u} \end{aligned}\right\} \tag{29}$$

Here z is a time advance operator, i.e. $z\mathbf{x}(t)=\mathbf{x}(t+T)$ where T is the sample period. A' and B' are calculated using the formulae :

$$A'=\exp(AT) \tag{30}$$

$$B'=[\exp(AI)-I]A^{-1}B \tag{31}$$

C and D remain unchanged.

For a sample period of 0·1 the resulting A' and B' are :

$$\left.\begin{aligned} A'&=\begin{bmatrix} 1\cdot105171 & 0\cdot0000000 & 0\cdot0000000 & 0\cdot0000000 \\ 0\cdot1909601 & 0\cdot8187307 & -0\cdot2124512 & 0\cdot0000000 \\ 0\cdot0000000 & 0\cdot0000000 & 1\cdot349859 & 0\cdot0000000 \\ 0\cdot0513138 & 0\cdot0000000 & 0\cdot0000000 & 0\cdot9512294 \end{bmatrix} \\ B'&=\begin{bmatrix} 0\cdot1051709 & 0\cdot00000000 \\ 0\cdot009690836 & -0\cdot01039399 \\ 0\cdot000000000 & 0\cdot1166196 \\ 0\cdot002543245 & 0\cdot00000000 \end{bmatrix} \end{aligned}\right\} \tag{32}$$

Figure 21 (a) shows the root loci on the z-plane for this sample period. The stable part of the complex z-plane is the inside of the unit circle, so the system can be made stable with diagonal scalar feedback since the root loci pass through this circle. The characteristic gain loci (Fig. 21 (c)) and the inverse characteristic gain loci (Fig. 21 (d)) show the stable range of feedback gains. Two axis crossings are required since the system has two open-loop unstable poles. The gain loci are evaluated round the unit circle in an anti-clockwise direction, to ensure that the unstable part of the plane is on the right-hand side of the loci.

Rather than show the inverse z-plane, one may use the result of a bilinear transform defining a b-plane (of Saucedo and Schiring 1968).

$$b=\frac{2}{T}\frac{(z-1)}{(z+1)} \tag{33}$$

Here T is the sample period. This transform has the property of mapping the unit circle in the z-plane to the imaginary axis in the b-plane, hence the stable part of the b-plane is the left-half plane. The scaling factor of $2/T$ is introduced to make distances near the origin of the b-plane the same as the corresponding distances on the s-plane. A comparison of Figs. 18 (a), (c), (d) and 21 (b), (c), (d) shows how the characteristic gain and frequency loci have been slightly affected by the introduction of the sampling.

The transition matrices for a sample period of 1/3 are :

$$\left.\begin{aligned} A'&=\begin{bmatrix} 1\cdot395612 & 0\cdot0000000 & 0\cdot0000000 & 0\cdot0000000 \\ 0\cdot5881302 & 0\cdot5134172 & -0\cdot8819457 & 0\cdot0000000 \\ 0\cdot0000000 & 0\cdot0000000 & 2\cdot718282 & 0\cdot0000000 \\ 0\cdot1830436 & 0\cdot0000000 & 0\cdot0000000 & 0\cdot8464817 \end{bmatrix} \\ B'&=\begin{bmatrix} 0\cdot3956124 & 0\cdot0000000 \\ 0\cdot1015473 & -0\cdot1317876 \\ 0\cdot00000000 & 0\cdot5727605 \\ 0\cdot02952531 & 0\cdot0000000 \end{bmatrix} \end{aligned}\right\} \tag{34}$$

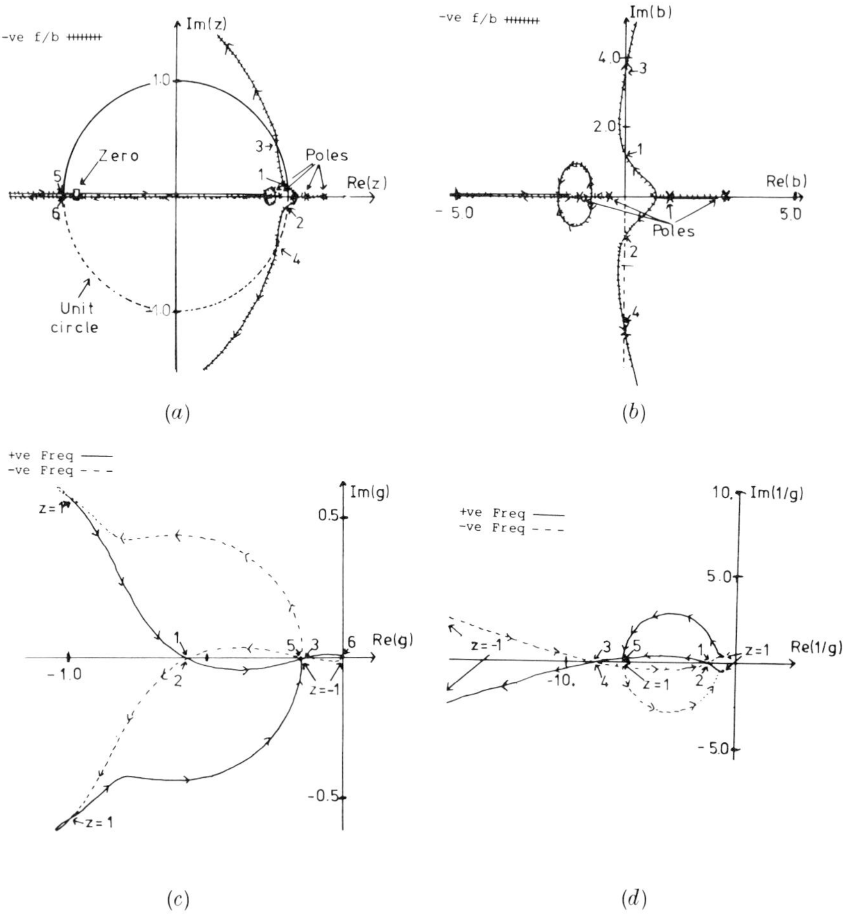

Figure 21. (*a*) Characteristic frequency loci on the z-plane for Example 7, with a sample period of 0·1. (*b*) Characteristic frequency loci on the b-plane for Example 7, with a sample period of 0·1. (*c*) Characteristic gain loci evaluated round the unit circle on the z-plane for Example 7, with a sample period of 0·1. (*d*) Inverse characteristic gain loci evaluated round the unit circle on the z-plane for Example 6, with a sample period of 0·1.

Figures 21 (*a*), (*b*), (*c*), (*d*) show the characteristic gain and frequency loci for a sample time of 1/3. It can be seen that the loci with the increase in sample time indicate that there is now no stable range of diagonal scalar feedback gain. The deterioration of the stability margins with increasing sample period can be expected generally, since, in the b-plane, the sampling tends to introduce zeros at $2/T$, and tends to move all the poles closer together and towards the origin. Both of these effects will move the root loci further right, and so decrease stability margins.

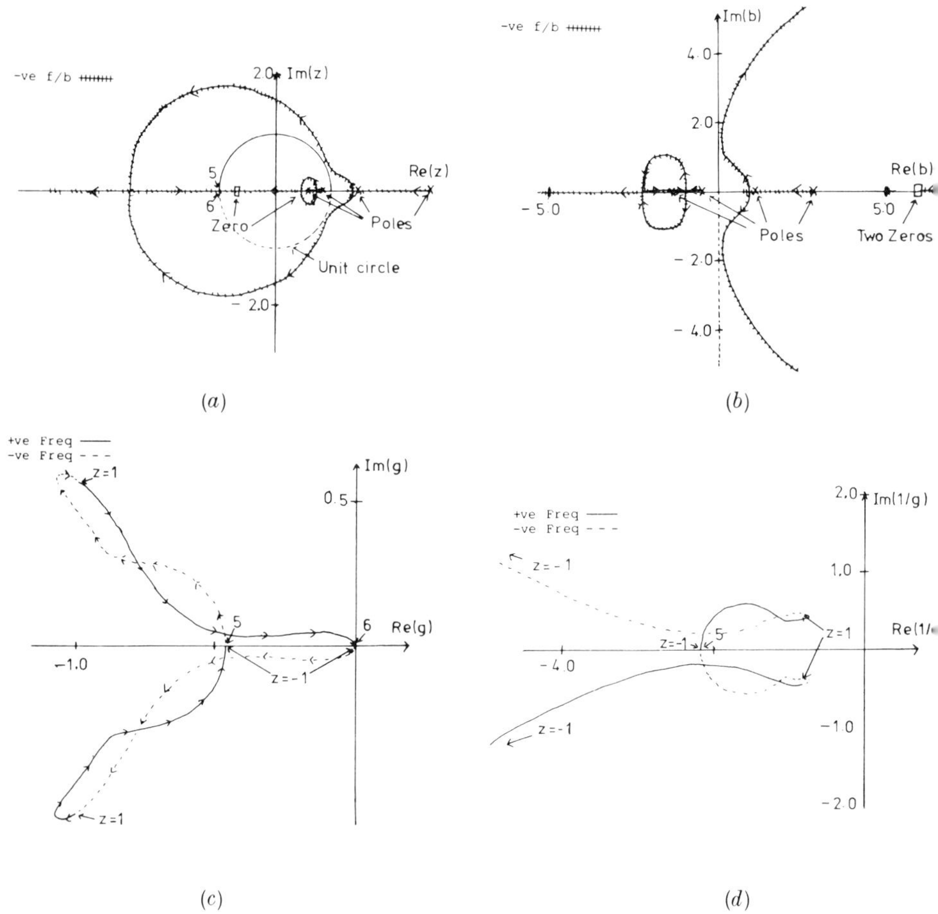

Figure 22. (*a*) Characteristic frequency loci on the z-plane for Example 7, with a sample period of 1/3. (*b*) Characteristic frequency loci on the b-plane for Example 7, with a sample period of 1/3. (*c*) Characteristic gain loci evaluated round the unit circle on the z-plane for Example 7, with a sample period of 1/3. (*d*) Inverse characteristic gain loci evaluated round the unit circle on the z-plane for Example 7, with a sample period of 1/3.

10. Choice of system description

It will have been noticed that in order to determine the stability of the closed-loop system, the number of open-loop unstable poles, or the number of closed-loop unstable poles for some given feedback gain must be known. The number of open-loop unstable poles can be easily counted if one is using state-space descriptions. However if rational polynomial Laplace transforms are used, there can be difficulties in determining this number of open-loop unstable poles. Difficulties arise if the system has an unstable state which is affected by more than one input, and which affects more than one output. It is then found that most small perturbations of the numerator polynomials of the elements of the transfer-function matrix which involve this state will require the introduction of a second unstable pole together with a zero which almost cancels it. This extra state can be almost unobservable, or almost uncontrollable. Similarly most small perturbations of one of the denominator polynomials of the elements of the transfer-function matrix which involve this state will require the introduction of a second unstable pole very close to the original unstable one. Since these new states are almost unobservable or almost uncontrollable they will not greatly affect the frequency response of the system ; however they will make it almost impossible to ensure that the closed-loop system is stable since they change the number of unstable poles for that system description. Therefore rational polynomial Laplace transfer functions should not be used when counting the number of open-loop unstable poles if it is thought that the system may have an unstable state which affects more than one output and is affected by more than one input. If they are used when there are unstable poles, the denominators should be kept in a factorized form, or a common form in order to avoid the introduction of extra states by small perturbations of the denominators.

10.1. *A simple example of the increase in number of states*

Suppose that we have a system with one state, two inputs and two outputs.

$$\dot{\mathbf{x}} = A\mathbf{x} + B\mathbf{u}$$
$$\mathbf{y} = C\mathbf{x}$$

$$A = [1], \quad B = [1 \quad 2], \quad C = \begin{bmatrix} 3 \\ 4 \end{bmatrix}$$

The transfer-function matrix $G(s)$ is :

$$\begin{aligned} G(s) &= C(sI - A)^{-1}B \\ &= \begin{bmatrix} 3 \\ 4 \end{bmatrix} \left[\frac{1}{s-1} \right] [1 \quad 2] \\ &= \frac{1}{s-1} \begin{bmatrix} 3 & 6 \\ 4 & 8 \end{bmatrix} \end{aligned} \tag{35}$$

Suppose we perturb $G(s)$ to

$$G'(s) = \frac{1}{s-1} \begin{bmatrix} 3 \cdot 1 & 6 \\ 4 & 8 \end{bmatrix} \tag{36}$$

There is now no state-space representation of $G'(s)$ with just one state. This is because such a representation would require a pair of vectors **C** and **B** such that

$$\mathbf{CB} = \begin{bmatrix} 3 \cdot 1 & 6 \\ 4 & 8 \end{bmatrix} \tag{37}$$

and there is no such pair of vectors. Therefore the state-space representation of this perturbed transfer function now requires two states.

Note that this example is a very simple one and would be very unlikely to occur, since there are less states than inputs. However, with more complicated systems transformations can be carried out on the A, B and C matrices to make the A matrix diagonal and leave the input–output relationships unchanged. The argument used in this example can then be used with the individual rows of the modified B matrix together with corresponding columns of the C matrix.

11. Conclusions

(1) If each of the elements of the frequency response of the system shown in Fig. 1 is a regular function of frequency, the loci of the closed-loop system poles are continuous, and each locus goes through a pole of one of the elements of the transfer-function matrix.

(2) Characteristic gain loci can be used to determine the numbers of right-half-plane characteristic frequencies for a wide range of closed-loop linear systems. The restriction required is that the system's characteristic frequency loci should be continuous ; in particular these loci can be used for systems with time delays.

(3) Characteristic gain loci can be used with systems which are made up of sub-systems connected in series, parallel or in a feedback configuration, subject to certain restrictions concerning the counting of the open-loop unstable poles.

(4) If the number of open-loop unstable poles for a system is not known the characteristic gain loci can still be used to indicate the only possible ranges of feedback gain for stability.

(5) Loci similar to the characteristic gain loci can be used to indicate the number of closed-loop poles in any given area of the frequency plane.

(6) For many real continuous-time systems the characteristic gain loci will indicate the stable ranges of feedback gain, if the number of open-loop unstable poles can be counted.

(7) Laplace and z-transfer functions are unsuitable basic or primary descriptions for use with multivariable systems which have an unstable state which is affected by more than one input, and which affects more than one output, since small variations in the coefficients of the polynomials of Laplace or z-transfer functions can result in extra unstable states in the system model. This suggests that :

(*a*) It may be best to use identification methods which produce models which are not liable to have the number of right-half-plane poles changed by small parameter variations. For example state-space models could be identified instead of transfer-function models.

(*b*) Minimal realization methods should not be needed since small changes of parameters should theoretically restore any state which the minimal realization has been able to remove.

(*c*) Simulation of time responses for open-loop unstable systems, which have unstable states which are affected by more than one input and affect more than one output, should be very difficult if not impossible to compute directly from a transfer function, since there will probably be extra unstable states added, which are almost uncontrollable or almost unobservable.

Appendix A

An alternative proof of the continuity of the characteristic frequency loci

If each of the elements of $G(s)$ has only simple isolated poles in the unstable region of the frequency plane and is analytic at all other points in the unstable region then, in any small region of this area, the elements can each be approximated arbitrarily well by finite-order rational polynomials in s. This implies that the characteristic gains as a function of frequency are approximated locally by an algebraic function, and so the characteristic gains at a point x can be expressed as :

$$\sum_{n=-m}^{\infty} C_n[1/z]^n$$

Where $z^p = x - s$, with p equal to one if x is not a branch point and equal to the number of sheets in the cycle (i.e. the number of sheets joined together at the branch point). If x is a branch point, m is finite. Suppose that the point x is on a characteristic frequency locus and that m is zero (i.e. this is not a pole or zero). Then near to this point on this set of sheets, the characteristic gains are approximately :

$$C_0 + C_m z^m$$

Here the m'-th term is the first non-zero term. Hence there are m values of z which make this function have a given small magnitude and be real and positive, and similarly m values of z which make this function have a given small magnitude and be real and negative. Therefore the frequency loci do not increase in number near this point x. Similarly if this point is a pole near to this point the gains are approximately :

$$C_{-m} z^{-m}$$

Hence there are the same number of values of z such that the gains are equal to a given large positive value or negative value. Therefore the frequency loci are continuous at all points with finite gain, and do not split at any points including those with infinite gain. Note that at poles the inverse frequency loci will be continuous.

Acknowledgments

I would like to thank Dr. I. Postlethwaite for many stimulating discussions on the subjects in this report, Dr. B. Kouvaritakis for his helpful criticisms, and Professor A. G. J. MacFarlane for his encouragement and guidance. This work was financed by the Science Research Council.

References

Bode, H. W., 1945, *Network Analysis and Feedback Amplifier Design* (Princeton, New Jersey: D. Van Nostrand Company, Inc.).

Callier, F. M., and Desoer, C. A., 1976, *Automatica*, **12,** 507.

Cook, P. A., 1973, *Proc. Instn elect. Engrs*, **120,** 1557.

Desoer, C. A., and Vidyasagar, M., 1975, *Feedback Systems Input–Output Properties* (New York: Academic Press).

Knopp, K., 1947, *Theory of Functions*, Part 2 (Chapter on Algebraic Functions) (New York: Dover Publications Inc.).

MacFarlane, A. G. J., and Postlethwaite, I., 1977, *Int. J. Control*, **26,** 265.

Saucedo, R., and Schiring, E. E., 1968, *Introduction to Continuous Digital Control Systems* (Macmillan).

Takahashi, Y., Rabins, M. J., and Auslander, D. M., 1970, *Control* (Reading, Mass.: Addison-Wesley).

III. MULTIVARIABLE ZEROS AND MULTIVARIABLE ROOT LOCI

A systematic attack on the problem of developing a frequency-response theory for multivariable systems was begun in a pioneering paper by Rosenbrock (1966), which ushered in a decade of increasing interest in a revitalized frequency-response approach. The definition and study of the poles and zeros of a rational transfer-function matrix is an essential first step in such an undertaking and Rosenbrock (1970) put forward the required extension in his important study of algebraic system theory. This approach to the definition of multivariable poles and zeros was based on McMillan's work on a canonical form for rational matrices (McMillan 1952). The McMillan form of a rational matrix had previously played an important role in Kalman's study of the realization problem (Kalman 1965).

Poles and zeros of multivariable systems

If the dynamical system associated with a transfer-function matrix $G(s)$ has at least as many output as inputs, and if the complex number v is neither a pole nor a zero of $G(s)$, then it can be shown (Desoer and Schulman 1974) that, for all non-zero vectors

$$k \in \mathbb{C}^l$$

where $\mathbb{C}^l$ denotes the space of l-dimensional complex vectors, there is a polynomial

$$m(s) = m_0 + m_1 s + m_2 s^2 + \dots$$

such that the input

$$u(t) = k \exp(vt) 1(t) + \sum_{\alpha} m_\alpha \delta^{(\alpha)}(t) \tag{3.1}$$

produces the zero-state response with exponential form

$$y(t) = G(v) k \exp(vt) \neq 0, \quad \text{for all } t > 0$$

Here $1(t)$ denotes a Heaviside unit step function and $\delta^{(\alpha)}(t)$ denotes a Dirac impulse function of order α. (The role of the generalized impulse functions in the input specified in eqn. (3.1) is to kick the dynamical system into an appropriate initial condition at the time $t = 0+$.) This shows that, assuming we have an appropriate set of initial conditions, the response to a vector exponential input of the form

$$u(t) = k \exp(vt)$$

can be made to be of the form

$$y(t) = G(v) k \exp(vt)$$

and is thus associated with a complex constant transmittance matrix found by substituting $s = v$ in the system transfer-function matrix $G(s)$. Hence the transfer-function matrix $G(s)$ may be regarded as defining the system's response to a vector of exponential inputs of exponent s, and thus the complex variable s can be regarded as a complex frequency variable as is familiar in the single-input, single-output case.

Poles are associated with specific values of complex frequency at which a transmittance becomes unbounded, and zeros with specific values of complex frequency at which a transmittance vanishes. Roughly speaking, a pole is associated with a characteristic frequency of a dynamical system. The set of system poles is a subset of the eigenvalues of the matrix A which appears in a state-space realization $\{A, B, C, D\}$ for $G(s)$ and the poles are related to characteristic modes which are both controllable and observable. Poles are thus characteristic of the ' internal dynamical machinery ' of a system. Zeros are characteristic of the ways in which this dynamical machinery is coupled to the environment in which the system is embedded. A survey of the various roles played by poles and zeros in linear multivariable systems has been given by MacFarlane and Karcanias (1976).

The generalization of the concepts of the poles and zeros of a scalar transfer function to the general matrix transfer-function case is done using a standard form for transfer-function matrices due to McMillan (1952); this is usually called the Smith–McMillan form since it is a direct consequence of the Smith form for polynomial matrices (Barnett 1971).

Smith form: If all the elements of a real-coefficient rational-function matrix $G(s)$ are polynomials in s it is called a polynomial matrix. Such an $m \times l$ matrix of normal rank r can be put into a standard form called the Smith form $S(s)$ by a series of elementary row and column operations:

$$S(s) = \begin{bmatrix} S^*(s)_{r,r} & 0_{r,\,l-r} \\ 0_{m-r,\,r} & 0_{m-r,\,l-r} \end{bmatrix}$$

where

$$S^*(s) = \text{diag}\,\{s_1(s), s_2(s), \ldots, s_r(s)\}$$

and

$$s_1(s)\,|\,s_2(s)\,|\,s_3(s)\,|\ldots|\,s_r(s)$$

where the notation $s_i(s)\,|\,s_j(s)$ denotes that the polynomial $s_i(s)$ divides the polynomial $s_j(s)$ without remainder.

Smith–McMillan form: Any rational function matrix $G(s)$ of order $m \times l$ and normal rank r can be written in the form

$$G(s) = \frac{1}{d(s)}\,N(s)$$

where $N(s)$ is a polynomial matrix and $d(s)$ is the least common denominator of all the elements of $G(s)$. If $N(s)$ has a Smith form $S(s)$ then

$$N(s) = L^{-1}(s)S(s)R^{-1}(s)$$

where $L^{-1}(s)$ and $R^{-1}(s)$ are appropriate unimodular matrices. (A unimodular matrix is one having a determinant independent of s.) The Smith–McMillan form of $G(s)$ is then defined as

$$M(s) = \frac{1}{d(s)}\,S(s)$$

where $M(s)$ has the form

$$M(s)=\begin{bmatrix} M^*(s)_{r,r} & 0_{r,\,l-r} \\ 0_{m-r,\,r} & 0_{m-r,\,l-r} \end{bmatrix}$$

with

$$M^*(s)=\operatorname{diag}\left[\frac{\epsilon_1(s)}{\psi_1(s)},\frac{\epsilon_2(s)}{\psi_2(s)},\ldots,\frac{\epsilon_r(s)}{\psi_r(s)}\right]$$

where each $\epsilon_i(s)$ divides all $\epsilon_{i+j}(s)$, and each $\psi_i(s)$ divides all $\psi_{i-j}(s)$. If

$$y(s)=G(s)u(s)$$

then it can be shown that, when $G(s)$ has a zero kernel (as a rational matrix) :

(i) $y(s)$ vanishes for $u(s)\neq 0$ and s finite if and only if some $\epsilon_i(s)$ is zero.
(ii) $\|y(s)\|\to\infty$ for $\|u(s)\|<\infty$ and s finite if and only if some $\psi_i(s)\to 0$.

These considerations led Rosenbrock (1970, 1973) to the following definitions :

Poles of $G(s)$. The poles of $G(s)$ are defined to be the set of all zeros of the set of polynomials $\{\psi_i(s) : i=1, 2, \ldots, r\}$.

Zeros of $G(s)$. The zeros of $G(s)$ are defined to be the set of all zeros of the set of polynomials $\{\epsilon_i(s) : i=1, 2, \ldots, r\}$.

It is not necessary to carry out a transformation to Smith–McMillan form in order to calculate the zero polynomial $z(s)$ and the pole polynomial $p(s)$ given by

$$z(s)=\prod_{i=1}^{r}\epsilon_i(s),\quad p(s)=\prod_{i=1}^{r}\psi_i(s)$$

Simple rules have been given for finding these polynomials (Kontakos 1973, MacFarlane and Karcanias 1976). The order $\delta[p(s)]$ of the pole polynomial $p(s)$ is the same as that of a minimal realization for $G(s)$ (Kalman 1965) : thus a completely controllable and completely observable realization for $G(s)$ will be stable if and only if none of the zeros of the pole polynomial $p(s)$ lie in the right-half complex plane.

The zeros defined above are what are usually called the transmission zeros for the system. There are many other definitions ; for further information the interested reader should consult Rosenbrock (1970, 1973, 1974), MacFarlane and Karcanias (1976, 1978), Davison and Wang (1974), Desoer and Schulman (1974) and Wolovich (1977).

When the poles and zeros of a tranfer-function matrix $G(s)$ have been defined and related to those of the characteristic gain function $g(s)$ (Postlethwaite and MacFarlane 1979), it is important to investigate how these relate to the parameters of a state-space realization for $G(s)$. Apart from the fact that such an investigation is of intrinsic interest in establishing links between state-space and frequency-response approaches, it is of great value for design studies since the parameters of state-space models are of direct physical significance.

The first two papers in the reprints presented in this section are concerned with the relationship between the zeros of $G(s)$ and the structure of state-space models. The third paper (by Kouvaritakis and Edmunds) is concerned with the relationships between state-space model parameters and the key features of multivariable root loci.

Finally, in the paper by Kouvaritakis, the asymptotic behaviour of the closed-loop characteristic frequencies of optimal feedback systems is examined.

References

Barnett, S., 1971, *Matrices in Control Theory* (London : Van Nostrand Reinhold).

Davison, E. J., and Wang, S. H., 1974, Properties and calculation of transmission zeros of linear multivariable systems. *Automatica*, **10,** 643–658.

Desoer, C. A., and Schulman, J. D., 1974, Zeros and poles of matrix transfer functions and their dynamical interpretation. *I.E.E.E. Trans. Circuits Syst.*, **21,** 3–8.

Kalman, R. E., 1965, Irreducible realizations and the degree of a rational matrix. *SIAM J. Control*, **13,** 520–544.

Kontakos, T., 1973, Ph.D. Thesis, University of Manchester.

MacFarlane, A. G. J., and Karcanias, N., 1976, Poles and zeros of linear multivariable systems : a survey of the algebraic, geometric and complex-variable theory. *Int. J. Control*, **24,** 33–74 ; 1978, Relationships between state-space and frequency-response concepts. *Preprints of Seventh IFAC World Congress*, Vol. 3, 1771–1779.

McMillan, B., 1952, Introduction to formal realizability theory. *Bell Syst. tech. J.*, **31,** 217–279, 541–600.

Postlethwaite, I., and MacFarlane, A. G. J., 1979, *A Complex Variable Approach to the Analysis of Linear Multivariable Feedback Systems*, Lecture Notes in Control and Information Sciences, Vol. 12 (Berlin : Springer-Verlag).

Rosenbrock, H. H., 1966, On the design of linear multivariable control systems. *Proceedings of the 3rd IFAC Congress*, London, Vol. 1, Book 1, Paper 1A (London : Institution of Mechanical Engineers) ; 1970, *State Space and Multivariable Theory* (London : Nelson) ; 1973, The zeros of a system. *Int. J. Control*, **18,** 297–299 ; 1974, Correction to the zeros of a system. *Ibid.*, **20,** 525–527.

Wolovich, W. A., 1977, Multivariable system zeros. *Control System Design by Pole–Zero Assignment*, edited by F. Fallside (London : Academic Press), pp. 226–236.

Geometric approach to analysis and synthesis of system zeros
Part 1. Square systems

B. KOUVARITAKIS and A. G. J. MACFARLANE

For square matrix-valued functions of a complex variable a method is given for computing the invariant zeros in terms of the parameters of the corresponding state-space model. The analysis presented gives an interesting insight into the geometrical relationships which underly the concept of invariant zeros. An algorithm is derived which should be useful for the direct calculation of zeros from state-space descriptions.

1. Introduction

The concepts of poles and zeros of a matrix-valued function of a complex variable (MacFarlane 1974) form a natural link between frequency-response and state-space approaches to the multivariable feedback problem. From a frequency-response point of view, the nature and location of the zeros and poles of a transfer function matrix, defining the transmission round a set of loops which is to be closed, determine the phase characteristics of an appropriate set of characteristic loci (MacFarlane 1974), and these loci in turn determine the stability of the closed-loop system. From a state-space point of view, it turns out that the location in the complex plane of both the poles and zeros of a transfer-function matrix can be determined in terms of geometrical relationships involving the standard state-space model-defining matrices A, B and C. For the poles, the geometrical relationships are the obvious ones associated with the eigenvectors of the matrix A and with the controllability and observability of the related modes. It is the purpose of this paper to give the geometrical interpretation of zero locations in terms of the matrices A, B and C.

Only the case of square matrix-valued functions of the complex variable will be considered here ; the general non-square case will be treated in a later paper. For convenience it could be assumed that all the state-space system models involved are completely controllable and completely observable. In such a case the invariant factors of the Rosenbrock system matrix and the invariant factors of the transfer-function matrix referred to as the invariant zeros and transmission zeros, respectively, are the same and hence there is no ambiguity involved in simply using the term zero, as appropriate, without further qualification. However, the treatment given here is general and encompasses all the zeros that belong to the set of invariant zeros, that is both transmission and decoupling zeros.

2. Definition and main properties of zeros

We will confine our investigations here mostly to the case of proper systems, and accordingly take the appropriate system matrix (Rosenbrock 1970) to be of the form

$$P(s)=\begin{bmatrix} sI-A & -B \\ C & 0_{m,m} \end{bmatrix} \tag{2.0, 1}$$

where A is $n \times n$, B is $n \times m$, C is $m \times n$, and $0_{m,m}$ is an $m \times m$ zero matrix. The matrix $P(s)$ will thus be a square polynomial matrix of order $(n+m)$ having a determinant

$$\det P(s)=|P(s)|=z(s) \tag{2.0, 2}$$

and we may accordingly define the zeros of the square system to be the roots of the polynomial $z(s)$, which will be conveniently called the *zero polynomial* for the system.

2.1. *Invariance of zeros*

Let z be one of the zeros of the system. Then we must have that

$$|P(z)|=\begin{vmatrix} zI-A & -B \\ C & 0_{m,m} \end{vmatrix}=0 \tag{2.1, 1}$$

Using elementary properties of determinants, it may be easily shown that we must also have that

$$|P(z)|=|P_1(z)|=\begin{vmatrix} zI-A & -BK \\ C & 0_{m,m} \end{vmatrix}=0 \tag{2.1, 2}$$

$$|P(z)|=|P_2(z)|=\begin{vmatrix} zI-A+BKL & -BK \\ C & 0_{m,m} \end{vmatrix}=0 \tag{2.1, 3}$$

$$|P(z)|=|P_3(z)|=\begin{vmatrix} zI-A+BKHC & -BK \\ C & 0_{m,m} \end{vmatrix}=0 \tag{2.1, 4}$$

Equation (2.1, 2) represents the defining equation for the zeros of a system after a pre-compensator represented by a matrix K has been added to the system to change its transfer-function matrix from $G(s)$ to $G(s)K$. It shows that the system zeros are invariant under such a modification.

Equations (2.1, 3) and (2.1, 4) show that the system zeros are invariant under a range of state and output-feedback modifications. One important conclusion to be drawn from these results is that once a system has acquired a set of invariant zeros as a result of a ' squaring-down ' output transformation, no further transformation of outputs can change the locations of the zeros so formed.

2.2. *Limiting positions of closed-loop characteristic frequencies under high-gain feedback*

Consider the standard multivariable feedback arrangement shown in Fig. 1. We propose to investigate the asymptotic behaviour of the closed-loop characteristic frequencies when the feedback gain matrix K has the specific form

$$\left.\begin{aligned} K &= kI \\ \text{and} \quad k &\to \infty \end{aligned}\right\} \tag{2.2, 1}$$

Since, as already demonstrated, the invariant zeros are unaffected by feedback we will denote the system zero polynomial by $z(s)$ for both the open-loop and closed-loop situations. Let

$p_o(s) =$ open-loop system characteristic polynomial
$p_c(s) =$ closed-loop system characteristic polynomial

Then a standard result (Hsu and Chen 1968) gives that

$$|I + kG(s)| = \frac{p_c(s)}{p_o(s)} \tag{2.2, 2}$$

where

$$G(s) = C(sI - A)^{-1}B \tag{2.2, 3}$$

is the transfer-function matrix of the plant. We also have that, using Schur's formula for partitioned determinants (Gantmacher 1959)

$$\begin{aligned} z(s) &= \begin{vmatrix} sI - A & -B \\ C & 0_{m,m} \end{vmatrix} \\ &= |sI - A|\,|C(sI - A)^{-1}B| \\ &= p_o(s)\,|G(s)| \end{aligned}$$

so that we get

$$|G(s)| = \frac{z(s)}{p_o(s)} \tag{2.2, 4}$$

The return-difference determinant on the left-hand side of eqn. (2.2, 2) can be expanded as

$$|I + kG(s)| = 1 + \text{trace}\,(kG(s)) + \ \dots\ + \det\,(kG(s)) \tag{2.2, 5}$$

Thus, using this with eqn. (2.2, 2), we have that

$$1 + \text{trace}\,(kG) + \ \dots\ + \det\,(kG) = \frac{p_c(s)}{p_o(s)} \tag{2.2, 6}$$

Therefore as $k \to \infty$ we will have that

$$\frac{p_c(s)}{p_o(s)} \to |kG(s)| = k^m |G(s)|$$

and so, using eqn. (2.2, 4), we have that

$$\frac{p_c(s)}{p_o(s)} \rightarrow \frac{k^m z(s)}{p_o(s)}$$

and so

$$p_c(s) \rightarrow k^m z(s) \tag{2.2, 7}$$

as $k \rightarrow \infty$. From this simple form of argument we conclude that, as the gain parameter k becomes arbitrarily large, m of the closed-loop characteristic frequencies become arbitrarily large and the remaining $(n-m)$ tend aribtrarily closely to the roots of the equation

$$z(s) = 0$$

that is to the transmission zeros of the system. This gives the basic idea behind the technique which we shall use to compute the system zeros : they are those limiting values of closed-loop characteristic frequency which tend to finite locations in the complex plane as $k \rightarrow \infty$.

3. Spectral decomposition of matrix *BC*

We assume that both the input matrix B and the output matrix C have full rank m. Such an assumption is reasonable since it has the physical significance that the system has m independent inputs and m independent outputs. We further assume to begin with that the rank of BC is the same as that of B or C, namely m ; the case in which the rank of BC is less than m will be dealt with later. The matrix BC will thus have m non-zero eigenvalues which we will denote by $\{\rho_i | i = 1, 2, \ldots, m\}$. The remaining eigenvalues of BC will all be zero. For the moment we will assume that the non-zero eigenvalues ρ_i are all distinct ; the case of repeated eigenvalues will be considered separately.

The matrix BC will thus have a spectral decomposition of the form

$$\begin{aligned} BC &= U_{BC} \Lambda_{BC} V_{BC} \\ &= \underset{\updownarrow n}{\overset{\leftarrow m \rightarrow \leftarrow n-m \rightarrow}{[\, W \,\vdots\, M \,]}} \left[\begin{array}{c|c} \Lambda_\rho & 0_{m,\, n-m} \\ \hline 0_{n-m,\, m} & 0_{n-m,\, n-m} \end{array} \right] \overset{\leftarrow n \rightarrow}{\left[\begin{array}{c} Z \\ \hline N \end{array} \right]} \begin{array}{l} \updownarrow m \\ \updownarrow n-m \end{array} \end{aligned} \tag{3.0, 1}$$

where Λ_{BC} is a diagonal matrix of non-zero eigenvalues of BC, and U_{BC} and V_{BC} have as their columns and rows respectively the appropriate eigenvectors and reciprocal eigenvectors of BC. The sub-matrices W and Z have as columns and rows respectively those eigenvectors and reciprocal eigenvectors of BC which correspond to the non-zero eigenvalues ρ_i. The sub-matrices M and N are respectively right and left annihilators of BC, since they contain those eigenvectors and reciprocal eigenvectors of BC which correspond to the set of zero eigenvalues. Thus we have that

$$(BC)M = 0_{n,\, n-m} \tag{3.0, 2}$$

$$N(BC) = 0_{n-m,\, n} \tag{3.0, 3}$$

Clearly the matrices M and N are not unique. If eqn. (3.0, 2) is multiplied through on the left by $B^\dagger$, a left-inverse of B, we get that

$$CM = 0_{m,\, n-m} \tag{3.0, 4}$$

so that M is also a right annihilator of the output matrix C. If eqn. (3.0, 3) is multiplied through on the right by $C^\dagger$, a right inverse of C, we get that

$$NB = 0_{n-m,\, m} \tag{3.0, 5}$$

so that N is also a left annihilator of B. We must also have that

$$NM = I_{n-m} \tag{3.0, 6}$$

where I_{n-m} is a unit matrix of order $(n-m)$; this follows directly from the facts that

$$U_{BC} V_{BC} = I_m \tag{3.0, 7}$$

and that N and M are appropriate sub-matrices of V_{BC} and U_{BC} respectively. It is shown in Appendix 1 that the set of conditions specified in eqns. (3.0, 4), (3.0, 5) and (3.0, 6) are both necessary and sufficient for the determination of pairs of matrices M and N that satisfy the spectral decomposition of eqn. (3.0, 1).

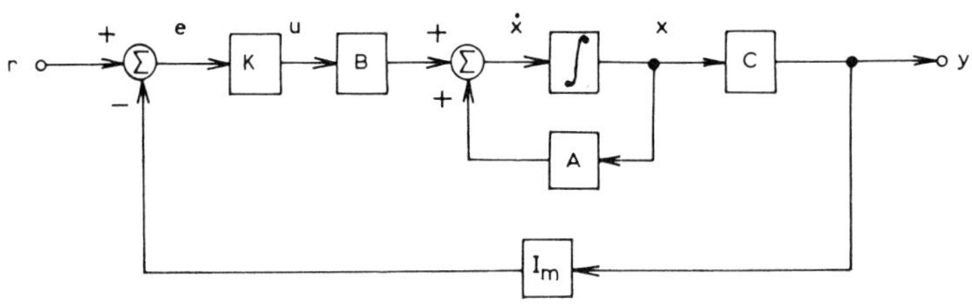

Figure 1.

4. Calculation of invariant zeros from state-space parameters

Consider the feedback arrangement shown in Fig. 1 and let the feedback gain matrix be

$$K = kI_m \tag{4.0, 1}$$

The closed-loop system dynamical transformation matrix will then be of the form

$$A_{\mathrm{cl}} = A - kBC \tag{4.0, 2}$$

Suppose we now carry out a change of basis in the state space, taking as the new set of basis vectors the eigenvectors of BC ; these are assumed to be a linearly independent spanning set for the state space. The eigenvalues of A_{cl} are invariant under this change of basis, and the effect of the gain parameter k

on the eigenvalues of $(A+kBC)$ is best determined by inspecting the transformed form of $(A+kBC)$ appropriate to this change of basis, which is given by

$$\begin{aligned}\tilde{A}_{\mathrm{cl}} &= V_{BC}(A-kBC)U_{BC}\\ &= V_{BC}AU_{BC}-k\Lambda_{BC}\\ &= \begin{bmatrix} Z \\ N \end{bmatrix} A[W \;\; M] - k\begin{bmatrix} \Lambda_\rho & 0_{m,\,n-m} \\ 0_{n-m,\,m} & 0_{n-m,\,n-m} \end{bmatrix}\\ &= \begin{bmatrix} (ZAW-k\Lambda_\rho) & ZAM \\ NAW & NAM \end{bmatrix} \end{aligned} \qquad (4.0,\ 3)$$

The closed-loop system characteristic frequencies are thus determined from a closed-loop characteristic polynomial given by

$$|sI_n-\tilde{A}_{\mathrm{cl}}| = \begin{vmatrix} (sI_m-ZAW+k\Lambda_\rho) & -ZAM \\ -NAW & (sI_{n-m}-NAM) \end{vmatrix}$$

Applying Schur's formula (Gantmacher 1959) to this partitioned determinant we then have that the closed-loop system characteristic frequencies are determined from

$$|sI_m-ZAW+k\Lambda_\rho||-NAW(sI_m-ZAW+k\Lambda_\rho)^{-1}ZAM + sI_{n-m}-NAM|=0 \qquad (4.0,\ 4)$$

where $|sI_m-ZAW+k\Lambda_\rho|$ is assumed to be non-zero and

$$|sI_{n-m}-NAM||-ZAM(sI-NAM)^{-1}NAW + sI_m-ZAW+k\Lambda_\rho|=0 \qquad (4.0,\ 5)$$

where $|sI_{n-m}-NAM|$ is assumed to be non-zero.

If we now allow k to become arbitrarily large, this pair of equations may be replaced by the simpler forms

$$|-NAW(sI_m+k\Lambda_\rho)^{-1}ZAM+sI_{n-m}-NAM|=0 \qquad (4.0,\ 6)$$

and

$$|sI_m+k\Lambda_\rho|=0 \qquad (4.0,\ 7)$$

The m solutions of eqn. (4.0, 7) which are of the form

$$s=-k\rho_i \qquad (4.0,\ 8)$$

correspond to that set of m closed-loop characteristic frequencies which tend to infinity with k tending to infinity. The remaining $(n-m)$ closed-loop characteristic frequencies, given by eqn. (4.0, 6), remain finite as k tends to infinity ; these therefore define the zeros of the system. Now for finite values of s and infinite values of k, eqn. (4.0, 6) may be replaced by the simple equation

$$|sI_{n-m}-NAM|=0 \qquad (4.0,\ 9)$$

since each diagonal entry of the diagonal matrix $(sI_m-k\Lambda_\rho)^{-1}$ which occurs in eqn. (4.0, 7) is of the form $(s+k\rho_i)^{-1}$ and will thus tend to zero as k tends to infinity.

We therefore reach the extremely useful and important conclusion that the zeros of the system are given by the eigenvalues of the matrix NAM, since these are the solutions of eqn. (4.0, 9). This fact leads immediately to a simple algorithm for determining the zeros of a system directly from its state-space model-defining matrices A, B and C. Before describing this algorithm however we must consider the implications of the non-uniqueness of the matrices M and N. It is shown in Appendices 1 and 2 that any appropriate pair of matrices M and N may be used to calculate the system zeros.

4.1. *Algorithm for calculation of zeros*

The above results lead to the following algorithm for the calculation of the transmission zeros of a system from its state-space parameters (Kouvaritakis 1974).

(i) For the output matrix C, compute any right annihilator M whose columns are linearly independent (see Appendix 1).

(ii) For the input matrix B, compute any left annihilator N whose rows are linearly independent (see Appendix 1).

(iii) Form the matrix NM and then pre-multiply N by $(NM)^{-1}$. This makes the resulting modified matrix $\bar{N}$, derived from N in this way, a left inverse of M.

(iv) Form the matrix $\bar{N}AM$.

(v) Find the invariant zeros of the system by calculating the eigenvalues of the matrix $\bar{N}AM$.

4.2. *Example of use of algorithm to calculate zeros*

This technique of zero calculation is now illustrated by means of an example for a system having two inputs, four states and two outputs. The corresponding state-space model-defining matrices are

$$A=\begin{bmatrix} -1 & 1 & 3 & -2 \\ 0 & -1 & -1 & -1 \\ 0 & 1 & -3 & -1 \\ 0 & 3 & -1 & -5 \end{bmatrix}, \quad B=\begin{bmatrix} 1 & 0 \\ 1 & 2 \\ 1 & 1 \\ 2 & 2 \end{bmatrix}, \quad C=\begin{bmatrix} 1 & -1 & 3 & 0 \\ 0 & -1 & -3 & 2 \end{bmatrix}$$

If the corresponding transfer function matrix $G(s)$ or the system matrix $P(s)$ is calculated, then an application of any of the standard techniques will show that the system has two zeros located at

$$z_1=-1 \quad \text{and} \quad z_2=-2$$

We now show that determining the matrix $\bar{N}AM$ and finding its eigenvalues gives the same result.

(i) and (ii). Following the procedure described in Appendix 1 we set

$$M=\begin{bmatrix} M' \\ I_2 \end{bmatrix} \quad \text{and} \quad N=[N' \ \ I_2]$$

where M' and N' are obtained as

$$M' = \begin{bmatrix} -6 & 2 \\ -3 & 2 \end{bmatrix} \quad \text{and} \quad N' = \tfrac{1}{2} \begin{bmatrix} -1 & -1 \\ -2 & -2 \end{bmatrix}$$

After substitution and appropriate scaling we then obtain

$$M = \begin{bmatrix} -6 & 2 \\ -3 & 2 \\ 1 & 0 \\ 0 & 1 \end{bmatrix}, \quad N = \begin{bmatrix} 1 & 1 & -2 & 0 \\ 2 & 2 & 0 & -2 \end{bmatrix}$$

(iii) In order to suitably adjust the rows of N we pre-multiply it by $(NM)^{-1}$ where

$$(NM)^{-1} = \begin{bmatrix} -11 & 4 \\ -18 & 6 \end{bmatrix}^{-1} = \tfrac{1}{6} \begin{bmatrix} 6 & -4 \\ 18 & -11 \end{bmatrix}$$

giving

$$\bar{N} = \tfrac{1}{6} \begin{bmatrix} 6 & -4 \\ 18 & -11 \end{bmatrix} \begin{bmatrix} 1 & 1 & -2 & 0 \\ 2 & 2 & 0 & -2 \end{bmatrix} = \tfrac{1}{3} \begin{bmatrix} -1 & -1 & -6 & 4 \\ -2 & -2 & -18 & 11 \end{bmatrix}$$

(iv) The matrix product $\bar{N}AM$ is then given by

$$\bar{N}AM = \tfrac{1}{3} \begin{bmatrix} -1 & -1 & -6 & 4 \\ -2 & -2 & -18 & 11 \end{bmatrix} \begin{bmatrix} -1 & 1 & 3 & -2 \\ 0 & -1 & -1 & -1 \\ 0 & 1 & -3 & -1 \\ 0 & 3 & -1 & -5 \end{bmatrix} \begin{bmatrix} -6 & 2 \\ -3 & 2 \\ 1 & 0 \\ 0 & 1 \end{bmatrix}$$

$$= \begin{bmatrix} -4 & 1 \\ -6 & 1 \end{bmatrix}$$

(v) This has as characteristic equation

$$\lambda^2 + 3\lambda + 2 = 0$$

so that the eigenvalues of $\bar{N}AM$ are

$$\lambda_1 = -1 \quad \text{and} \quad \lambda_2 = -2$$

which are indeed the system zeros.

4.3. *The case of repeated roots*

In the derivation of the algorithm for computing zeros given above it was assumed that the matrix BC could be expressed in the form shown in eqn. (3.0, 1); this is not possible if BC has a set of eigenvalues which are not all distinct. This difficulty can however be easily circumvented by then using the standard Jordan form decomposition for BC. Thus, if BC has repeated non-zero eigenvalues, we may express it in the form

$$BC = [W \;\; M] \begin{bmatrix} J & 0_{m,\,n-m} \\ 0_{n-m,\,m} & 0_{n-m,\,n-m} \end{bmatrix} \begin{bmatrix} Z \\ N \end{bmatrix} \qquad (4.3, 1)$$

where J now has the form appropriate to a sum of companion forms corresponding to the invariant polynomials of BC. If we again take a closed-loop system with gain matrix of the form

$$K = kI_m$$

then the closed-loop characteristic frequencies will be the roots of the pair of equations

$$|sI_m - ZAW + kJ| = 0 \tag{4.3, 2}$$

and

$$|-NAW(sI_m - ZAW + kJ)^{-1} + sI_{n-m} - NAM| = 0 \tag{4.3, 3}$$

Following a similar line of argument to that adopted for the case of distinct non-zero eigenvalues of BC we replace this pair of equations, as k tends to infinity, by the pair

$$|sI_m + kJ| = 0 \tag{4.3, 4}$$

and

$$|-NAW(sI_m + kJ)^{-1}ZAM + sI_{n-m} - NAM| = 0 \tag{4.3, 5}$$

The matrix J is of triangular form, having as diagonal elements the set of non-zero eigenvalues of BC, which we will again denote by ρ_i. As sI_m is diagonal, the matrix $sI_m + kJ$ will have triangular form and hence the m closed-loop poles that go to infinity with k are given by

$$s_i = -k\rho_i$$

For finite s and arbitrarily large k we will have that

$$|sI_m + kJ| \simeq k^m \prod_{i=1}^{m} \rho_i$$

and that

$$(sI_m + kJ)^{-1} \simeq \begin{vmatrix} \frac{1}{k\rho_1} & \times & \times & \times \cdots & \times \\ & \frac{1}{k\rho_2} & \times & \times \cdots & \times \\ & & \ddots & & \vdots \\ & & & & \frac{1}{k\rho_m} \end{vmatrix}$$

where the non-zero off-diagonal elements (denoted by a $\times$) will have the form of constants divided by powers of k. We will thus have that

$$(sI_m + kJ)^{-1} \rightarrow 0_{m,m}$$

as $k \to \infty$. Therefore, for finite values of s and arbitrarily large values of k, eqn. (4.3, 5) reduces to

$$|sI_{n-m} - NAM| = 0 \tag{4.3, 6}$$

which establishes that the algorithm previously given remains valid when BC has repeated non-zero eigenvalues.

5. Case when *CB* does not have full rank : zeros at infinity

Since we may assume that all the elements of the matrices N, A and M are finite, the existence of a full rank CB will guarantee the existence of $(n-m)$ finite zeros. It is however possible to have a system for which the degree of the zero polynomial $z(s)$ is less than $(n-m)$. Since it is conventional in complex variable work to regard a function as having an equal number of poles and zeros when the zeros at infinity are included, we will regard such systems as having 'extra' zeros at infinity. It should be clear, from the nature of the arguments deployed in previous sections, that this can only happen if CB is not of full rank. It is the purpose of this section to examine the implications of the situation which arises when

$$|CB| = 0$$

We first observe that if BC has more than $(n-m)$ zero eigenvalues, it is no longer possible to express the matrix product BC in the form shown in eqn. (3.0, 1). The reason for this is that the order of the left and right null spaces of B and C respectively, and hence of BC, is $(n-m)$ so that only $(n-m)$ independent eigenvectors may be determined for the zero eigenvalues. A recourse to the Jordan form of BC will not help either, since zero elements will now appear along the principal diagonal of the matrix J and the reduction of eqn. (4.3, 3) to eqn. (4.3, 5) will no longer be possible. Further difficulty arises from the fact that, as will be proved shortly, the singularity of CB implies that NM will also be singular ; this means that we will no longer be able to choose an N and M such that

$$NM = I_{n-m}$$

It will now be shown that, in the general case where CB may not have full rank, the zeros of a proper system $S(A, B, C)$ are the solutions of the equation

$$|sNM - NAM| = 0 \tag{5.0, 1}$$

To establish this, we again use the fact that the finite zeros are the limiting finite positions of the closed-loop characteristic frequencies as $k \to \infty$ with $K = kI$. The closed-loop characteristic frequencies are the solutions of

$$|sI - A + kBC| = 0 \tag{5.0, 2}$$

Let the pair of matrices N and M be again defined as right and left annihilators of C and B respectively, and introduce a pair of matrices L and R such that

$$L = \begin{bmatrix} N \\ N' \end{bmatrix} \quad \text{and} \quad R = [M \quad M']$$

where N' and M' are any constant matrices such that the columns of R and the rows of L span the state space. L and R are then non-singular and we may multiply the left-hand side of eqn. (5.0, 2) on the left by L, and on the right by R and then take determinants to get that

$$|L||sI-A+kBC||R| = \begin{vmatrix} N \\ N' \end{vmatrix} |sI-A+kBC| \; |M \;\; M'|$$

$$= \begin{vmatrix} (sNM-NAM) & (sNM'-NAM') \\ (sN'M-N'AM) & (sN'M'-N'AM'+kN'BCM') \end{vmatrix}$$

$$=0$$

Using Schur's formula for partitioned determinants then gives that

$$|-(sNM'-NAM')(sN'M'-N'AM'-kN'BCM')^{-1}(sN'M-N'AM) + sNM-NAM||sN'M'-N'AM'+kN'BCM'|=0 \qquad (5.0, 3)$$

Since, as usual, we are only concerned with the situation for which s is finite and k is infinite, this may be reduced to

$$|sNM-NAM||kN'BCM'|^{-1}=0 \qquad (5.0, 4)$$

It follows directly from the way in which the matrices N' and M' were introduced that their rows and columns are independent of the rows and columns of N and M respectively. Hence the product matrix $N'BCM'$ will have full rank. It therefore follows that those closed-loop characteristic frequencies of the feedback system which remain finite as $k\to\infty$, and hence the required zeros, are the roots of the equation

$$|sNM-NAM|=0 \qquad (5.0, 5)$$

5.1. *Rank deficiency of NM*

Before proceeding to a revised form of algorithm for calculating the zeros in the case where CB does not have full rank, we must relate the rank deficiency of CB to the rank deficiency of NM. Specifically, we shall show that the degree of rank deficiency of CB is equal to the degree of rank deficiency of NM.

Since it has been assumed that both the matrices B and C have full rank, and since we can always carry out an appropriate re-numbering of the system inputs and outputs, we may partition the matrices B and C to get

$$B = \overset{\leftarrow\; m\; \rightarrow}{\begin{bmatrix} B_1 \\ \cdots \\ B_2 \end{bmatrix}} \begin{matrix} \updownarrow n-m \\ \updownarrow m \end{matrix}$$

and

$$C = \underset{\updownarrow}{m} \overset{\leftarrow n-m \rightarrow \leftarrow m \rightarrow}{[\; C_1 \;\vdots\; C_2 \;]}$$

where both B_2 and C_2 are non-singular. We may then, as shown in Appendix 1, express the matrices M and N in the form

$$M = \begin{bmatrix} I_{n-m} \\ -C_2^{-1}C_1 \end{bmatrix} K_m$$

and

$$N = K_n[I_{n-m} \quad -B_1B_2^{-1}]$$

where K_m and K_n are appropriate $(n-m)\times(n-m)$ constant matrices of full rank. It follows that

$$\text{rank of } NM = \text{rank of } K_n[I_{n-m} \quad -B_1B_2^{-1}]\begin{bmatrix} I_{n-m} \\ -C_2^{-1}C_1 \end{bmatrix} K_m$$

$$= \text{rank of } (I_{n-m} + B_1B_2^{-1}C_2^{-1}C_1) \qquad (5.1,\ 1)$$

CB however does not have full rank and so we will have that

$$\begin{aligned} |CB| &= |C_1B_1 + C_2B_2| \\ &= |C_2||B_2||I_m + C_2^{-1}C_1B_1B_2^{-1}| \\ &= 0 \end{aligned}$$

so that we have

$$|I_m + C_2^{-1}C_1B_1B_2^{-1}| = 0 \qquad (5.1,\ 2)$$

since C_2 and B_2 are taken to be non-singular.

Now the eigenvalue-shift theorem tells us that when a unit matrix is added to some given matrix then unity is added to all its eigenvalues. Thus eqn. (5.1, 2) implies that some of the eigenvalues of the matrix $C_2^{-1}C_1B_1B_2^{-1}$ must be -1. In particular, if the rank deficiency of CB is d, then d of the eigenvalues of $C_2^{-1}C_1B_1B_2^{-1}$ will be equal to -1. A standard result of matrix theory states that the non-zero eigenvalues of the product of two matrices are the same irrespective of the order in which they are multiplied. It thus follows that the matrix $B_1B_2^{-1}C_2^{-1}C_1$ will also have d of its eigenvalues equal to -1. Therefore eqn. (5.1, 1) gives that the rank deficiency of NM is also d, which is the result we wished to establish.

5.2. *Degree of zero polynomial*

We are now in a position to prove that the degree of the zero polynomial of the system $S(A, B, C)$, when CB has rank deficiency d, is reduced from $(n-m)$ to $(n-m-d)$. To do this we first note that since the product NM has rank $(n-m-d)$ we may select K_m and K_n to be such that

$$NM = \begin{bmatrix} I_{n-m-d} & 0_{n-m-d,\ d} \\ 0_{d,\ n-m-d} & 0_{d,\ d} \end{bmatrix} \qquad (5.2,\ 1)$$

The system zero polynomial can therefore be written in the form

$$z(s) = \left| \begin{bmatrix} sI_{n-m-d} & 0 \\ 0 & 0 \end{bmatrix} - NAM \right|$$

Now let the matrix NAM be partitioned in the following way :

$$NAM = \left[\begin{array}{c:c} A_0 & B_0 \\ \hdashline -C_0 & -D_0 \end{array}\right] \begin{array}{l} \updownarrow n-m-d \\ \updownarrow d \end{array}$$

(column widths: $\leftarrow n-m-d \rightarrow \leftarrow d \rightarrow$)

so that the zero polynomial assumes the form

$$z(s) = \begin{vmatrix} (sI_{n-m-d} - A_0) & -B_0 \\ C_0 & D_0 \end{vmatrix} \tag{5.2, 2}$$

This shows that the degree of $z(s)$ is in general $(n-m-d)$ (and never more than $(n-m-d)$). Thus we can consider d of the zeros of the system $S(A, B, C)$ to have vanished to infinity, and can determine the remaining $(n-m-d)$ finite zeros as the zeros for the associated reduced non-proper system $S(A_0, B_0, C_0, D_0)$ whose system matrix determinant is exhibited in eqn. (5.2, 2).

5.3. *Calculation of zeros for associated reduced non-proper systems*

To complete our investigation and compute the zeros of a system for which CB is rank deficient, we need to know how to locate the zeros of the non-proper $S(A_0, B_0, C_0, D_0)$. To do this we again investigate the limiting positions of an appropriate set of closed-loop eigenvalues with

$$K = kI_{n-m-d} \quad \text{and} \quad k \to \infty$$

The appropriate closed-loop dynamical transformation matrix now takes the form

$$\bar{A}_{\rm cl} = A_0 - B_0 K(I + D_0 K)^{-1} C_0 \tag{5.3, 1}$$

which for our particular form of K becomes

$$\bar{A}_{\rm cl} = A_0 - B_0 \left(\frac{1}{k} I + D_0\right)^{-1} C_0 \tag{5.3, 2}$$

As $k \to \infty$ we thus have that

$$\bar{A}_{\rm cl} \to A_0 - B_0 D_0^{-1} C_0 \tag{5.3, 3}$$

if we assume that D_0 is non-singular. Since A_0 has dimension $(n-m-d)$ we have that the $(n-m-d)$ limiting closed-loop characteristic frequencies, and hence the required zeros, are the eigenvalues of the matrix $(A_0 - B_0 D_0^{-1} C_0)$.

Finally we consider what to do should D_0 be singular. Suppose then that D_0 is rank deficient by f degrees. The zeros are determined from the zero polynomial

$$z(s) = \begin{vmatrix} sI_{n-m-d} - A_0 & -B_0 \\ C_0 & D_0 \end{vmatrix} \tag{5.3, 4}$$

which, after a series of appropriate right and left unimodular transformations, may be replaced by

$$z'(s) = \begin{vmatrix} sI_{n-m-d} - A & -B_1' & -B_2' \\ C_1' & 0_{f,f} & 0_{f,m-f} \\ C_2' & 0_{m-f,f} & I_{m-f} \end{vmatrix} \tag{5.3, 5}$$

and so by

$$z'(s) = \alpha \begin{vmatrix} sI_{n-m-d-f} - A_1' & -B_1' \\ C_1' & 0_{f,f} \end{vmatrix} \tag{5.3, 6}$$

where α is an appropriate constant and

$$A_1' = A_0 - B_2'C_2' \tag{5.3, 7}$$

The problem has thus been reduced to that of finding the zeros of the system $S(A_1', B_1', C_1')$ which may be done by an application of the standard algorithm.

5.4. *General form of algorithm for determination of zeros*

The above results lead to the following modified form of algorithm which may be used to compute the finite zeros of a system $S(A, B, C)$ for which the product CB does not have full rank.

(i) Find d, the rank deficiency of CB.

(ii) Compute a pair of matrices N and M by the procedure discussed in § 5.1 and such that

$$NM = \begin{bmatrix} I_{n-m-d} & 0 \\ 0 & 0_{d,d} \end{bmatrix} \tag{5.4, 1}$$

(iii) Calculate NAM and partition it to get

$$NAM = \begin{bmatrix} A_0 & B_0 \\ -C_0 & -D_0 \end{bmatrix} \tag{5.4, 2}$$

(iv) If D_0 is non-singular, then the required zeros are the eigenvalues of the matrix $(A_0 - B_0D_0^{-1}C_0)$. If D_0 is not of full rank then form the appropriately partitioned polynomial matrix

$$|sNM - NAM|$$

and apply unimodular transformations to bring $-D_0$ into diagonal form and having only 0's and 1's along its principal diagonal. By collecting together those columns of B_0 and rows of C_0 which are associated with the zero entries, form from them the pairs of matrices B_1', C_1' and B_2', C_2' respectively. The zeros of the original system are then found by applying the standard 'NAM algorithm' to the system having

$$A \to A_0 - B_2'C_2'$$

$$B \to B_1'$$

and

$$C \to C_1'$$

6. Conclusions

It has been shown how the invariant zeros of a system can be calculated directly from its state-space model-defining matrices A, B and C. The relationships involved give a most useful and interesting insight into the relationships between state-space and frequency-response descriptions of system behaviour. In particular they show clearly that the zeros are characteristic of the way in which the internal dynamical mechanism of the system (associated with the matrix A) is coupled to the environment (this coupling represented by the matrices B and C).

Appendix 1

This Appendix is largely concerned with the implications of the non-uniqueness of the matrices M and N. The effects of this are examined by deriving a pair of uniquely specified matrices in terms of which all other admissible candidates may be expressed.

Let M_0 be a solution to eqn. (3.0, 4) so that

$$CM_0 = 0_{m,\, n-m}$$

This form of relationship with C will obviously still hold if some of the columns of the matrix M_0 are replaced by linear combinations of its columns. It follows that all other solutions to eqn. (3.0, 4) can be written in the form

$$M = M_0 K_m \tag{A 1, 1}$$

where K_m is an $(n-m)\times(n-m)$ matrix of constants. By a similar form of argument, applied to rows, we see that if N_0 is a left annihilator of B all other left annihilators can be expressed in the form

$$N = K_n N_0 \tag{A 1, 2}$$

where K_n is again an $(n-m)\times(n-m)$ matrix of constants.

Consideration of the dimensions and assumed ranks of the matrices B and C shows that there will exist $(n-m)$ linearly independent column vectors μ_i and $(n-m)$ linearly independent row vectors $v_i{}^t$ which satisfy the equations

$$C\mu_i = 0_{m,\,1}$$

and

$$v_i{}^t B = 0_{1,\, m}$$

Clearly any linear combinations of these column vectors μ_i and row vectors $v_i{}^t$ will also satisfy these equations. It thus follows that, in selecting any particular μ_i and $v_i{}^t$, any $(n-m)$ of their entries may be specified arbitrarily. It is therefore possible for us to select right annihilators of C and left annihilators of B of the form

$$M_0 = \begin{bmatrix} M_0{}' \\ \cdots \\ I_{n-m} \end{bmatrix} \begin{matrix} \updownarrow m \\ \updownarrow n-m \end{matrix}$$

and

$$N_0 = [\overset{\leftarrow m \rightarrow}{N_0{}'} \;\vdots\; \overset{\leftarrow n-m \rightarrow}{I_{n-m}}]$$

in which all columns of M_0 and all rows of N_0 are linearly independent. It is now shown that such an M_0 and N_0 can be uniquely specified. To see this we insert these forms of expression for M_0 and N_0 into eqns. (3.0, 4) and (3.0, 5) and get, after appropriate partitioning of the matrices C and B respectively, that

$$CM = [\overset{\leftarrow m \rightarrow}{C_1} \;\vdots\; \overset{\leftarrow n-m \rightarrow}{C_2}] \begin{bmatrix} M_0' \\ \hdashline I_{n-m} \end{bmatrix} = 0_{m,\, n-m}$$

from which we have

$$M_0' = -C_1^{-1}C_2 \tag{A 1, 3}$$

and also that

$$NB = [N_0' \vdots I_{n-m}] \begin{bmatrix} B_1 \\ \hdashline B_2 \end{bmatrix} \begin{matrix} \updownarrow m \\ \updownarrow n-m \end{matrix}$$

from which we have

$$N_0' = -B_2B_1^{-1} \tag{A 1, 4}$$

Naturally the exercise of choice involved in assigning the $(n-m)$ elements of the vectors μ_i and v_i' which is involved in carrying out this procedure must be done in such a way that neither B_1 nor C_1 is singular.

We can now examine the implications of the condition

$$NM = I_m$$

by substituting into it the general forms of M and N given by eqns. (A 1, 1) and (A 1, 2). This gives

$$K_nN_0M_0K_m = I_m$$

from which

$$K_n = K_m^{-1}(N_0M_0)^{-1}$$

We thus see that the pair of matrices

$$M = M_0K_m \tag{A 1, 5}$$

and

$$N = K_m^{-1}(N_0M_0)^{-1}N_0 \tag{A 1, 6}$$

will satisfy all three conditions stipulated by eqns. (3.0, 4), (3.0, 5) and (3.0, 6).

To complete our investigation we must now show that any pair of matrices M and N, given by eqns. (A 1, 5) and (A 1, 6), will satisfy the spectral decomposition shown in eqn. (3.0, 1). To do this we first note that the matrix product BC is independent of both M and N, being given by

$$\begin{aligned} BC &= [W \quad M] \begin{bmatrix} \Lambda_\rho & 0_{m,\, n-m} \\ 0_{n-m,\, m} & 0_{n-m,\, n-m} \end{bmatrix} \begin{bmatrix} Z \\ N \end{bmatrix} \\ &= W\Lambda_\rho Z \end{aligned}$$

All that is required to be shown therefore is that such an M and N ensure that

$$\begin{bmatrix} Z \\ N \end{bmatrix} [W \quad M] = I_n$$

that is that the rows of N and the columns of M have a suitably orthonormal relationship. Written out in more detail this last set of relationships requires that

$$ZM = 0_{m,\, n-m} \tag{A 1, 7}$$

$$NW = 0_{n-m,\, m} \tag{A 1, 8}$$

$$NM = I_{n-m} \tag{A 1, 9}$$

Now N and M have already been chosen such that (A 1, 9) is satisfied. To show that they also satisfy conditions (A 1, 7) and (A 1, 8) we pre-multiply and post-multiply equations

$$(BC)W = W\Lambda_\rho$$

and

$$Z(BC) = \Lambda_\rho Z$$

by N and M respectively to obtain

$$\begin{aligned} N(BC)W &= (NB)(CW) \\ &= 0_{n-m,\, m} \\ &= NW\Lambda_\rho \end{aligned}$$

and

$$\begin{aligned} Z(BC)M &= ZB(CM) \\ &= 0_{m,\, n-m} \\ &= \Lambda_\rho ZM \end{aligned}$$

Finally, since Λ_ρ is a diagonal matrix of non-zero eigenvalues of BC, we can conclude that these show that eqns. (A 1, 8) and (A 1, 7) must be satisfied.

Appendix 2

The purpose of this Appendix is to consider the effect of different choices of K_m. In the previous Appendix we have derived the following expressions for M and N :

$$M = M_0 K_m$$

and

$$N = K_m^{-1}(N_0 M_0)^{-1} N_0$$

Clearly different choices of K_m will lead to different annihilators M and N. The key matrix NAM whose eigenvalues are the system zeros is given by

$$NAM = K_m^{-1}(N_0 M_0)^{-1} N_0 A M_0 K_m \tag{A 2, 1}$$

This is obviously a similarity transformation of

$$N'AM' = (N_0M_0)^{-1}N_0AM_0 \tag{A 2, 2}$$

and therefore has the same eigenvalues as $N'AM'$, which is independent of K_m. It follows that the eigenvalues of NAM, and hence the zeros as computed by the algorithm, will be independent of the choice of K_m.

For the purposes of zero determination therefore, we may choose for K_m any matrix of full rank. In particular we can choose

$$K_m = I_{n-m}$$

and M and N then assume the simple forms

$$M = M_0 \tag{A 2, 3}$$

and

$$N = (N_0M_0)^{-1}N_0 \tag{A 2, 4}$$

References

Gantmacher, F. R., 1959, *Theory of Matrices*, Vol. 1 (New York : Chelsea).
Hsu, C. H., and Chen, C. T., 1968, *Proc. I.E.E.E.*, **56,** 2061.
Kouvaritakis, B., 1974, Ph.D. Thesis, University of Manchester.
MacFarlane, A. G. J., 1974, *I.F.A.C. Symposium on Multivariable Technological Systems*, Manchester.
Rosenbrock, H. H., 1970, *State Space and Multivariable Theory* (London : Nelson).

Geometric approach to analysis and synthesis of system zeros

Part 2. Non-square systems

B. KOUVARITAKIS and A. G. J. MacFARLANE

A method is derived for computing the invariant zeros of a general non-square matrix-valued function of a complex variable in terms of the parameters of the corresponding state-space system model. Based on this a method is developed for choosing a ' squaring down ' output matrix to give a desired set of zero locations for a system whose states are all accessible. This technique is then extended to deal with the situation when all the states are not accessible.

1. Introduction

The preceding paper (Kouvaritakis and MacFarlane 1976) has described a geometrical approach to the calculation of the invariant zeros of a given linear dynamical system in terms of its state-space model-defining parameters. The treatment given there was however confined to systems having the same number of inputs and outputs, and thus to square matrix-valued functions of a complex variable. Since zeros can be defined for systems with arbitrary (and in particular unequal) numbers of inputs and outputs (MacFarlane 1974) it is important that such a geometrical approach be extended to the general case, and this extension is given here. The theory developed gives an insight into the now well-established fact (Davison and Wang 1974) that non-square transfer-function matrices only have zeros under very special circumstances. Roughly speaking, a general *non-square* transfer-function matrix almost always has no transmission zeros ; this fact becomes obvious in terms of the geometrical relationships developed here.

2. Zeros of a non-square system

Let $G(s)$ be an $m \times l$ matrix-valued function of the complex variable s. Suppose $l \neq m$ and assume that $m > l$. Let ρ_G be the normal rank of $G(s)$, and assume that $\rho_G = l$. Then $z(s)$, the zero polynomial of $G(s)$, is the greatest common divisor of all minors of $G(s)$ of order l *provided that* each minor has been adjusted to have the system open-loop characteristic polynomial $p(s)$ as its common denominator (Kontakos 1973). The purpose of the denominator adjustment is to ' put back ' any zeros cancelled by coincident poles in evaluating the determinants involved. An alternative way of looking at the zeros in terms of state-space parameters is as those values of s for which the $(n+m) \times (n+l)$ system matrix (for a proper system)

$$P(s) = \begin{bmatrix} sI - A & -B \\ C & 0 \end{bmatrix} \tag{2.0, 1}$$

loses rank. Such values can be found by determining, for a $P(s)$ of normal rank $[n+\min(l, m)]$ what values of s make all minors of order $[n+\min(l, m)]$ vanish. Consideration of the forms of such minors shows that they will contain among them ones whose value is a polynomial of degree $[n-\max(l, m)]$ (corresponding to those minors in which the whole of the zero matrix in the bottom right-hand corner of $P(s)$ is incorporated). We thus note that the *maximum* degree of $z(s)$ cannot be more than $[n-\max(l, m)]$. Under our assumptions, then $z(s)$ cannot have degree greater than $(n-m)$.

The selection of minors of $G(s)$ of order l corresponds to finding the determinants of the transfer-function matrices of the l-input l-output subsystems formed from the given system by selecting sets of l outputs. This involves selecting l rows from $G(s)$ or, alternatively, the formation of new systems from the original system $S(A, B, C)$ which generated $G(s)$ by selecting l rows of C, to get new $l \times n$ output matrices C_i, and computing the determinants of the transfer functions for the corresponding set of systems $S_i(A, B, C_i)$. It follows from the fact that $z(s)$ is the greatest common divisor of all the (adjusted) zero polynomials of the set of $l \times l$ subsystems which may be formed from the original system, that the zeros of the given non-square system are the *intersection* of the sets of zeros for all these $l \times l$ subsystems.

This way of looking at the zeros of a non-square system, in terms of the intersection of all the sets of zeros calculated for all square subsystems of order l which can be formed from it by different selections of l outputs, could obviously be made the basis of an algorithm for calculating zeros which used the technique previously developed for square systems. We wish however to develop a more direct geometrical approach to the problem of finding zeros for non-square systems, and this we now consider.

3. Geometric method for calculation of zeros from state-space model parameters

As in the previously discussed case of systems having the same number of inputs and outputs, the key to the geometrical approach developed here is the introduction of a pair of matrices N and M which are left and right annihilators of B and C respectively, that is such that

$$NB = 0_{n-l,\, l} \tag{3.0, 1}$$

$$CM = 0_{m,\, n-m} \tag{3.0, 2}$$

It will now be shown that, if the system is such that CB has full rank, and if the number of outputs is greater than or equal to the number of inputs $(m \geqslant l)$, then the system matrix

$$P(s) = \begin{bmatrix} sI - A & -B \\ C & 0_{m,\, l} \end{bmatrix} \tag{3.0, 3}$$

loses rank if and only if the matrix

$$\begin{bmatrix} sI - A_1 \\ -C_1 \end{bmatrix}$$

loses rank, where A_1 and C_1 are given by

$$A_1=(NM)^\dagger NAM \qquad (3.0,\ 4)$$

$$C_1=(NM)^0 NAM \qquad (3.0,\ 5)$$

Here $(NM)^\dagger$ is a left inverse of NM and $(NM)^0$ is a left annihilator of NM so that

$$(NM)^\dagger NM=I_{n-m} \qquad (3.0,\ 6)$$

and

$$(NM)^0 NM=0_{m-l,\ n-m} \qquad (3.0,\ 7)$$

It is shown in the following section that NM will have full rank when CB has full rank.

This result turns the problem of calculating the zeros of a non-square system with more outputs than inputs into an equivalent problem of finding the output decoupling zeros (Rosenbrock 1970) of a related lower-order system. The case of systems having more inputs than outputs is treated by deriving a dual result which turns the zero determination problem into one of finding the *input* decoupling zeros of a related lower-order system.

3.1. *Generation of system whose output decoupling zeros are transmission zeros of given non-square system*

Let $M^\dagger$ be a right inverse of C and $N^\dagger$ a left inverse of B, and define a pair of matrices L and R by

$$L=\begin{bmatrix} N & 0_{n-l,\ m} \\ N^\dagger & 0_{l,\ m} \\ 0_{m,\ n} & I_m \end{bmatrix} \qquad (3.1,\ 1)$$

and

$$R=\begin{bmatrix} M & M^\dagger & 0_{n,\ l} \\ & & \\ 0_{l,\ n-m} & 0_{l,\ m} & I_l \end{bmatrix} \qquad (3.1,\ 2)$$

If the system matrix $P(s)$ is then multiplied on the left by L and on the right by R, we obtain a matrix $P'(s)$ given by

$$\begin{aligned} P'(s)&=LP(s)R \\ &=\begin{bmatrix} (sNM-NAM) & (sNM^\dagger-NAM^\dagger) & 0_{n-l,\ l} \\ (sN^\dagger M-N^\dagger AM) & (sN^\dagger M^\dagger-N^\dagger AM^\dagger) & -I_l \\ 0_{m,\ n-m} & I_m & 0_{m,\ l} \end{bmatrix} \end{aligned} \qquad (3.1,\ 3)$$

Since both the matrices L and R are square and of full rank, the normal ranks of the matrices $P(s)$ and $P'(s)$ will be the same. Furthermore, the special form of the matrix $P'(s)$ is such that it will be rank deficient if and only if the matrix

$$Z(s)=sNM-NAM \qquad (3.1,\ 4)$$

loses rank.

Now let a further transforming matrix L_1 be defined by

$$L_1 = \begin{bmatrix} (NM)^\dagger \\ \\ (NM)^0 \end{bmatrix} \tag{3.1, 5}$$

where, as indicated in eqns. (3.0, 6) and (3.0, 7) respectively, $(NM)^\dagger$ is a left inverse of NM and $(NM)^0$ is a left annihilator of NM. If the matrix $Z(s)$ is multiplied on the left by the matrix L_1, then we will obtain another matrix, $Z_1(s)$, having the same rank as $Z(s)$ and given by

$$Z_1(s) = L_1 Z(s) = \begin{bmatrix} (NM)^\dagger \\ \\ (NM)^0 \end{bmatrix} [sNM - NAM]$$

so that

$$Z_1(s) = \begin{matrix} \scriptstyle n-m \\ \\ \scriptstyle m-l \end{matrix} \overset{\longleftarrow \quad n-m \quad \longrightarrow}{\begin{bmatrix} sI_{n-m} & -(NM)^\dagger NAM \\ \\ & -(NM)^0 NAM \end{bmatrix}} \tag{3.1, 6}$$

and thus

$$Z_1(s) = \begin{bmatrix} sI & -A_1 \\ \\ & -C_1 \end{bmatrix} \tag{3.1, 7}$$

where A_1 and C_1 are as defined in eqns. (3.0, 4) and (3.0, 5) respectively.

Thus the system matrix $P(s)$ given by eqn. (3.0, 3) will lose rank if and only if the matrix $Z_1(s)$ loses rank, which is the case if and only if s is an output decoupling zero for the system with state-space matrix pair A_1 and C_1 as dynamical transformation matrix and output coupling matrix respectively. Hence the problem of finding the zeros of a non-square system with more outputs than inputs is equivalent to finding the output decoupling zeros of a related, lower-order, dynamical system.

3.2. *Generation of system whose input decoupling zeros are transmission zeros of given non-square system*

If the given system has more inputs than outputs, then we can find for NM a right inverse and a right annihilator. An argument, similar to that just given, can then be constructed to show that the problem of finding the transmission zeros of the given system is equivalent to finding the input decoupling zeros of a related, lower-order, dynamical system with a dynamical transformation matrix and an input coupling matrix given by

$$A_1 = NAM(NM)^\dagger \tag{3.2, 1}$$

and

$$B_1 = NAM(NM)^0 \tag{3.2, 2}$$

where $(NM)^\dagger$ is a right inverse of NM, and $(NM)^0$ is a right annihilator of NM.

3.3. *Illustrative example*

Consider the system whose state-space model-defining matrices are

$$A=\begin{bmatrix} -2 & -6 & 3 & -7 & 6 \\ 0 & -5 & 4 & -4 & 8 \\ 0 & 2 & 0 & 2 & -2 \\ 0 & 6 & -3 & 5 & -6 \\ 0 & 2 & 2 & -2 & 5 \end{bmatrix}, \quad B=\begin{bmatrix} -2 & 7 \\ -8 & -5 \\ -3 & 0 \\ 1 & 5 \\ -8 & 0 \end{bmatrix}$$

$$C=\begin{bmatrix} 0 & -1 & 2 & -1 & -1 \\ 1 & 1 & 1 & 0 & -1 \\ 0 & 3 & -2 & 3 & -1 \end{bmatrix}$$

A suitable pair of matrices N and M is found to be

$$N=\begin{bmatrix} -5 & -7 & 22 & 0 & 0 \\ -35 & 17 & 0 & 66 & 0 \\ -20 & -28 & 0 & 0 & 33 \end{bmatrix}$$

$$M=\begin{bmatrix} 1 & -1 \\ -1 & 1 \\ 0 & 1 \\ 1 & 0 \\ 0 & 1 \end{bmatrix}$$

This gives the matrix NAM to be

$$NAM=\begin{bmatrix} 8 & -74 \\ 56 & -254 \\ 32 & -131 \end{bmatrix}$$

Finally the matrices A_1 and C_1 are found as

$$\begin{aligned} A_1 &= (NM)^{\dagger} NAM \\ &= \frac{-1}{110}\begin{bmatrix} 25 & 0 & -20 \\ -8 & 0 & 2 \end{bmatrix}\begin{bmatrix} 8 & -74 \\ 56 & -254 \\ 32 & -131 \end{bmatrix} \\ &= \begin{bmatrix} 4 & 7 \\ 0 & -3 \end{bmatrix} \\ C_1 &= (NM)^0 NAM \\ &= [3,\ -5,\ 8]\begin{bmatrix} 8 & -74 \\ 56 & -254 \\ 32 & -131 \end{bmatrix} = [0,\ 0] \end{aligned}$$

Since C_1 is a zero matrix, the output decoupling zeros corresponding to the pair (A_1, C_1) will simply be the eigenvalues of A_1 which are

$$z_1 = 4 \quad \text{and} \quad z_2 = -3$$

It can easily be verified that $P(s)$ loses rank for each of these values of the complex variable s. Furthermore, since the maximum number of invariant zeros which the given system can have is $(n-m)=5-3=2$, the method has found all the invariant zeros.

3.4. *Remarks on the zeros of non-square systems*

Reflection on the nature of the arguments just presented will show that a system having different numbers of inputs and outputs will only have invariant zeros in very special circumstances. The simplest way to see why any arbitrarily chosen system with unequal numbers of inputs and outputs will almost always not have any transmission zeros is to note that in general there is no reason why any of the eigenvectors of the matrix A_1, given by eqn. (3.0, 4), should lie in the kernel of the matrix, C_1, given by eqn. (3.0, 5). Furthermore, the maximum number of transmission zeros which a non-square system may have will depend on the rank of the appropriate related matrix C_1. If C_1 has full rank, then the maximum number of output decoupling zeros which the pair (A_1, C_1) may have is

$$n_z = (n-m) - (m-l) = n + l - 2m$$

As C_1 loses rank, the value of the integer variable n_z will increase, and each additional degree of rank deficiency of C_1 will imply the possibility of one more zero. The integer n_z will attain a maximum value of $(n-m)$ when C_1 becomes a zero matrix, and in this case the $(n-m)$ invariant zeros may be obtained as the $(n-m)$ eigenvalues of the matrix A_1.

We thus reach the important conclusion that zeros are ' naturally ' associated with square transfer-function matrices. This is intuitively satisfying since it can be shown that *square* matrix-valued functions of a complex variable can be related to the standard concepts used in the classical theory of functions of a complex variable (MacFarlane 1974).

4. Zeros at infinity : case when NM does not have full rank

We now consider the situation when NM is deficient in rank. As in the square case, it turns out that this is so if and only if CB is deficient in rank, and furthermore, the degrees of deficiency of rank of CB and of NM are the same. To show this we argue in the following way.

If CB has full rank, then a selection of any l of the m rows of C, to give a matrix C_1 say, will result in a product C_1B which will also have full rank. If C_1B has full rank, and M_1 is a right annihilator of C_1, then NM_1 will also have full rank since C_1B will be a square non-singular matrix. But if NM_1 has full rank, so must NM and we thus conclude that NM will have full rank if and only if CB has full rank.

Now suppose that the rank deficiency of CB is of degree one. Then there will exist a non-zero vector, v_1 say, such that

$$CBv_1 = 0$$

Since M is a right annihilator of C we can pick a non-zero vector v_2 such that

$$CMv_2 = 0$$

and so

$$CBv_1 = CMv_2$$

from which we deduce

$$Bv_1 = Mv_2$$

and thus

$$NBv_1 = NMv_2$$

which implies that NM is deficient in rank by at least one degree. To see that it can only be deficient in rank by one degree and no more, assume that there existed a non-zero vector $v_3 \neq v_2$ such that

$$NMv_3 = 0$$

We could then find another vector $v_4 \neq v_1$ such that

$$Mv_3 = Bv_4$$

which would in turn imply that

$$CBv_4 = CMv_3 = 0$$

where

$$v_4 \neq v_1$$

This however contradicts our assumption that CB is only deficient in rank by one degree and we therefore conclude that the degree of deficiency in rank of NM is one, the same as that of CB. Similar arguments may be constructed for CB rank deficient by two, three, etc. degrees.

4.1. *Zero locations when NM is not of full rank*

As in the analysis of § 3.1 we have that the system zeros are those values of s for which the matrix

$$Z(s) = sNM - NAM \tag{4.1, 1}$$

loses rank. Suppose CB, and therefore also NM, is rank deficient by d degrees. In this case we may always adjust the rows of N and the columns of M in such a way that

$$NM = \begin{bmatrix} I_{n-m-d} & 0_{n-m-d,\, d} \\ 0_{d,\, n-m-d} & 0_{d,\, d} \\ 0_{m-l,\, n-m-d} & 0_{m-l,\, d} \end{bmatrix} \tag{4.1, 2}$$

The matrix $Z(s)$, as given by eqn. (4.1, 1), may then be partitioned comformably with the partitioning of NM specified in eqn. (4.1, 2) to give a matrix of the form

$$Z_1(s) = \begin{bmatrix} (sI_{n-m-d} - A_1) & -B_1 \\ C_1 & -D_1 \end{bmatrix} \tag{4.1, 3}$$

This may be operated on from the left by a set of rank-preserving transformations to generate a matrix

$$Z_2(s) = \begin{bmatrix} sI_{n-m-d} - A_1 & -B_1 \\ -C_2 & I_d \\ -C_3 & 0 \end{bmatrix} \tag{4.1, 4}$$

This in turn may be operated on by rank-preserving transformations from the left to remove B_1, and from the right to remove C_2, to generate the matrix

$$Z_3(s)=\begin{bmatrix} sI_{n-m-d}-A_1+B_1C_2 & 0 \\ 0 & I_d \\ -C_3 & 0 \end{bmatrix} \tag{4.1, 5}$$

An inspection of the form of the matrix $Z_2(s)$ now enables us to conclude that the matrix $Z(s)$, of eqn. (4.1, 1), will lose rank if and only if the matrix $Z_3(s)$, of eqn. (4.1, 5), loses rank. Thus the matrix $Z(s)$ will lose rank if and only if s is an output decoupling zero corresponding to the pair (A_3, C_3) where

$$A_3=A_1-B_1C_2 \tag{4.1, 6}$$

The maximum number of such output decoupling zeros, corresponding to the case when C_3 has full rank, is

$$\begin{aligned} n_z' &= n-m-d-(m-l) \\ &= n_z-d \end{aligned}$$

and the absolute maximum number of transmission zeros which $S(A, B, C)$ may have (which will occur when $C_3=0$) is $n-m-d$. We thus conclude that the rank deficiency of CB by d degrees has moved d of the invariant zeros to infinity. This view is reinforced by noting that if CB loses rank by d degrees, then so will every possible square system which may be formed by selecting any l of the m rows of C. Hence each subsystem so created will be such that d of its invariant zeros will have been moved to infinity. We already know however that the zeros of the non-square system are the intersection of the sets of the zeros of all the square $l\times l$ subsystems we can create from it. This intersection set will thus contain among its members those d zeros which have moved to infinity.

5. Dependence of zero locations on output matrix C

Suppose we were interested in designing a disturbance-rejection regulator for a system with more outputs than inputs. In forming the feedback connections from output to input via the regulator, a ‘ squaring down ’ process would be involved which would introduce zeros into the corresponding loop-transmission transfer-function matrix. The location of these zeros would vitally affect the stability characteristics of the closed-loop system. The following question then naturally arises. What control do we have over the placing of invariant zeros when we make combinations of output variables in this way ? In most practical situations the available system outputs will already be linear combinations of system states and there will therefore be less freedom in creating and placing zeros than if all the system states were accessible. Nevertheless it is logical to first consider the case when we have complete freedom over the choice of the output coupling matrix C and we will accordingly do so.

Before giving a general treatment to this, however, it is useful to note at this point that there is one fairly obvious way to choose an output matrix C to get a required set of invariant zeros $\{z_1, \ldots, z_{n-l}\}$. This applies in the situation where we can assume that the corresponding system transfer-function matrix loses rank at each zero, which implies that there is no coincidence between poles and zeros. All that is required is to choose $(n-l)$ of the elements of the matrix C such that the set of matrices $G(z_i)$ given by

$$\begin{aligned} G(z_1) &= C(z_1 I - A)^{-1}B \\ G(z_2) &= C(z_2 I - A)^{-1}B \\ &\vdots \\ G(z_{n-l}) &= C(z_{n-l} I - A)^{-1}B \end{aligned}$$

all have zero determinant.

5.1. *Necessary and sufficient set of conditions for output matrix C to result in stipulated set of zero locations*

Suppose we have a linear dynamical system with l inputs, and with all state variables directly accessible for measurement. Let the corresponding A and B matrices of the system state-space model be known, and let N be a left annihilator of B so that

$$NB = 0_{n-l,\, l} \tag{5.1, 1}$$

Let $\{z_1, z_2, \ldots, z_{n-l}\}$ be a set of complex numbers (with non-wholly-real numbers in complex-conjugate pairs) which it is desired to be the set of invariant zeros of the system resulting from an appropriate choice of an output matrix C. Then it is necessary and sufficient for an $l \times n$ matrix C to result in these zeros being formed that every right annihilator M of C is also a right annihilator of the matrix C_0 given by

$$C_0 = \Lambda_z PN - PNA \tag{5.1, 2}$$

where

$$\Lambda_z = \text{diag}\ (z_1, z_2, \ldots, z_{n-l}) \tag{5.1, 3}$$

and P is an arbitrary non-singular matrix of dimension $(n-l) \times (n-l)$.

To establish the necessity of condition (5.1, 2), assume that C is such that the set of zeros is as required, and let M be a corresponding right annihilator of C so that

$$CM = 0_{l,\, n-l} \tag{5.1, 4}$$

Since, by assumption, the system is such that it has $(n-l)$ zeros CB will have full rank, and therefore NM will have full rank. Hence we can form a matrix M' which is also a right annihilator of C and given by

$$M' = M(NM)^{-1} \tag{5.1, 5}$$

This right annihilator of C will be such that it is related to the left annihilator of B by

$$NM' = I_{n-l} \tag{5.1, 6}$$

It then follows immediately that the stipulated set of zeros $\{z_1, z_2, \ldots, z_{n-l}\}$ must be the eigenvalues of the matrix NAM'. We must therefore have

$$NAM' = U_z \Lambda_z V_z \tag{5.1, 7}$$

where Λ_z is as defined by eqn. (5.1, 3) and $U_z = V_z^{-1}$ is a matrix whose columns are a set of eigenvectors for NAM'. Substituting for M' from (5.1, 5) into (5.1, 7) we thus get

$$NAM(NM)^{-1} = U_z \Lambda_z V_z \tag{5.1, 8}$$

Multiplying from the right by NM and the left by V_z and rearranging then gives that

$$(\Lambda_z V_z N - V_z NA)M = 0 \tag{5.1, 9}$$

Finally, since we can choose the eigenvectors of NAM' in an arbitrary manner, we must have that

$$(\Lambda_z PN - PNA)M = 0 \tag{5.1, 10}$$

where P is an arbitrary $(n-l) \times (n-l)$ non-singular matrix.

Before proceeding further let us examine the dimensions of the various matrices involved. We have that

C_0 is of dimension $(n-l) \times n$

C is of dimension $l \times n$

M is of dimension $n \times (n-l)$

We will assume for the moment that

$$l \geqslant (n-l)$$

The case when the number of outputs formed is less than n minus the number of inputs will be considered later. We note that

$$\text{dimension Ker }(C) = n - l$$

and

$$\text{dimension Ker }(C_0) = l$$

so that

$$\text{dimension Ker }(C_0) \geqslant \text{dimension Ker }(C)$$

The necessary condition which we have just derived therefore may be replaced by the alternative condition

$$\text{Ker }(\Lambda_z PN - PNA) \supseteq \text{Ker } C \tag{5.1, 11}$$

To show that this condition is also sufficient, suppose that

$$(\Lambda_z V_z N - V_z NA)M = 0_{l,\, n-l} \tag{5.1, 12}$$

and that

$$CM = 0_{n-l,\, n-l} \tag{5.1, 13}$$

Then after appropriate manipulations we will have that

$$NAM(NM)^{-1} = U_z \Lambda_z V_z$$

where

$$U_z = V_z^{-1}$$

and thus we have that

$$NAM' = U_z \Lambda_z V_z \tag{5.1, 14}$$

where

$$M' = M(NM)^{-1} \tag{5.1, 15}$$

Clearly if the columns of M lie in the subspace Ker (C) then so will the columns of M'. Furthermore, M' is such that

$$NM' = I_{n-l} \tag{5.1, 16}$$

and hence the zeros of $S(A, B, C)$ must be the eigenvalues of NAM', which are of course the stipulated set of zeros $\{z_1, z_2, \ldots, z_{n-l}\}$.

To complete this discussion of the dependence of system invariant zeros on the output matrix C, we naturally seek an expression from which an appropriate C matrix may be directly obtained. We start by noting that if the kernel space of C is contained in the kernel space of C_0, then this is equivalent to saying that

$$C_0 = RC \tag{5.1, 17}$$

where R is an $(n-l) \times l$ constant matrix of full rank. To verify this, suppose the kernel space of C_0 is spanned by the columns of some matrix M_0 and further suppose, for convenience, that the first l rows of M_0 are linearly independent. We may then partition M_0 as

$$M_0 = \begin{matrix} \updownarrow l \\ \updownarrow n-l \end{matrix} \begin{bmatrix} M_0' \\ \hdashline M_0'' \end{bmatrix} \tag{5.1, 18}$$

and hence derive a general expression for C_0 in terms of M_0' and M_0'' as

$$C_0 = K_0[-M_0''(M_0')^{-1} \vdots I_{n-l}] \tag{5.1, 19}$$

where K_0 is a non-singular $(n-l) \times (n-l)$ constant matrix. Now suppose that C has a right-hand annihilator M. Then from (5.1, 11) we have that the columns of M must be linear combinations of the columns of M_0 so that

$$M = M_0K \tag{5.1, 20}$$

where K is an $l \times (n-l)$ matrix of full rank. This enables us to derive an expression for C in terms of M_0' and M_0'' as

$$C = K_{sq}[I_l \vdots M_0'K(M_0''K)^{-1}] \tag{5.1, 21}$$

where K_{sq} is a non-singular matrix of dimension $l \times l$. Comparing eqns. (5.1, 19) and (5.1, 21) gives that

$$\begin{aligned} C_0 &= K_0M_0''(M_0')^{-1}K_{sq}^{-1}C \\ &= RC \end{aligned} \tag{5.1, 22}$$

where

$$R = K_0M_0''(M_0')^{-1}K_{sq}^{-1} \tag{5.1, 23}$$

is an $(n-l) \times l$ matrix corresponding to the values of K_0 and K_{sq}.

Thus we see that if C is to result in a system having its invariant zeros in the stipulated locations then it must be such that the rows of the matrix C_0 must be expressible as linear combinations of the rows of C. Further insight into the reasons why this should be so can be seen by considering the rows of C_0.

Let v_i^t be the ith row of P (that is the ith row of an equivalent V_z). Then the ith row of C_0 will be given by

$$C_{0,i}{}^t = v_i{}^t N(z_i I - A) \tag{5.1, 24}$$

However, $N(z_i I - A)$ must be a left annihilator of $(z_i I - A)^{-1}B$. Furthermore, if

$$G(s) = C(sI - A)^{-1}B \tag{5.1, 25}$$

is to lose rank at $s = z_i$ then its rows must be linearly dependent so that some vector, α^t say, will exist such that

$$\alpha^t C(z_i I - A)^{-1}B = 0 \tag{5.1, 26}$$

This shows that some linear combination of the rows of C, here denoted by $\alpha^t C$, must lie in the left null space of $(z_i I - A)^{-1}B$ and so an α will exist such that

$$\alpha^t C = v_i{}^t N(z_i I - A)^{-1} = C_{0,i}{}^t \tag{5.1, 27}$$

exactly as required by the geometrical interpretation we have given for eqn. (5.1, 17).

Thus the problem of choosing an output matrix C so as to place the invariant zeros of a state-accessible system in a set of required locations in the complex plane reduces to solving the equation

$$RC = \Lambda_z PN - PNA \tag{5.1, 28}$$

The above discussion specifically assumed that

$$l \geqslant (n - l)$$

and it is obvious that eqn. (5.1, 28) cannot hold if this condition is violated. What must then be done is that the dimension of the kernel of C_0 must be suitably augmented. This can be done by sacrificing some of the freedom of choice involved in selecting the matrix P, that is in selecting the eigenvector set of the appropriate matrix NAM', whose eigenvalues are the required set of zeros, in such a way as to reduce the rank of C_0.

5.2. *Example of zero placement for a state-accessible system*

Consider a state-accessible system having four states and two inputs and with

$$A = \begin{bmatrix} -1 & 1 & 3 & -2 \\ 0 & -1 & -1 & -1 \\ 0 & 1 & -3 & -1 \\ 0 & 3 & -1 & -5 \end{bmatrix}, \quad B = \begin{bmatrix} 1 & 0 \\ 1 & 2 \\ 1 & 1 \\ 2 & 2 \end{bmatrix}$$

and suppose we wished two transmission zeros to be located at

$$z_1 = -1 \quad \text{and} \quad z_2 = -2$$

so that

$$\Lambda_z = \begin{bmatrix} -1 & 0 \\ 0 & -2 \end{bmatrix}$$

An appropriate left annihilator of B is found as

$$N=\begin{bmatrix}1 & 1 & -2 & 0\\ 2 & 2 & 0 & -2\end{bmatrix}$$

and selecting

$$R=P=V_z=I_2$$

we get that

$$C=C_0=\Lambda_z N-NA$$

so that

$$C=\begin{bmatrix}-1 & 0\\ 0 & -2\end{bmatrix}\begin{bmatrix}1 & 1 & -2 & 0\\ 2 & 2 & 0 & -2\end{bmatrix}-\begin{bmatrix}1 & 1 & -2 & 0\\ 2 & 2 & 0 & -2\end{bmatrix}\begin{bmatrix}-1 & 1 & 3 & -2\\ 0 & -1 & -1 & -1\\ 0 & 1 & -3 & -1\\ 0 & 3 & -1 & -5\end{bmatrix}$$

$$=\begin{bmatrix}0 & 1 & -6 & 1\\ -2 & 2 & -6 & 0\end{bmatrix}$$

Calculating the appropriate transfer-function matrix $G(s)$ and finding its determinant (or finding the zeros in any other way) gives

$$\det\,[G(s)]=\frac{6(s+1)(s+2)}{(s+1)(s+2)(s+3)(s+4)}$$

and verifies that the required placement has been achieved.

6. Zero placement for system whose states are not all accessible

In most practical situations we do not have all the states of a system accessible. It does however often arise that we have more outputs than inputs, so that we can consider 'squaring down' from the set of outputs to form a set of regulatory feedback connections to a smaller number of inputs. If the existing system $S(A, B, C)$ already has n_z zeros, these will not be affected by any 'squaring down' operation so that we will only be able to place $(n-l-n_z)$ of the $(n-l)$ zeros which the l-input l-output system formed may have. Since $n_z \leqslant (n-m)$, the number of zeros we will be able to place will always be at least $(m-l)$.

Suppose then we have a system $S(A, B, C)$ with $m>l$ and follow the existing set of outputs by a further set of connections represented by an $l\times m$ post-compensator real constant matrix L, to get a new state-to-output matrix given by

$$C_{\text{sq}}=LC \tag{6.0, 1}$$

All the analysis of the previous section applies to the choice of the matrix C_{sq} so that, if the overall system is to have a set of zeros $\{z_1, z_2, \ldots, z_{n-l}\}$ which are the elements of a diagonal matrix Λ_z, we will have that L must satisfy the equation

$$RC_{\text{sq}}=RLC=\Lambda_z PN-PNA \tag{6.0, 2}$$

where, as before, we have that (i) P is an arbitrary $(n-l)\times(n-l)$ non-singular constant matrix, (ii) N is a left annihilator of B and (iii) R is an $(n-l)\times l$ constant matrix of full rank. In this case we must observe the following important constraints in solving eqn. (6.0, 2) to find a suitable post-compensating matrix L.

(i) The matrix Λ_z must contain among its diagonal elements all the existing zeros of the system $S(A, B, C)$ whose outputs are being squared down.

(ii) No more than $(n-l-n_z)$ new zeros should be specified, as more than this number would lead to a set of inconsistent equations.

The procedure is an obviously straightforward extension of that developed for the state-accessible case. It will now be illustrated by a simple example. Suppose we have a system $S(A, B, C)$ with

$$A=\begin{bmatrix}-1 & 1 & 3 & -2\\ 0 & -1 & -1 & -1\\ 0 & 1 & -3 & -1\\ 0 & 3 & -1 & -5\end{bmatrix},\quad B=\begin{bmatrix}1 & 0\\ 1 & 2\\ 1 & 1\\ 2 & 2\end{bmatrix}$$

$$C=\begin{bmatrix}0 & 1 & -6 & 1\\ 1 & -1 & 3 & 0\\ 2 & 4 & -3 & -3\end{bmatrix}$$

This system already has a zero at

$$z_1=-1$$

Thus one of the entries to Λ_z must be -1. Suppose we wished to connect the three outputs together to form two modified outputs in such a way as to place a further zero at -3. We would then have

$$\Lambda_z=\begin{bmatrix}-1 & 0\\ 0 & -3\end{bmatrix}$$

and taking $P=I_2$ and using the value of N found in the example of the previous section we find that

$$\Lambda_z PN-PNA=\begin{bmatrix}0 & 1 & -6 & 1\\ -4 & 0 & -6 & 2\end{bmatrix}$$

Inserting this into eqn. (6.0, 2) we get

$$RL\begin{bmatrix}0 & 1 & -6 & 1\\ 1 & -1 & 3 & 0\\ 2 & 4 & -3 & -3\end{bmatrix}=\begin{bmatrix}0 & 1 & -6 & 1\\ -4 & 0 & -6 & 2\end{bmatrix}$$

Setting $R=I_2$ and solving for L gives

$$L=\begin{bmatrix}1 & 0 & 0\\ 0 & -\frac{8}{3} & -\frac{2}{3}\end{bmatrix}$$

and, after scaling the second row, we get

$$L' = \begin{bmatrix} 1 & 0 & 0 \\ 0 & -4 & -1 \end{bmatrix}$$

as a suitable post-compensating matrix. The new composite system output matrix is

$$C_{\mathrm{sq}} = LC = \begin{bmatrix} 0 & 1 & -6 & 1 \\ -6 & 0 & -9 & 3 \end{bmatrix}$$

Forming the overall system's transfer-function matrix

$$G_{\mathrm{sq}}(s) = C_{\mathrm{sq}}(sI - A)^{-1}B$$

and taking its determinant gives

$$\det\left[G_{\mathrm{sq}}(s)\right] = \frac{-9(s+1)(s+3)}{(s+1)(s+2)(s+3)(s+4)}$$

which shows that we have indeed obtained a post-compensator L giving a pair of zeros at -1 and -3 for the overall system.

References

Davison, E. J., and Wang, S. H., 1974, *Automatica*, **10,** No. 6, 643.

Kontakos, T., 1973, Ph.D. Thesis, University of Manchester.

Kouvaritakis, B., and MacFarlane, A. G. J., 1976, *Int. J. Control*, **23,** 149.

MacFarlane, A. G. J., 1974, *I.F.A.C. Symposium on Multivariable Technological Systems*, Manchester.

Rosenbrock, H. H., 1970, *State Space and Multivariable Theory* (London: Nelson).

Multivariable root loci: a unified approach to finite and infinite zeros

B. KOUVARITAKIS and J. M. EDMUNDS

A MacLaurin's series expansion is used to approximate the individual sheets of the characteristic gain function of a linear multivariable system in terms of its state space matrices, at the point at infinity on the frequency plane for small values of characteristic gain. This yields the asymptotic root locus directions for systems with one or more sets of closed loop poles going to infinity for large feedback gains. A bilinear transformation on the frequency variable is used to allow the approximation at any point on the frequency plane. This gives the angles of approach to finite zeros in terms of the state space matrices, and gives a simple way of calculating zeros for square systems. A bilinear transformation on the gain variable is then introduced allowing the characteristic gain function to be approximated at any values of frequency and gain. This enables the directions of departure from the open-loop poles to be determined in terms of the state space matrices of the system.

1. Introduction

Recent research (Kouvaritakis and Shaked 1976, Kouvaritakis 1978 c, and MacFarlane and Postlethwaite 1977 b) has shown that the work of Evans (1948) on the root locus of single-input single-output systems may be generalized to multivariable systems. The natural medium for this generalization is provided by the theory of algebraic functions of a complex variable through the appropriate definition of the characteristic gain and characteristic frequency functions. Thus the root loci (RL) of multivariable systems emerge as the loci on the complex plane for which the gain function becomes real and negative. It is not unreasonable therefore to expect that the set of properties of the RL of scalar systems form a special set of the properties of the RL of multivariable systems. Particular areas where this intuitive conjecture has been proved to hold true concern the behaviour of the RL when the feedback gain parameter, k, assumes either arbitrarily high or low values. The study of the latter yields properties such as the ' angles of departure ' from open-loop poles (Shaked 1976) while the study of the first yields properties such as the ' angles of approach ' to the system zeros (Shaked 1976) and the ' asymptotic root locus behaviour ' (ARLB) of those closed-loop poles that vanish to the point at infinity.

The results presented to date are not in their most general form and the rectification of this forms one of the aims of the present paper. A further objective of this work is to propose a method of analysis which enables a unified approach to the study of the aforementioned aspects of RL behaviour. The treatment given here is based on the series expansion of the appropriate transfer function matrix, $G(s)$, about a general point in the complex plane. Use of such an expansion about the point at infinity enables the investigation of the infinite terminal locations of the RL as $k \to \infty$, called the infinite zeros (IZ), and hence yields the asymptotic directions of the RL. The application of the series expansion about the finite points of the complex plane leads to the study of the approach of the RL to their finite terminal locations, the zeros of $G(s)$, which for convenience are referred to as the finite zeros (FZ). A more penetrating study of the relation between the expansion about the point at infinity and the expansion about a finite point reveals that the latter corresponds to a new transfer function matrix $\tilde{G}(p)$ which may be obtained from $G(s)$ by means of a bilinear transformation of the complex variable s. Such a transformation results in a conformal map of the s-plane onto a new plane, the p-plane. The matrices $G(s)$ and $\tilde{G}(p)$ however share the same Riemann surface so that the RL properties of $G(s)$ may be obtained from the RL of $\tilde{G}(p)$ by a back transformation from p to s.

The arguments above readily highlight the unity between the concepts of FZ and IZ. In particular it is shown that it is possible to arrange for $\tilde{G}(p)$ to have FZ only, so that both the FZ and IZ of $G(s)$ are mapped onto FZ. The transformation of variable therefore may be exploited in order to enable the adoption of a common methodology for the analysis of the behaviour of approach of the RL to both the FZ and IZ. This approach not only yields the means of computing the 'angles of approach' to FZ but it also suggests a simple algorithm for the calculation of the FZ.

The mathematical tool of the bilinear transformation may also be applied to the characteristic gain variable and by an appropriate interpretation one may set up a duality between poles and zeros. A consequence of this is that the study of the behaviour of departure from system poles may be unified to that of the behaviour of approach to FZ and IZ.

The material of the paper is arranged as follows. The ARLB results presented by Kouvaritakis and Shaked (1976) hold true for proper systems which exhibit some block triangularizability properties. The treatment contained in § 2 extends these results to the general case and deals with proper as well as non-proper systems. Further, a method for the systematic adjustment of the ARLB is discussed. The general formulae for the computation of the 'pivots', that is the points of radiation of the asymptotes, are obtained in § 3, while the corrected formulae for the angles of approach are derived in § 4. Here it is shown that it is always possible to adjust these angles by means of input compensation and a systematic procedure is outlined. In addition a simple algorithm for the computation of the FZ is proposed. For the sake of completeness the topic of the 'angles of departure' is dealt with in § 5 where the general case of multiple open loop poles is considered. The paper is concluded by § 6, where a numerical example is examined for illustrative purposes.

2. Infinite zeros and asymptotes

2.1. *Introduction*

Consider the system described by the set of state-space equations

$$\left.\begin{aligned} \dot{\mathbf{x}} &= A\mathbf{x} + B\mathbf{u} \\ \mathbf{y} &= C\mathbf{x} + D\mathbf{u} \end{aligned}\right\} \qquad (2.1)$$

where **x**, **u** and **y** are the n, l and m dimensional state, input and output vectors and where A, B, C and D are matrices of constants of appropriate dimensions. Then an alternative representation of such a system is given in terms of the transfer function matrix

$$G(s) = C(sI - A)^{-1}B + D \tag{2.2}$$

where s denotes the Laplace transform complex variable. Next assume the system $S = \mathscr{S}(A, B, C, D)$ to be square, that is assume the outputs of S to be equal in number to the inputs ($l = m$) and as such to be available for the direct feedback to the inputs as indicated in Fig. 1. Note that a scalar gain operator, kI, where k is the gain parameter and I denotes the identity matrix, has been inserted in the feedback loop.

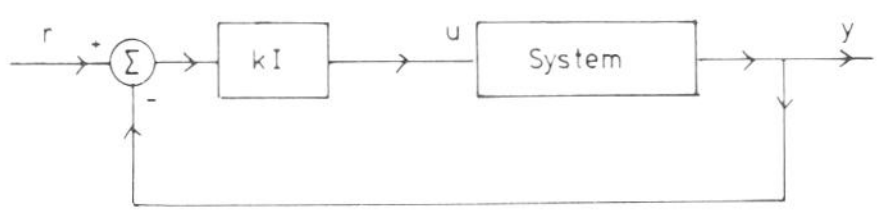

Figure 1.

We define the multivariable root loci (RL) as the loci traced by the closed-loop poles of the system in Fig. 1 when the gain k assumes values from 0 to some arbitrarily large positive number. The RL therefore can be obtained from the equation

$$f(s, k) = |F(s)| = |I + kG(s)| = 0 \tag{2.3}$$

where $F(s)$ denotes the return difference matrix of S. Note that special attention is required at values of k for which $F(s)$ becomes singular, a condition only possible for systems with a direct pass operator D which has negative eigenvalues (Kouvaritakis 1978 c, and Edmunds 1978). The multivariable RL defined by means of eqn. (2.3) can be seen to form the generalization of the classical concept of RL for systems with one input and one output. Thus for both the classical and the multivariable case the RL are the loci on the complex plane where the 'characteristic gain function' (MacFarlane and Postlethwaite 1977 b) becomes real and negative. In the scalar case, this function is the rational transfer function and has as its domain of definition a single copy of the complex plane, while in the multivariable case the characteristic gain function is the algebraic function $g(s)$ defined by MacFarlane and Postlethwaite (1977 a)

$$|gI - G(s)| = 0 \tag{2.4}$$

and has as its domain of definition m copies of the complex plane joined together to form the appropriate Riemann surface. Intuitively therefore one expects the properties of the RL of scalar systems to form special cases of these for multivariable systems. Indeed recent research (Kouvaritakis and

Shaked 1976, Kouvaritakis 1978 c, and MacFarlane and Postlethwaite 1977 b) has proven this conjecture to hold true. In the present section we examine aspects of the 'asymptotic root locus behaviour' (ARLB) of multivariable systems, that is the behaviour of the multivariable RL as the gain k assumes arbitrarily large values.

An inspection of the form of the return difference matrix, $F(s)$ in eqn. (2.3), establishes that for $k \to \infty$ eqn. (2.3) requires that at least one of the branches of the eigenfunctions of $G(s)$, that is the branches of the characteristic gain function, $g(s)$, must vanish. It follows therefore that the multivariable RL terminate at the zeros of the system, namely the points at which $g(s)$ becomes zero. Some of these points exist at finite location and for this reason are called the finite zeros (FZ) (Kouvaritakis and Shaked 1976, Kouvaritakis and MacFarlane 1976), while the remainder are located at the point at infinity and hence are called infinite zeros (IZ) (Kouvaritakis and Shaked 1976). Note that the set of FZ is coincident with the set of transmission zeros (Desoer and Schulman 1974, and MacFarlane and Karcanias 1976). From a RL point of view therefore there appears to be no qualitative difference between FZ and IZ, and IZ may be simply thought of as FZ that have been moved to the point of infinity. This assertion will be further justified in a later section when use of a transformation of variable is made in order to show both FZ and IZ may be obtained as maps of the FZ only of a different system. The implication of this is that a unified approach may be employed to examine the properties of both FZ and IZ. In this section we develop the mathematical technique for dealing with the IZ and this is subsequently modified to give a technique suitable for the treatment of the FZ in a later section.

2.2. *The first order IZ*

For simplicity of presentation the case of proper systems is investigated first. Hence we consider the systems $S_1 = \mathscr{S}(A_1, B_1, C_1)$ where the direct pass operator D_1 is assumed to be zero and is therefore omitted. The corresponding transfer function matrix, $G(s)$, by the suitable application of the Taylor series expansion about the point at infinity, may be written as

$$G_1(s) = \sum_{i=1}^{\infty} \frac{1}{s^i} C_1 A_1^{i-1} B_1$$

$$= \frac{1}{s} G_1 + P_1(s) \tag{2.5}$$

where

$$G_1 = C_1 B_1 \tag{2.6}$$

and

$$P_1(s) = \sum_{i=2}^{\infty} \frac{1}{s^i} C_1 A_1^{i-1} B_1 \tag{2.7}$$

For reasons to become apparent later the first Markov parameter (MP), G_1, will also be called the first projected Markov parameter (PMP). Now the spectral decomposition of G_1 may in general be written as

$$G_1=[U_1 \,\vdots\, M_1]\left[\begin{array}{c:c} \Lambda_1 & 0 \\ \hdashline 0 & 0_{d_1} \end{array}\right]\left[\begin{array}{c} V_1 \\ \hdashline N_1 \end{array}\right] \tag{2.8}$$

where Λ_1 consists of the Jordan canonical blocks of G_1 that correspond to the non-zero eigenvalues of G_1. The associated eigenvectors and dual eigenvectors appear as column and row vectors in the matrices U_1 and V_1, respectively. The matrices M_1 and N_1 contain the eigenvectors corresponding to the d_1 zero eigenvalues. Note that G_1 is assumed to have simple null structure (SNS), that is to say that the zero eigenvalues of G_1 are assumed to appear in simple canonical blocks. The case of non-simple null structure (NSNS) is somewhat involved mathematically and does not offer any further insight into the study of the ARLB. Furthermore, it can be shown (Kouvaritakis 1978 a) that the condition of NSNS is undesirable from a design point of view and that generically almost any input compensator commutes a NSNS to a SNS. For these reasons the case of NSNS will not be treated in the present paper.

Substituting $G(s)$ in eqn. (2.3) by the transfer function matrix $G_1(s)$ of eqn. (2.5) we obtain

$$\left| I+\frac{k}{s}\,G_1+kP_1(s)\right| =0 \tag{2.9}$$

which after the diagonalization of G_1 in accordance with eqn. (2.8) becomes

$$\begin{bmatrix} I+\frac{k}{s}\,\Lambda_1+k\,V_1P_1(s)U_1 & k\,V_1P_1M_1 \\ \\ kN_1P_1(s)U_1 & I+kN_1P_1(s)M_1 \end{bmatrix}=0 \tag{2.10}$$

This in turn by Schur's formula for determinants of partitioned matrices (Gantmacher 1959) yields

$$\left| I+\frac{k}{s}\,\Lambda_1+\,V_1P_1(s)U_1\right| =0 \tag{2.11}$$

or

$$\Delta_2(s,k)=\left| I+kN_1P_1(s)M_1-k^2N_1P_1(s)U_1 \times\left[I+\frac{k}{s}\,\Lambda_1+k\,V_1P_1(s)U_1\right]^{-1} V_1P_1(s)M_1\right| =0 \tag{2.12}$$

Equation (2.11) now describes the behaviour of all the RL that vanish to the IZ according to the first order law

$$s=\lambda k \tag{2.13}$$

Indeed, substitution of the above into eqn. (2.11) readily yields that solutions for λ exist and are given by

$$\lambda=-\lambda_j^{(1)} \quad j=1,2,\ldots,m-d_1 \tag{2.14}$$

where $\lambda_j^{(1)}$ denotes the jth non-zero eigenvalue of G_1. We are therefore led to the following result (Kouvaritakis and Shaked 1976).

Theorem 2.1

A system $S_1 = \mathscr{S}(A_1, B_1, C_1)$ with m inputs and outputs for which the first MP, G_1, has SNS with rank defect d_1, has $m-d_1$ first order IZ. The associated asymptotes make with the negative real axis angles equal to the phase of the non-zero eigenvalues of G_1.

Clearly first order type of solutions are the only type of solutions that can be obtained from eqn. (2.11).

2.3. *The second order IZ*

All orders of asymptotic behaviour higher than 1 are described by eqn. (2.12). In order to investigate these an equivalent system $S_2 = \mathscr{S}(A_2, B_2, C_2)$ with no first order behaviour but otherwise identical ARLB is derived and Theorem 2.1 is suitably extended to deal with the second order IZ. The arguments employed in the derivation of the system S_2 from S_1 may be repeated to obtain further equivalent systems and thus describe the ARLB of all orders. The derived results form generalizations of the results presented by Kouvaritakis and Shaked (1976) since, as will be discussed later, they deal with systems which may or may not be block triangularizable and furthermore do not require all orders of IZ from the lowest to the highest to be present (Owens 1977 and Kouvaritakis 1978 b).

Thus consider the form of eqn. (2.12) as $k \to \infty$. Then whether s remains finite or whether it tends to the point at infinity in such a way that $s/k \to 0$, it can be shown (Appendix A) that eqn. (2.12) reduces to

$$|I + kN_1 P_1(s)[I + sG_1^{\dagger} P_1(s)]^{-1} M_1| = 0 \tag{2.15}$$

where $G_1^{\dagger}$ denotes the Penrose commuting g_2-inverse (Pringle and Rayner 1971) of G_1 and is given by

$$G_1^{\dagger} = U_1 \Lambda_1^{-1} V_1 \tag{2.16}$$

Now, let us define $G_2(s)$ to be the transfer function of the system $P_1(s)$ with the feedback operator $sG_1^{\dagger}$ around it after appropriate projection onto the right and left null space of G_1 such that

$$G_2(s) \triangleq N_1(P_1(s)[I + sG_1^{\dagger} P_1(s)]^{-1}) M_1 \tag{2.17}$$

Then eqn. (2.15) becomes

$$f_2(s, k) = |I + kG_2(s)| = 0 \tag{2.18}$$

and as such can be seen to have the same form as eqn. (2.3). The significance of this course is that if S_2 denotes the system described by $G_2(s)$ then S_1 and S_2 share the same ARLB except that S_2 does not exhibit any first order behaviour.

Further insight into the equivalence between S_1 and S_2 may be gained by the schematic representation of S_2 depicted in Fig. 2.

Note that use of eqn. (2.5) has been made in order to replace $P_1(s)$ by $G_1(s) - (1/s)G_1$ in the above diagram. We may now recapitulate as follows.

Lemma 2.1

Given the system of S_1 $\mathscr{S}(A_1, B_1, C_1)$ for which the first MP, G_1, has the decomposition of eqn. (2.8), a new system S_2 with no first order IZ but with an

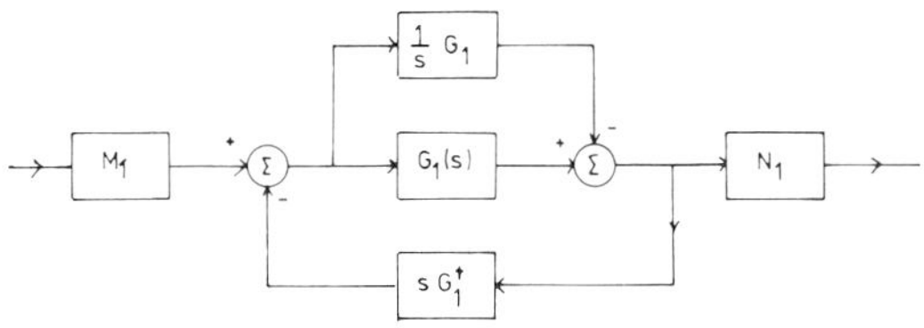

Figure 2.

otherwise identical ARLB to that of S_1 may be derived. S_2 is obtained by removing the first order behaviour of S_1, then correcting the remainder by applying feedback through the operator $sG_1{}^{\dagger}$ and finally projecting the resulting system onto the right and left null space of G_1.

An alternative representation of S_2 is given by the state space model derived in Appendix B and described below.

Lemma 2.2

Let $S_1 = \mathscr{S}(A_1, B_1, C_1)$ have a first MP, G_1, with the decomposition of eqn. (2.8). Then the transformed operators

$$A_2 = T_1 A_1, \quad T_1 = I - B_1 G_1{}^{\dagger} C_1, \quad B_2 = B_1 M_1 \quad \text{and} \quad C_2 = N_1 C_1 \tag{2.19}$$

define a new system $S_2 = \mathscr{S}(A_2, B_2, C_2)$ which has no first order IZ but has an otherwise identical ARLB with S_1.

Having established the equivalence between S_1 and S_2 we may now repeat the procedure of § 2.2 on S_2 in order to derive the second order type of ARLB described by the equation

$$s^2 = \lambda k \tag{2.20}$$

Thus consider the Taylor series of $G_2(s)$ about the point at infinity which may be written as

$$G_2(s) = \sum_{i=1}^{\infty} \frac{1}{s^i} C_2 A_2{}^{i-1} B_2 = \frac{1}{s^2} G_2 + P_2(s) \tag{2.21}$$

where G_2 is the second MP of $G_2(s)$

$$G_2 = C_2 A_2 B_2 = N_1 C_1 A_1 B_1 M_1 \tag{2.22}$$

and

$$P_2(s) = \sum_{i=3}^{\infty} \frac{1}{s^i} C_2 A_2{}^{i-1} B_2 \tag{2.23}$$

Note that by eqn. (2.22) G_2 emerges as the projection of the second MP of S_1 onto the null space of G_1 and for this reason G_2 will be called the second PMP

Next we assume the spectral decomposition of G_2 to be

$$G_2 = [U_2 \;\; M_2] \begin{bmatrix} \Lambda_2 & 0 \\ 0 & 0_{d_2} \end{bmatrix} \begin{bmatrix} V_2 \\ N_2 \end{bmatrix} \tag{2.24}$$

and we substitute this together with eqn. (2.21) into eqn. (2.18) in order to obtain

$$\left| I + \frac{k}{s^2} \Lambda_2 + V_2 P_2(s) U_2 \right| = 0 \tag{2.25}$$

or

$$\Delta_3(s, k) = \left| I + kN_2 P_2(s) M_2 - k^2 N_2 P_2(s) V_2 \times \left[I + \frac{k}{s^2} \Lambda_2 + k V_2 P_2(s) U_2 \right]^{-1} V_2 P_2(s) M_2 \right| = 0 \tag{2.26}$$

Upon substitution of eqn. (2.20) into (2.25) we deduce that solutions for λ exist and are given by

$$\lambda = -\lambda_j^{(2)}, \quad j = 1, 2, \ldots, d_1 - d_2 \tag{2.27}$$

where $\lambda_j^{(2)}$ denotes the jth non-zero eigenvalue of G_2. Thus we have that :

Theorem 2.2

The system $S_1 = \mathscr{S}(A_1, B_1, C_1)$ of Theorem 2.1, for which the second PMP, G_2, has SNS with rank defect d_2, has $d_1 - d_2$ second order IZ.

The associated asymptotes appear in Butterworth patterns in pairs, one such pattern corresponding to each non-zero eigenvalue of G_2. The angles $\alpha_j^{(2)}$ that these asymptotes make with the positive real axis are given by

$$\alpha_j^{(2)} = \tfrac{1}{2}[\angle -\lambda_j^{(2)} + 2t\pi], \quad j = 1, 2, \ldots, d_1 - d_2, \quad t = 0, 1 \tag{2.28}$$

Equation (2.25) yields second order solutions only and the high order ARLB is contained in eqn. (2.26). The methodology developed in §§ 2.2 and 2.3 can be repeated to generate solutions for the ARLB of all orders and this is undertaken in the following section. Here we conclude the discussion by noting the two systems S_1 and S_2 not only exhibit an equivalence in ARLB for all orders greater than one but they also share common FZ as proved in Appendix C and as stated below.

Theorem 2.3

The system S_1 of Theorem 2.1 and the system S_2 of Lemma 2.2 have an identical set of IZ and FZ with the only difference that S_2 has no first order IZ ; the first order IZ in S_2 have been replaced by an equal number of FZ located at the origin.

2.4. *The higher order IZ*

The steps taken to reduce eqn. (2.12) to eqn. (2.15) may be duplicated in order to reduce eqn. (2.26) to

$$|I + kG_3(s)| = 0 \tag{2.29}$$

where

$$G_3(s) = N_2 P_2(s)[I + s^2 G_2^{\dagger} P_2(s)]^{-1} M_2 \tag{2.30}$$

and where $G_2^{\dagger}$ is the Penrose commuting g_2-inverse of G_2. This of course could have been otherwise obtained by writing the transfer function matrix $G_2(s)$ as

$$G_2(s) = \frac{1}{s} G'_2(s) \tag{2.31}$$

where $G'_2(s)$ now corresponds to a system with first order IZ. The two transfer function matrices $G_2(s)$ and $G'_2(s)$ are related by a simple scalar multiplication so that their respective characteristic gain functions are also related by the same scalar multiplication and share the same Riemann surface as their domain of definition. It follows then that the third order ARLB of $G_2(s)$ may be extracted from the second order ARLB of $G'_2(s)$ after suitable adjustment of the relevant power of s. Thus in order to obtain the configuration which yields the third order asymptotes and has identical ARLB with $G_2(s)$ (and therefore with S_1) for all orders of IZ higher than 2 we derive the configuration which has an identical ARLB with $G'_2(s)$ for all orders of IZ higher than 1 and cascade the result with the scalar operator $(1/s)I$. The transfer function matrix of the overall configuration obtained in this manner may be readily shown to be equal to $G_3(s)$ of eqn. (2.30). The third order ARLB of S_1 is dictated by the non-zero eigenvalues of the third MP of $G_3(s)$.

The arguments used above are general and may be used to obtain the system that yields the $(i+1)$th order IZ of S_1 and has an identical ARLB with S_1 for all orders of IZ higher than i, when a transfer function matrix $G_i(s)$, with no IZ of order less than i but with an otherwise identical ARLB to that of S_1, is given. Thus we write $G_i(s)$ as

$$G_i(s) = \frac{1}{s^{i-1}} G'_i(s) \tag{2.32}$$

where $G'_i(s)$ corresponds to a system with first order IZ.

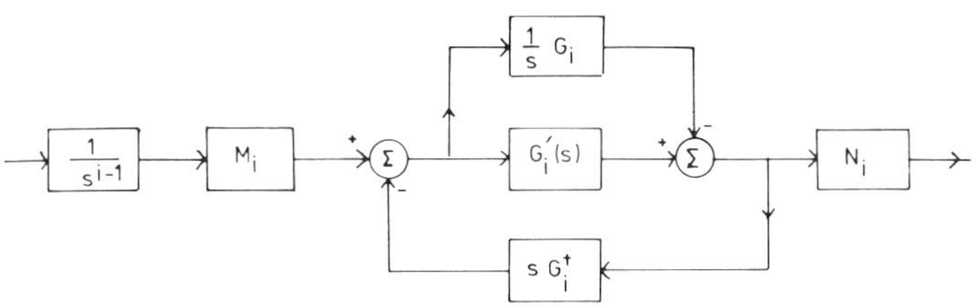

Figure 3.

The $(i+1)$th order IZ may now be obtained from the configuration of Fig. 3 where G_i is the ith MP of $G_i(s)$, $G_i^{\dagger}$ is the Penrose commuting g_2-inverse of G_i and N_i and M_i are the left and right annihilators of G_i. The system of Fig. 3 is of course equivalent to that shown in Fig. 4. Thus we are led to the following result.

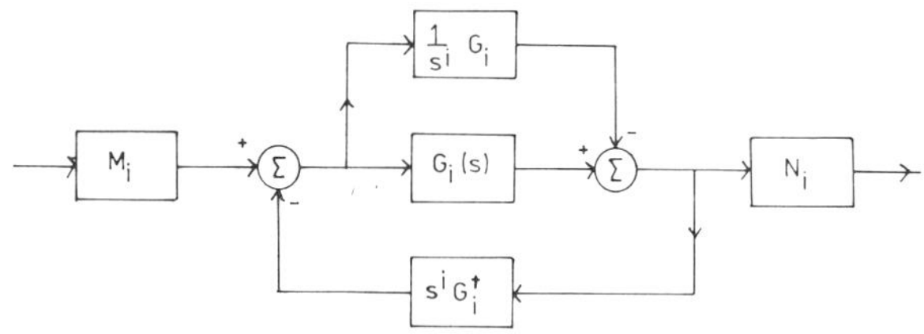

Figure 4.

Theorem 2.4

The asymptotes of the RL of a system described by the transfer function matrix $G_1(s)$ appear in Butterworth patterns described by the equation

$$s=[-\lambda_j^{(i)}k]^{1/i}, \quad i=1, 2, \ldots, \nu, \quad j=1, 2, \ldots, d_{i-1}-d_i \tag{2.33}$$

and make angles $\alpha_j^{(i)}$ with the positive real axis given by

$$\alpha_j^{(i)}=\frac{1}{i}[\angle-\lambda_j^{(i)}+2t\pi], \quad i=1, 2, \ldots, \nu, \quad j=1, 2, \ldots, d_{i-1}-d_i, \quad t=0, 1, \ldots, i-1 \tag{2.34}$$

where $\lambda_j^{(i)}$ is the jth non-zero eigenvalue of the ith PMP, G_i, d_i is the rank defect of G_i with $d_0=m$ and ν is the smallest i for which G_i is full rank. The ith PMP is defined as the ith MP of $G_i(s)$ where $G_i(s)$ describes the transference from the input $u_i(s)$ to the output $y_i(s)$ of the nested configuration in Fig. 5. In the special case of a PMP, G_j, which is zero, the Penrose commuting g_2-inverse must be replaced by zero and the corresponding annihilators N_j and M_j are replaced by the identity operator. All the PMP, G_i, $i=1, 2, \ldots, \nu-1$ are assumed to have SNS.

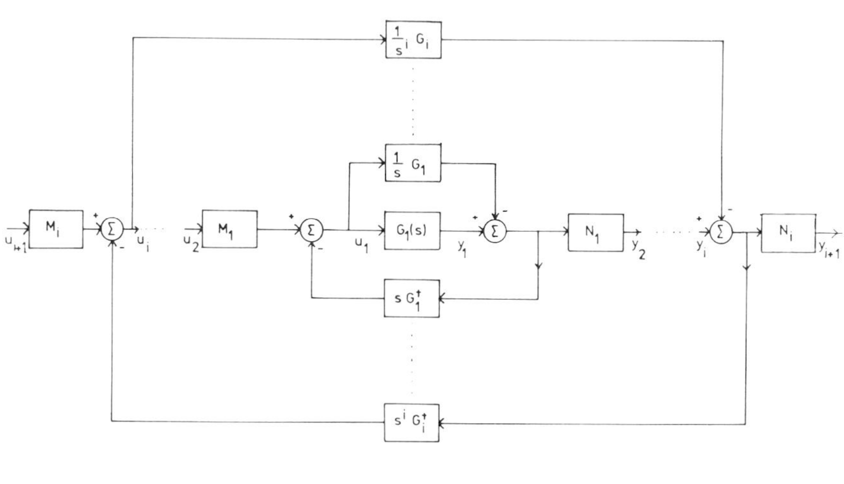

Figure 5.

Similar arguments may be employed in order to derive a state-space representation $\mathscr{S}(A_i, B_i, C_i)$ for the nested configurations of Fig. 5 that correspond to the transfer function matrices, $G_i(s)$. Furthermore this procedure could be used to relate the PMP to the original state-space matrices A_1, B_1 and C_1. Thus by Appendix B we have that the state-space representation $S_2 = \mathscr{S}(A_2, B_2, C_2)$ of $G_2(s)$ is as given by eqn. (2.19). Now by definition $C_2B_2 = 0$ so that S_2 may be equivalently represented by an integration operator, $\int (\cdot)\, dt \times I$, in cascade with the system

$$S'_2 = \mathscr{S}(A_2, B_2, C'_2) \quad \text{where} \quad C'_2 = C_2A_2 \tag{2.35 a, b}$$

S'_2 however has first order IZ and may be treated therefore in the same way as S_1 in order to obtain the higher order IZ. Then by Lemma 2.2 we derive the system

$$S'_3 = \mathscr{S}(A_3, B_3, C'_3) \tag{2.36 a}$$

with

$$A_3 = T_2A_2, \quad T_2 = I - B_2G_2^{\dagger}\, C'_2, \quad G_2 = C'_2B_2 \tag{2.36 b–d}$$

$$B_3 = B_2M_2, \quad C'_3 = C_3A_2 \quad \text{and} \quad C_3 = N_2C_2 \tag{2.36 e–g}$$

Now by the definition of the relevant parameters we have that

$$C_3A_2 = C_3A_3 \tag{2.37}$$

so that the system S'_3 may be written as

$$S'_3 = \mathscr{S}(A_3, B_3, C_3A_3) \tag{2.38}$$

Reversing previous arguments we may combine S'_3 together with the integration operator that exists in cascade with S'_3 in order to obtain the system S_3

$$S_3 = \mathscr{S}(A_3, B_3, C_3) \tag{2.39}$$

which has no first and second order IZ but otherwise exhibits an identical ARLB with S_1. Upon reflection on the manner in which S_3 was derived it becomes apparent that $\mathscr{S}(A_3, B_3, C_3)$ is the state-space representation of $G_3(s)$.

The above procedure is general and may be used to generate the state-space representation, $S_{i+1} = \mathscr{S}(A_{i+1}, B_{i+1}, C_{i+1})$, of $G_{i+1}(s)$ in terms of that of $G_i(s)$. It may be readily verified that the parameters A_{i+1}, B_{i+1}, C_{i+1} obey the conditions

$$A_{i+1} = T_iA_i, \quad T_i = I - B_iG_i^{\dagger}\, C_iA_i^{i-1} \tag{2.40 a, b}$$

$$G_{i+1} = C_{i+1}A_{i+1}^{\,i}\, B_{i+1}, \quad B_{i+1} = B_iM_i, \quad C_{i+1} = N_iC_i \qquad i = 1, 2, \ldots, \nu - 1 \tag{2.40 c–e}$$

These equations define explicit formulae for the computation of the PMP in terms of the matrices A_1, B_1 and C_1 of S_1. The following procedure may be pursued.

Step 1. Given $S_1 = \mathscr{S}(A_1, B_1, C_1)$ form the first PMP, $G_1 = C_1B_1$ and test its rank defect d_1. If $d_1 = 0$ stop otherwise compute the annihilators N_1 and M_1 of G_1.

Step 2. Form $B_2 = B_1 M_1$, $C_2 = N_1 C_1$ and $A_2 = T_1 A_1$ where $T_1 = I - B_1 G_1^{\dagger} C_1$. Compute the second PMP as the second MP of $S_2 = \mathscr{S}(A_2, B_2, C_2)$, $G_2 = C_2 A_2 B_2$ and test its rank defect d_2. If $d_2 = 0$ stop otherwise calculate the annihilators of G_2, N_2 and M_2.

Step 3. Form $B_3 = B_2 M_2$, $C_3 = N_2 C_2$ and $A_3 = T_2 A_2$ where $T_2 = I - B_2 G_2^{\dagger} C_2 A_2$. Then the third PMP may be computed as the third MP of $S_3 = \mathscr{S}(A_3, B_3, C_3)$, $G_3 = C_3 A_3^2 B_3$.

Steps 4–5... Proceed as above until a non-singular PMP, G_ν, is defined. The nested projection procedure is terminated at the νth step.

Note that the nested projection of the input and output maps, B_1 and C_1 implied by

$$B_{i+1} = B_1 M_1 M_2 \ldots M_i, \quad C_{i+1} = N_i N_{i-1} \ldots N_1 C_1 \tag{2.41}$$

together with the fact that by eqn. (2.40 *c*) G_i emerges as the ith MP of S_i account for the term ' projected Markov parameters ' used to describe the set of parameters G_i.

It was observed in § 2.3 that the systems S_1 and S_2 not only share a common ARLB for all orders of IZ higher than 1 but they also have a common set of FZ. This property carries over to the case of S_i and S_{i+1} as stated in the theorem below :

Theorem 2.5

The systems S_i and S_{i+1} of eqn. (2.40) have an identical set of FZ and IZ with the only difference that S_{i+1} has no ith order IZ ; the ith order IZ of S_{i+1} have been replaced by an equal number of FZ located at the origin.

The proof of this theorem is the same as that of Theorem 2.3 ; naturally one has to make the necessary adjustments of superscripts and subscripts. The present section is now brought to a close by the statement of the following result (Kouvaritakis and Shaked 1976) :

Corollary 2.5.1

The system $S_1 = \mathscr{S}(A_1, B_1, C_1)$ with PMP, G_i, $i = 1, 2, \ldots, \nu$ which have SNS with rank defect d_i has exactly $n_z = n - m - \sum_{i=1}^{\nu} d_i$ FZ and $m + \sum_{i=1}^{\nu} d_i$ IZ; of the IZ, $m - d_1$ are first order, $2(d_1 - d_2)$ are second and generally $i(d_{i-1} - d_i)$ are ith order for all $i = 1, 2, \ldots, \nu$.

2.5. *The case of non-proper systems*

The method of analysis developed in the preceding section, although derived for the case of proper systems, can be suitably extended to deal with the case of non-proper systems. It is shown here that subject to appropriate augmentation of the MP chain to include the non-proper part of the system the results remain the same. Furthermore it is shown that the new set of PMP can be generated by the same procedure discussed earlier. Thus consider the non-proper system

$$S_0 = \mathscr{S}(A_0, B_0, C_0, D_0) \tag{2.42}$$

whose transfer function matrix is given as

$$G_0(s)=C_0(sI-A_0)^{-1}B_0+D_0 \tag{2.43}$$

Then $G_0(s)$ may be written as

$$G_0(s)=sG'_0(s) \tag{2.44}$$

where $G'_0(s)$ is now proper and is related to $G_0(s)$ by a simple scalar multiplication. A system S'_1 therefore with no first order IZ but with an otherwise identical ARLB to that of $G'_0(s)$ for all orders of IZ higher than 1 can be derived as before and this is displayed by means of the diagram in Fig. 6, where

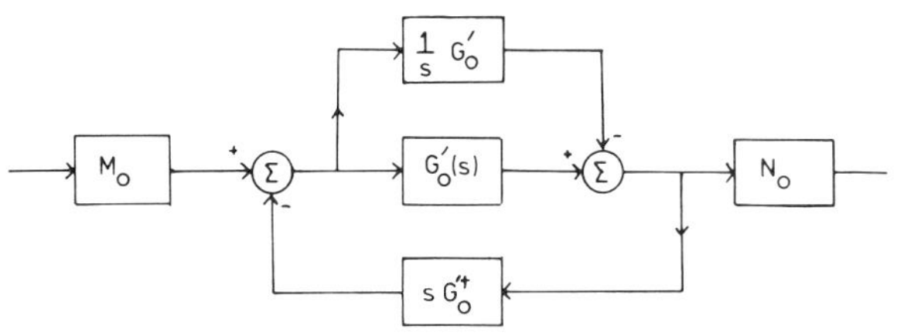

Figure 6.

$G'_0=D_0$ is the first non-zero MP of $G'_0(s)$. It follows then that the system S_1 with an ARLB which is identical to that of $G_0(s)$ for all orders of IZ higher or equal to 1 is given by the configuration of Fig. 7, where $G_0=G'_0$. The structure of this diagram is consistent with that of Fig. 5 so we may embed the case of non-proper systems into the statement of Theorem 2.4. simply by replacing $G_1(s)$ of Fig. 5 by the configuration of Fig. 7. Note that $d_0\neq m$ any longer, instead d_0 must denote the rank defect of G_0. The state space representation of S_1, from Fig. 7, may be deduced to be

$$S_1=\mathscr{S}(A_1,\,B_1,\,C_1) \tag{2.45 a}$$

where

$$A_1=T_0A_0,\quad T_0=I-B_0G_0{}^\dagger\,C_0A_0{}^{-1} \tag{2.45 b, c}$$

$$B_1=B_0M_0,\quad C_1=N_0C_0 \tag{2.45 d, e}$$

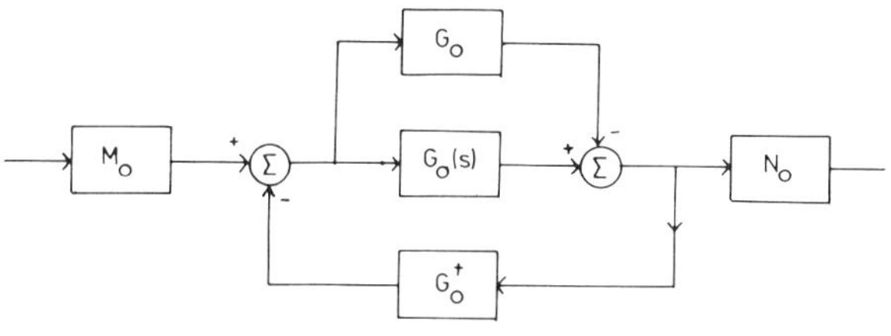

Figure 7.

Comparison of the above with eqn. (2.40) shows that the transition from the non-proper system S_0 to the proper system S_1 with the identical ARLB can be effected by the procedure of § 2.4 which obtains the general system S_{i+1} from S_i. The only modification required is that i now must be allowed to assume the values $i=0, 1, 2, \ldots, \nu$ and G_0 must be defined to be

$$G_0 = D_0 \tag{2.46}$$

2.6. *Discussions on the ARLB*

The analysis undertaken thus far has led to a number of results concerning the properties of the ARLB of both proper and non-proper systems. Here we relate these results to the relevant results presented by Kouvaritakis and Shaked (1976). Further, we discuss some of the properties of the PMP under feedback, input and output compensation. Finally, a design procedure for the adjustment of the ARLB is proposed.

The complexity of the formulae for the computation of the PMP of eqn. (2.40) derives from the fact that in general the equation defining the ' characteristic gain function ', $g(s)$, is irreducible in $g(s)$. As a result the branches of $g(s)$ are irrational functions of the complex variable s and therefore $G(s)$ is non-diagonalizable in the field of rational functions. A weaker condition to diagonalizability however is that of block triangularizability and this may reduce the complexity of the expressions for the PMP. The simplicity of the results of Kouvaritakis and Shaked (1976) emanates from assumptions of this kind. First of all we note that the first two PMP defined by Kouvaritakis and Shaked (1976) are identical to these defined in §§ 2.2 and 2.3. Thus the two approaches agree for all systems with first and second IZ only. It should be noted further that this agreement is also valid for all systems with any two consecutive orders of IZ only as for example second and third orders only or third and fourth orders only etc. The two sets of results deviate beyond a third order of IZ because of error terms which are propagated through off diagonal blocks and which are not taken into account by Kouvaritakis and Shaked (1976). Indeed it is these error terms that are responsible for the presence of the feedback terms $sG_1^\dagger$, $s^2G_2^\dagger$, ... in the configurations of Fig. 5. The nested projection of Theorem 2.4 is otherwise identical to that given by Kouvaritakis and Shaked (1976). The results presented in the latter hold true for all orders of IZ for the case of a $G(s)$ which is block triangularizable and the subsystems corresponding to each diagonal block have at the most two orders of IZ. Under these assumptions the PMP take the simple form of the nested projection of the MP of the given systems. The results discussed in this paper, though more complicated on account of the correction terms, deal with the general case. A further degree of generality of the present treatment derives from the fact that it is no longer necessary to assume any of the PMP to be non-zero (other than G_ν of course (Owens 1977 and Kouvaritakis 1978 b). The presence of a zero PMP does not affect the manner in which subsequent PMP are defined. Thus if for example the lth PMP is given as

$$G_l = 0 \tag{2.47}$$

then clearly there will exist no lth order IZ while the higher order IZ may be obtained by computing the PMP, G_i for $i > l$ using the procedure of § 2.4. We do this by setting

$$N_l = I, \quad M_l = I, \quad T_l = I \tag{2.48}$$

and introducing these parameters in eqn. (2.40). A further topic of interest concerns the effect that feedback has on the PMP. We have the following result.

Theorem 2.6

The first two PMP of a system $\mathscr{S}(A, B, C)$, G_1 and G_2 remain invariant under state and output feedback while the third PMP, G_3 remains invariant under output feedback only.

The proof of the above is simple and consists merely of deriving expressions for the PMP in terms of the closed-loop state-space matrices. The implication of the theorem is important from a design point of view. To see this consider the case of a system with first and second order IZ only. Suppose that either one or both the first and/or second order Butterworth patterns of asymptotes contain members that lie on the positive half of the real axis. Then a system would display poor gain margins and no amount of output or state feedback would be capable of improving these margins.

Fortunately the situation is different with regard to input or output compensation. Because of the symmetry of the expressions involved we shall only consider the use of the input compensation as provided by means of a precompensator of constant gains, K, introduced in the feedback loop in the manner indicated in Fig. 8.

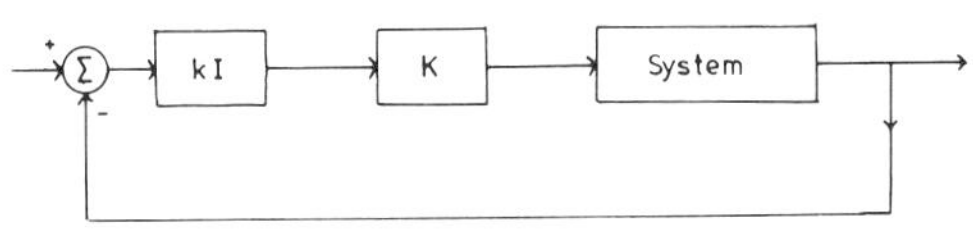

Figure 8.

Suppose now we make the particular choice for K

$$K = U_1K_1V_1 + M_1U_2K_2V_2N_1 + M_1M_2U_3K_3V_3N_2N_1 + \dots \\ + M_1M_2 \dots M_{\nu-1}U_\nu K_\nu V_\nu N_{\nu-1} \dots N_2N_1 \tag{2.49}$$

Then it is easy to show the set of PMP for the compensated open-loop system is simply given by

$$G'_i = G_iK_i \quad \text{for} \quad i = 1, 2, \dots, \nu \tag{2.50}$$

It follows therefore that the controller of eqn. (2.49) offers the facility for the individual adjustment of the PMP. One could further make use of the ' nested commutativity ' in order to gain immediate access to the non-zero eigenvalues of the PMP by setting.

$$K_i = U_i\Lambda_{K_i}V_i \tag{2.51}$$

where Λ_{K_i} is some diagonal matrix of constants. Upon substitution of the above into eqn. (2.50) we obtain

$$G'_i = U_i \Lambda_i \Lambda_{K_i} V_i \tag{2.52}$$

The eigenvalues of the modified PMP can now be seen to be equal to their original value multiplied by the corresponding diagonal element of Λ_{K_i}. Note that for the case of non-proper systems i must assume the values $i=0, 1, \ldots, \nu$. Input compensation therefore may be used in order to manipulate the eigenvalues of the PMP individually and hence to adjust the orientation of the associated asymptotes. In the case of the hypothetical example discussed earlier in the present section this means that all the first order asymptotes could be placed on the negative real axis while the second order asymptotes could be arranged so as to make 90° angles, with the real axis. Naturally compensation of this kind would result in a substantial improvement of the system gain margins.

Now, depending on whether the second order asymptotes above lie in the left or the right half of the complex plane, the compensated system may or may not have infinite gain margins. The reason for this is that the orientation of the asymptotes alone does not completely define the Butterworth patterns, instead one requires further information about the location of the patterns relative to the origin of the complex plane. This information is provided by the ' pivots ', that is the points of asymptote radiation shared in common by all asymptotes which belong to one and the same Butterworth pattern (Kouvaritakis and Shaked 1976). Multivariable pivots form the generalization of the asymptote intercepts with the real axis for scalar systems. Unlike the case of the latter, multivariable systems may have more than one pivot just as they may have more than one Butterworth pattern of asymptotes. Furthermore the pivots need not lie on the real axis but may appear in complex conjugate pairs.

3. Multivarible pivots

3.1. *The location of pivots*

In order to compute the location of the pivots we write the equation of the asymptote to the IZ of order i, associated with the jth non-zero eigenvalue of G_i in the following manner

$$s = [-\lambda_j^{(i)} k]^{1/i} + \eta_j^{(i)} \quad \text{for} \quad i = 1, 2, \ldots, \nu, \quad j = 1, 2, \ldots, d_{i-1} - d_i \tag{3.1}$$

where $\eta_j^{(i)}$ is some constant. The value of $\eta_j^{(i)}$, whether real or complex, gives the location of the relevant pivot since $\eta_j^{(i)}$ is taken to be common to all the i asymptotes associated with the ith roots of $-\lambda_j^{(i)}$. Now the ith order ARLB is obtained from the characteristic equation for the system S_i given by

$$|I + kG_i(s)| = \left| I + \frac{k}{s^i} G_i + P_i(s) \right| = 0 \tag{3.2}$$

where

$$P_i(s) = G_i(s) - \frac{1}{s^i} G_i \tag{3.3}$$

Following the procedure of §§ 2.2 and 2.3 we diagonalize G_i next and subsequently apply Schur's formula in order to derive the ARLB from the equation

$$\left| I + \frac{k}{s^i} \Lambda_i + k V_i P_i(s) U_i \right| = 0 \tag{3.4}$$

or after retention of the most significant term, P_i, of the Taylor series expansion of $P_i(s)$ only

$$\left| I + \frac{k}{s^i} \Lambda_i + k V_i P_i U_i \right| = 0 \tag{3.5}$$

where $P_i = C_i A_i^{\,i} B_i$.

Finally upon substitution of eqns. (3.1) into the above and upon consideration of the most significant terms we may readily deduce that

$$\eta_j^{(i)} = \frac{[\mathbf{v}_j^{(i)}]^t P_i \mathbf{u}_j^{(i)}}{i \lambda_j^{(i)}} \tag{3.6}$$

where $\mathbf{u}_j^{(i)}$ and $\mathbf{v}_j^{(i)}$ denote the eigenvector and reciprocal eigenvector of G_i associated with $\lambda_j^{(i)}$ respectively. Equation (3.6) therefore provides the means of computing the location of pivots.

3.2. *Pivot placement*

We examine now the possibility of manipulating the location of the pivots. To do this let us once again consider the use of the precompensator of eqn. (2.47) for which eqn. (3.2) becomes

$$\left| I + \frac{k}{s^i} \hat{G}_i + k \hat{P}_i \right| = 0 \tag{3.7}$$

where

$$\hat{G}_i = G_i K_i \quad \text{and} \quad \hat{P}_i = P_i K_i \tag{3.8}$$

If the spectral decomposition of $\hat{G}_i$ is given by

$$\hat{G}_i = [\hat{U}_i \mid \hat{M}_i] \begin{bmatrix} \hat{\Lambda}_i & 0 \\ 0 & 0_{d_i} \end{bmatrix} \begin{bmatrix} V_i \\ N_i \end{bmatrix} \tag{3.9}$$

then the new locations of the pivots are obtained by

$$\hat{\eta}_j^{(i)} = \frac{[\hat{\mathbf{v}}_j^{(i)}]_t P_i K_i \hat{\mathbf{u}}_j^{(i)}}{i \lambda_j^{(i)}} \tag{3.10}$$

where $\hat{\lambda}_j^{(i)}$ is the jth non-zero eigenvalue of $\hat{G}_i$ and $\hat{\mathbf{u}}_j^{(i)}$ and $\hat{\mathbf{v}}_j^{(i)}$ are the corresponding eigenvector and dual eigenvector.

Clearly then the introduction of K into the feedback loop, in general, alters both the eigenvalues and eigenvectors of the PMP and therefore by eqn. (3.10) the methodical placement of the pivots for a general K could be cumbersome. To circumvent this difficulty we may consider the case of a controller K which leaves the eigenvalues of the PMP unaltered. Such a K may be constructed from eqn. (2.49) by selecting

$$K_i = G_i^{\dagger} U'_i \Lambda_i V'_i \tag{3.11}$$

with

$$U'_i = U_i R_i \quad \text{and} \quad V'_i = R_i^{-1} V_i \tag{3.12}$$

where R_i is a full rank square matrix of constants. By substitution of the above into eqn. (3.10) it may be readily deduced that the new pivots are given by the diagonal elements of the matrix $(1/i) V'_i P_i G_i^{\dagger} U'_i$. Hence by an appropriate choice of the matrix R_i the pivots may be placed at desired locations.

4. The approach to the finite zeros

In an attempt to unify the concepts of FZ and IZ we develop here an approach to the analysis and computation of the zeros which is common to both FZ and IZ. Thus once again we consider the behaviour of the closed-loop system poles for $k \to \infty$ as obtained by the characteristic eqn. (2.3), only now we shall be interested in those poles that remain finite. Hence we shall no longer be concerned with the Taylor series expansion of $G_1(s)$ about the point at infinity but instead we shall examine the expansion of $G_1(s)$ about a finite test point s_0 as dictated by the appropriate MacLaurin series. The approach adopted leads to a bilinear transformation of the complex variable s to a new variable p, through which the FZ and IZ emerge as unified concepts. The treatment undertaken culminates in a new algorithm for the computation of zeros which though related to the algorithm of Kouvaritakis (1974) and Kouvaritakis and MacFarlane (1976) is easier to implement. Further, the unified treatment adopted enables the derivation of the angles of approach by the same mechanism that is employed for the calculation of the ARLB.

4.1. *The finite zeros*

Consider the MacLaurin series expansion of the transfer function matrix $G_0(s)$ of eqn. (2.30) about the test point s_0 on the complex plane, given as

$$G_0(s) = \sum_{i=0}^{\infty} (s - s_0)^i \frac{d^i}{ds^i} [G_0(s)]_{s=s_0} \tag{4.1}$$

where of course

$$\frac{d^i}{ds^i} [G_0(s)]_{s=s_0} = (-1)^i C_0 (s_0 I - A_0)^{-(i+1)} B_0 \tag{4.2}$$

Note that for the inverse of $(s_0 I - A_0)$ to exist we must assume that s_0 does not coincide with any of the eigenvalues of A_0. For the moment we shall be interested in the solutions of eqn. (2.3) in s which for $k \to \infty$ tend to finite points like s_0. Furthermore we wish to be able to obtain these solutions by means of the analysis used in earlier sections. Consideration of these requirements suggests the use of the bilinear transformation defined in terms of three finite complex numbers s_0, p_0 and ρ as

$$s - s_0 = \frac{\rho}{p - p_0} \tag{4.3}$$

where the role of the inversion implicit in eqn. (4.3) is to ensure that as $k \to \infty$ and $s \to s_0$ the complex quantity $p - p_0$ and therefore p tends to infinity. Upon substitution of eqns. (4.2) and (4.3) into (4.1) we derive that

$$G_0(s) = D_0 + \sum_{i=0}^{\infty} \frac{(-1)^i \rho^i}{(p-p_0)} C_0(s_0 I - A_0)^{-1} B_0 \tag{4.4}$$

which may now be written as a new transformed transfer function matrix in the variable p

$$G_0(s) = \tilde{G}_0(p) = \tilde{C}_0(pI - \tilde{A}_0)^{-1}\tilde{B}_0 + \tilde{D}_0 \tag{4.5}$$

where the transformed state-space matrices $\tilde{A}_0$, $\tilde{B}_0$, $\tilde{C}_0$ and $\tilde{D}_0$ may be obtained as functions of the original matrices A_0, B_0, C_0, D_0 and the parameters s_0, p_0 and ρ. A more general form of the transformation of eqn. (4.3) is given by

$$s = \frac{f'_1(p)}{f'_2(p)} = \frac{\alpha p + \beta}{\gamma p + \delta} \tag{4.6}$$

where it is assumed that the parameters α, β, γ and δ are such that the determinant of the transformation, $\Delta = \alpha\delta - \beta\gamma$, does not vanish. The transformations of eqns. (4.3) and (4.6) can be shown to be equivalent under the further assumption that $\gamma \neq 0$, since then the latter may be written as

$$s - \frac{\alpha}{\gamma} = \frac{(\Delta/\gamma^2)}{p + \dfrac{\gamma}{\delta}} \tag{4.7}$$

which by the substitutions

$$s_0 = \frac{\alpha}{\gamma}, \quad p_0 = -\frac{\delta}{\gamma} \quad \text{and} \quad \rho = \frac{\Delta}{\gamma^2} \tag{4.8}$$

reduces to eqn. (4.3).

Solving eqn. (4.6) for p we obtain the dual bilinear transformation

$$p = \frac{f_1(s)}{f_2(s)} = \frac{-\delta s + \beta}{\gamma s - \alpha} \tag{4.9}$$

Either of the two forms of the bilinear transformation sets up a one–one correspondence between s and p and therefore defines a map between two copies of the complex plane, the s-plane and the p-plane, which is conformal everywhere except at the critical points $s = s_0$, $s = \infty$ and $p = p_0$, $p = \infty$. To overcome this difficulty we make use of the standard extension of the definition of conformal representation (Phillips 1940). Thus we say that $s = f(p)$ transforms the neighbourhood of p_0 conformally into a neighbourhood of $s = \infty$ if the function $t = 1/f(p)$ transforms the neighbourhood of p_0 into a neighbourhood of $t = 0$. Also we say that $s = f(p)$ transforms the neighbourhood of $p = \infty$ conformally into a neighbourhood of s_0 if $s = \phi(p) = f(1/p)$ transforms the neighbourhood of $p = 0$ conformally into a neighbourhood of s_0. Then the bilinear transformation gives a one-one conformal representation of the whole closed p-plane on the whole closed s-plane and hence it gives a biuniform mapping for the complete s- and p-plane.

The state-space matrices $\tilde{A}_0$, $\tilde{B}_0$, $\tilde{C}_0$, $\tilde{D}_0$, for the transformed system of eqn. (4.5) under the transformation of eqn. (4.6) may be derived (Appendix D) in a manner independent of the series expansion of eqn. (4.1) or (4.4) so that eqn. (4.5) may be proved to hold true for all appropriate pairs (s, p) whether these exist in the vicinity of (s_0, p_0) or not. The matrices $\tilde{A}_0$, $\tilde{B}_0$, $\tilde{C}_0$ and $\tilde{D}_0$ are given by

$$\tilde{A}_0 = [f_1(A_0)][f_2(A_0)]^{-1} = [f_2(A_0)]^{-1}[f_1(A_0)] \tag{4.10 a}$$

$$\tilde{B}_0 = [f_2(A_0)]^{-1}B_0, \quad \tilde{C}_0 = \Delta C[f_2(A_0)]^{-1} \tag{4.10 b, c}$$

$$\tilde{D}_0 = -\gamma C_0[f_2(A_0)]^{-1}B_0 + D_0 = G_0(s_0) \tag{4.10 d}$$

By substitution and using eqn. (4.9) these become

$$\tilde{A}_0 = (-\delta A_0 + \beta I)(\gamma A_0 - \alpha I)^{-1} \tag{4.10 e}$$

$$\tilde{B}_0 = (\gamma A_0 - \alpha I)^{-1}B_0 \tag{4.10 f}$$

$$\tilde{C}_0 = (\alpha\delta - \beta\gamma)C_0(\gamma \tilde{A}_0 - \alpha)^{-1} \tag{4.10 g}$$

$$\tilde{D}_0 = -\gamma C_0(\gamma A_0 - \alpha)^{-1}B_0 + D_0 \tag{4.10 h}$$

We are therefore led to the following result.

Lemma 4.1

The transfer function matrices $G_0(s)$ and $\tilde{G}_0(p)$ of the systems $S_0 = \mathscr{S}(A_0, B_0, C_0, D_0)$ and $\tilde{S}_0 = \mathscr{S}(\tilde{A}_0, \tilde{B}_0, \tilde{C}_0, \tilde{D}_0)$ of eqn. (2.42) and (4.10) respectively satisfy the condition that

$$G_0(s) = \tilde{G}_0(p) \tag{4.11}$$

for any pair of complex variables s and p which are related by the bilinear transformation of eqn. (4.6) where the parameters α, β, γ, δ are some complex constants such that the critical point of the transformation at s_0 given by eqn. (4.8) does not coincide with a characteristic frequency of S_0.

Lemma 4.1 implies that the finite RL behaviour in s of S_0 for $k \to \infty$ may be derived from the ARLB in p of $\tilde{S}_0$. Thus RL properties of the FZ such as the ‘ angles of approach ’ and the ‘ attractivity ’ (Shaked 1976) may be investigated using the analysis already developed in § 2 ; this task is performed in the following section. Presently we establish a simple algorithm for the calculation of the FZ. The existence of IZ for a non-proper system by the results of § 2.5, implies the existence of a PMP chain and this in turn requires the direct pass operator of the system to be rank deficient. Hence $\tilde{S}_0$ will have IZ if and only if

$$|\tilde{D}_0| = |C_0(s_0 I - A_0)^{-1}B_0 + D_0| = 0 \tag{4.12}$$

Hence the only possible finite solutions s_0 to which poles of S_0 may tend as $k \to \infty$ are given by the set of FZ $\{z_i\}$ of S_0 namely the values of s for which $G(s)$ loses rank. By Lemma 4.1, however, we have that if $G_0(s)$ loses rank at $s = z_i$ then $\tilde{G}_0(p)$ also loses rank at

$$p = \tilde{z}_i = \frac{-\delta z_i + \beta}{\gamma z_i - \alpha} \tag{4.13}$$

The locations of the FZ of S_0 therefore may be obtained from those of the FZ of $\tilde{S}_0$ as

$$z_i = \frac{\alpha \tilde{z}_i + \beta}{\gamma \tilde{z}_i + \delta} \tag{4.14}$$

The equation leads to a convenient way of calculating the FZ of a system in the following manner. The FZ of a system $\mathscr{S}(A_0, B_0, C_0, D_0)$ for which D_0 is full rank are given by the eigenvalues of the matrix $A_0 - B_0 D_0^{-1} C_0$ (Kouvaritakis 1974). This simple method for the computation of the FZ unfortunately breaks down for systems with a rank deficient D_0 operator and the modified algorithm of Kouvaritakis and MacFarlane (1976) must be pursued in cases like this. Working with $\tilde{S}_0$ however we gain the advantage that we can always arrange for D_0 to be full rank by choosing $\alpha/\gamma = s_0$ to not coincide with a zero of S_0. Note that for $\tilde{D}_0$ full rank the system $\tilde{S}_0$ will have n FZ, while the systems S_0 with D_0 rank deficient will have FZ as well as IZ. Equation (4.14) may be used to obtain the IZ of S_0 also since it may be readily proved that for $|D_0| = 0$, $\tilde{S}_0$ has some of its FZ located at p_0. We are therefore led to the following theorem.

Theorem 4.1

The bilinear transformation of eqn. (4.6) sets a one-to-one correspondence between the set of FZ $\{z_i\}$ and IZ of a given system $S_0 = \mathscr{S}(A_0, B_0, C_0, D_0)$ and the set of FZ $\{\tilde{z}_i\}$ and IZ of the transformed system $\tilde{S}_0 = \mathscr{S}(\tilde{A}_0, \tilde{B}_0, \tilde{C}_0, \tilde{D}_0)$ of eqn. (4.5). For $\alpha/\gamma = s_0 \neq z_i$ (but $|D_0| = 0$) the system $\tilde{S}_0$ has no IZ so that both the FZ and the IZ of S_0 are mapped by eqn. (4.9) onto the FZ of $\tilde{S}_0$; the maps of the IZ are all located at $-\delta/\gamma = p_0$. Conversely if $-\delta/\gamma = p_0 \neq \tilde{z}_i$ (but $|\tilde{D}_0| = 0$) the FZ and IZ of $\tilde{S}_0$ are all mapped by eqn. (4.6) onto the zeros of S_0 which are all finite; the maps of the IZ of $\tilde{S}_0$ are located at $\alpha/\gamma = s_0$. If s_0 and p_0 can be chosen such that $s_0 \neq z_i$ and $p_0 \neq \tilde{z}_i$ then neither S_0 nor $\tilde{S}_0$ have any IZ.

Corollary 4.1

The FZ of a system $S_0 = \mathscr{S}(A_0, B_0, C_0, D_0)$ for $|D_0| \neq 0$ are given by the eigenvalues of $A_0 - B_0 D_0^{-1} C_0$. If $|D_0| = 0$ then the FZ of S_0 may be obtained from the FZ of the transformed system $\tilde{S}_0 = \mathscr{S}(\tilde{A}_0, \tilde{B}_0, \tilde{C}_0, \tilde{D}_0)$ for which $|\tilde{D}_0| \neq 0$. This system is given by eqn. (4.10) for a value of $\alpha/\gamma = s_0$ which does not coincide with any of the FZ of S_0. Then the FZ of S_0 are given by the transforms under eqn. (4.6) of all the FZ of $\tilde{S}_0$ which are not equal to $-\delta/\gamma = p_0$.

Note that under the particular set of conditions of $\beta = 1$, $\gamma = 1$, $\delta = 0$ the method for the zero calculation of the corollary above takes the special form of the method described by Patel (1976).

Theorem 4.1 highlights the close connection between the concepts of FZ and IZ. Now because of the regular nature of the function that relates the two complex variables s and p the transformation from the s-plane to the p-plane is conformal. From an input/output description point of view the two systems of the theorem above are indistinguishable so that the RL properties on the s-plane for the one system may be extracted from the RL properties on the p-plane for the other. The unification of the concept of FZ and IZ is further exploited in the following section to study this manner in which closed-loop poles approach the FZ with the increasing gain.

4.2. *Angles of approach*

The angles of approach of multivariable systems were first studied by Shaked (1976) who examined the case of the first two orders of approach for proper systems only. Due to some inadvertent error in the formulae developed it was suggested that the angles of approach are invariant under input compensation. Contrary to this, we give here the general treatment for all orders of approach which is common to proper and non-proper systems and from the corrected formulae derived we show that the angles of approach may be adjusted systematically through the use of pre-compensation.

We investigate the manner in which poles approach the FZ of S_0 through the ARLB of $\tilde{S}_0$. As mentioned earlier $\tilde{S}_0$ has IZ if, and only if $\tilde{D}_0$ loses rank, a condition secured by setting

$$\alpha/\gamma = s_0 = z_l, \quad l = 1, 2, \ldots, n_z \tag{4.15}$$

This particular choice of s_0 causes all the FZ of S_0 at z_l to be mapped onto the point at infinity on the p-plane. The number of IZ created in this way depends on the rank defects $\tilde{d}_i$, $i = 1, 2, \ldots, \tilde{\nu}$ of the resulting PMP of $\tilde{S}_0$. Now from eqn. (4.10 *d*) we have that the rank defect of $\tilde{D}_0$, the first parameter in the augmented MP chain for non-proper systems, depends on the multiplicity of the zero at z_l. Note that once again we assume all the relevant PMP to have SNS and for this reason any multiple zeros considered are assumed not to be branch points of the characteristic gain function of S_0.

Let the PMP of $\tilde{S}_0$ be denoted by $\tilde{G}_i$ and let their respective non-zero eigenvalues be represented by $\tilde{\lambda}_j^{(i)}$ for $j = 1, 2, \ldots, \tilde{d}_{i-1} - \tilde{d}_i$ and for $i = 1, 2, \ldots, \tilde{\nu}$ where $\tilde{\nu}$ is defined as the first integer i for which $\tilde{d}_i = 0$. Then, applying Theorem 2.4 to system $\tilde{S}_0$ we derive the ARLB on the p-plane as follows.

Lemma 4.2

For $k \to \infty$, the closed-loop poles, p, of $\tilde{S}_0 = \mathscr{S}(\tilde{A}_0, \tilde{B}_0, \tilde{C}_0, \tilde{D}_0)$ vanish at infinity in accordance to the law

$$p = [-\tilde{\lambda}_j^{(i)} k]^{1/i}, \quad i = 1, 2, \ldots, \tilde{\nu}, \quad j = 1, 2, \ldots, \tilde{d}_{i-1} - \tilde{d}_i \tag{4.16}$$

Hence, the ith order IZ of $\tilde{S}_0$ are reached along asymptotes grouped in Butterworth patterns; each pattern is associated with one eigenvalue $\tilde{\lambda}_j^{(i)}$, and has i asymptotes whose angles with the positive real axis are given by the phase of the ith roots of $-\tilde{\lambda}_j^{(i)}$.

Now by Theorem 4.1 we have that the transformation of eqn. (4.9) maps the FZ, of S_0, z_l, onto a FZ of $\tilde{S}_0$ which is not located at p_0. As we let s_0 however, go to z_l, this FZ of $\tilde{S}_0$ is removed to the point at infinity so that the approach of s to z_l for $k \to \infty$ may be derived from the approach of p to the corresponding IZ of $\tilde{S}_0$. Thus substituting eqn. (4.16) into (4.9) we obtain

$$s = \frac{1}{[-\tilde{\lambda}_j^{(i)} k]^{1/i}} + z_l, \quad l = 1, 2, \ldots, n_z, \quad i = 1, 2, \ldots, \tilde{\nu}$$
$$j = 1, 2, \ldots, \tilde{d}_{i-1} - \tilde{d}_i \tag{4.17}$$

from which we are led to the following result.

Theorem 4.2

The FZ of a system $S_0 = \mathscr{S}(A_0, B_0, C_0, D_0)$ as $k \to \infty$ are approached by closed-loop poles along directions which make with the positive real axis angles given by the phases of the ith roots of $[-1/\tilde{\lambda}_j^{(i)}]$. The approach is inversely proportional to the ith root of k and hence the order of convergence is said to be i.

Corollary 4.2.1

The order of convergence of poles to the distinct FZ of S_0 is 1 and the corresponding angles of approach are simply given by the phase of $[-1/\tilde{\lambda}_j^{(i)}]$.

Corollary 4.2.2

Multiple FZ are approached along directions which are grouped in the Butterworth patterns ; the number of directions in each pattern is equal to the order of convergence.

An inspection of eqn. (4.17) now reveals that the speed of convergence to the FZ does not only depend on the order of the $1/k$ term but it also depends on the scalar coefficient of this term which is related to the concept of ' attractivity '. We measure ' attractivity ' by introducing the parameter $\alpha_j^{(i)}$.

$$\alpha_j^{(i)} = -\frac{1}{i} \lim_{\substack{s \to z_i \\ k \to \infty}} k^{i+1} \left| \frac{ds}{dk} \right| \tag{4.18}$$

that quantifies the component of speed of approach to the FZ, z_i, which is independent of the order of convergence. Hence we may state the further result.

Corollary 4.2.3

The attractivity of the FZ, z_i, of the system S_0 is given by the reciprocal of modulus of the ith root of $\tilde{\lambda}_j^{(i)}$.

We discuss next the ability to adjust the angles and speed of approach. Consider the case of a proper system $S_1 = \mathscr{S}(A_1, B_1, C_1)$ with $|C_1 B_1| \neq 0$. Then the quantity $-1/\tilde{\lambda}_j^{(1)}$ that defines the angle and speed of approach to the distinct FZ, z_i of S_1 may be proved to be given by (Appendix E)

$$-\frac{1}{\tilde{\lambda}_j^{(i)}} = -\mathbf{v}_{z_i}{}^t N_1^{(1)} A_1 B_1 [C_1 B_1]^{-2} C_1 A_1 M_1^{(1)} \mathbf{u}_{z_i} \tag{4.19}$$

where $\mathbf{u}_{z_i}$ and $\mathbf{v}_{z_i}{}^t$ are the eigenvectors and dual eigenvectors of $N_1^{(1)} A_1 M_1^{(1)}$ corresponding to the eigenvalue z_i when the annihilators $N_1^{(1)}$ and $M_1^{(1)}$ are chosen such that $N_1^{(1)} M_1^{(1)} = I$. Equation (4.19) provides the corrected formula for the computation of the angles of approach to z_i and indicates that in general the use of a pre-compensator K (and by arguments of symmetry a post-compensator L) will alter the quantity $-1/\tilde{\lambda}_j^{(1)}$ and hence will modify the approach behaviour. This observation holds true for the case of multiple FZ of proper as well as non-proper systems. A simple way to establish this is to examine the effect K has on the state-space parameters of $\tilde{S}_0$ of eqn. (4.10). The new set $\tilde{A}'_0$, $\tilde{B}'_0$, $\tilde{C}'_0$ and $\tilde{D}'_0$ may be readily derived to be

$$\tilde{A}'_0 = A_0, \quad \tilde{B}'_0 = \tilde{B}_0 K, \quad \tilde{C}'_0 = \tilde{C}_0 \quad \text{and} \quad \tilde{D}'_0 = \tilde{D}_0 K \tag{4.20}$$

It follows then that K appears as a precompensator of the system $\tilde{S}_0$ as well as S_0. Hence K may be used to manipulate the eigenvalues of the PMP of $\tilde{S}_0$ precisely in the same manner employed in § 2.6 for S_0. The mechanism developed for the adjustment of the ARLB of S_0 when applied to $\tilde{S}_0$, however, produces the systematic means for the adjustment of the behaviour of approach to the FZ.

5. The angles of departure

The bilinear transformation applied to the frequency s leads to the definition of a new frequency p which tends to finite limits as s tends to infinity and as such enables the unification of the concepts of FZ and IZ. We may apply a bilinear transformation to the gain g in order to define a new gain h which tends to zero as g goes to infinity. Such an operation sets up a duality between the concept of poles and zeros and may be exploited to derive the general treatment of the study of the 'departure' behaviour of the RL from the open-loop poles. This subject was first touched upon by Shaked (1976) who undertook the study of the angles of departure from distinct poles. The treatment given here culminates in a method that deals with the general case of multiple poles. For the sake of completeness however we extend the results by means of the series expansion first which casts some light into the details of the behaviour of departure and we conclude the section by a brief description of the unified approach.

5.1. *The series expansion approach*

The application of the Laurent's series to the transfer function matrix $G_0(s)$ in the vicinity of a pole λ_i of $G_0(s)$ enables the study of the behaviour of s, as s tends to λ_i for $k \to 0$ to be obtained from an equation of the type

$$\left| I + \frac{k}{(s-\lambda)} R_1 + \frac{k}{(s-\lambda)^2} R_2 + \ldots + \frac{k}{(s-\lambda)^\mu} R_\mu \right| = 0 \tag{5.1}$$

Where μ is the multiplicity of λ and R_i for $i = 1, 2, \ldots, \mu$ denote the relevant residues of $G_0(s)$. The latter may be computed in terms of the state-space matrices A_0, B_0, C_0 of the system $S_0 = \mathscr{S}(A_0, B_0, C_0, D_0)$. Thus if λ_i for $i = 1, 2, \ldots, n$ denote the eigenvalues of A_0 and $\lambda_i = \lambda$ for $i = 1, 2, \ldots, \mu$, is a repeated pole appearing in simple Jordan canonical blocks with corresponding eigenvectors and dual eigenvectors denoted by $\mathbf{w}_i$ and $\mathbf{v}_i^t$ then λ will be associated with only one parameter R which is given as

$$R_1 = \sum_{i=1}^{\mu} C_0 \mathbf{w}_i \mathbf{v}_i^t B_0 \tag{5.2}$$

Appropriate expressions in terms of the generalized eigenvectors of A_0 may be derived for R_i in the case of eigenvalues which appear in non-simple Jordan blocks.

The behaviour of departure may now be obtained from eqn. (5.1) which for $k \to 0$ and s tending to λ in accordance with

$$s = [\alpha^{(\mu)} k]^{1/\mu} + \lambda \tag{5.3}$$

reduces to

$$\left| I + \frac{1}{\alpha^{(\mu)}} R_\mu \right| = 0 \tag{5.4}$$

Thus the angles of departure are given by the μth roots of the non-zero eigenvalues of R_μ, and the directions of departure can be seen to form Butterworth patterns. Note that if the repeated eigenvalue λ appears in more than one Jordan block then the associated residues R_i will vanish for all $i > \sigma$ where σ represents the largest dimension of all the Jordan blocks of λ. In a case like this the parameter μ of eqn. (5.3) and (5.4) must be replaced by σ. Furthermore note that if $\sigma \times \text{rank}\,(R_\sigma) < \mu$ then lower order solutions of the type

$$s = [\alpha^{(\sigma - j)} k]^{1/\sigma - j} + \lambda \tag{5.5}$$

for $j \geqslant 1$ will also exist. Such solutions may be retrieved by carrying out a nested projection on the null spaces of the residues R_i starting from the largest i and working backwards towards $i = 1$.

The detailed treatment of the above procedure is somewhat involved and will not be discussed here. Instead we conclude this section by giving a brief summary of the results in three simple special cases :

(1) λ has multiplicity μ and appears in one Jordan Block, then the angles of approach are given by the μth roots of the non-zero eigenvalue of R_μ, α_μ

$$\alpha_\mu = \mathbf{v}_\mu{}^t B_0 C_0 \mathbf{w}_\mu \tag{5.6}$$

(2) λ has multiplicity μ and appears in simple Jordan cannonical blocks, then the angles of approach are given by the phases of the μ non-zero eigenvalues of the residue R_μ of eqn. (5.2).

(3) λ is a distinct eigenvalue, then the angle of approach is given by the non-zero eigenvalue of the $R_1 = C_0 \mathbf{w}_1 \mathbf{v}_1{}^t B_0$, $\alpha^{(1)}$

$$\alpha^{(1)} = \mathbf{v}_1{}^t B_0 C_0 \mathbf{w}_1 \tag{5.7}$$

This last case is the case examined by Shaked (1976).

5.2. *The bilinear transformation approach*

The relationship between the closed-loop characteristic frequency s and the closed-loop characteristic gain g is described in a concise manner by the characteristic equation defined by the generalized system matrix, $P(s, g)$ (MacFarlane *et al.* 1977).

$$|P(s, g)| = \begin{vmatrix} sI - A_0 & B_0 \\ C_0 & gI - D_0 \end{vmatrix} = 0 \tag{5.8}$$

A change in the frequency variable according to the bilinear transformation of eqn. (4.9) converts the above to a transformed characteristic equation

$$|\tilde{P}(p, g)| = \begin{vmatrix} pI - \tilde{A}_0 & \tilde{B}_0 \\ \tilde{C}_0 & gI - \tilde{D}_0 \end{vmatrix} = 0 \tag{5.9}$$

where $\tilde{A}_0$, $\tilde{B}_0$, $\tilde{C}_0$ and $\tilde{D}_0$ are defined by eqn. (4.10). By arguments of symmetry however one may also apply a bilinear transformation on the gain variable and thus define the new gain parameter h to be

$$h = \frac{\phi_1(g)}{\phi_2(g)} = \frac{-dg+b}{cg-a} \tag{5.10}$$

The relationship between the new gain variable h and the frequency variable p is given by the modified characteristic equation

$$|\hat{P}(p, h)| = \begin{vmatrix} pI - \hat{A}_0 & \hat{B}_0 \\ \hat{C}_0 & hI - \hat{D}_0 \end{vmatrix} = 0 \tag{5.11}$$

where

$$\hat{D}_0 = [\phi_1(\tilde{D}_0)][\phi_2(\tilde{D}_0)]^{-1} = [\phi_2(\tilde{D}_0)]^{-1}[\phi_1(\tilde{D}_0)] \tag{5.12 a}$$

$$\hat{C}_0 = [\phi_2(\tilde{D}_0)]^{-1}\tilde{C}_0, \quad \hat{B}_0 = \Delta_\phi \tilde{B}_0[\phi_2(\tilde{D}_0)]^{-1} \tag{5.12 b, c}$$

$$\hat{A}_0 = -c\tilde{B}_0[\phi_2(\tilde{D}_0)]^{-1}\tilde{C}_0 + \tilde{A}_0 = \tilde{S}_0(g_0) \tag{5.12 d}$$

where Δ_ϕ and g_0 are the determinant and the finite critical point of the transformation of eqn. (5.10) and are therefore given as $\Delta_\phi = ad - bc$ and $g_0 = a/c$. The frequency matrix $\tilde{S}_0(g)$ is simply defined as the closed-loop state matrix of the system $\tilde{S}_0 = \mathscr{S}(\tilde{A}_0, \tilde{B}_0, \tilde{C}_0, \tilde{D}_0)$ when a scalar feedback gain $k = -1/g$ is applied around $\tilde{S}_0$.

We may now exploit the bilinear transformation for the purpose of unifying the problem of 'departure' with the problem of 'approach'. Clearly in order to be able to do that we must set up a duality between poles and zeros and this may be easily achieved by selecting the parameters of eqn. (5.10) to be $a=0$, $b=1$, $c=1$, $d=0$ for which h becomes

$$h = \frac{1}{g} \tag{5.13}$$

Under this definition of the gain h we have that the zeros of $\tilde{S}_0$, which of course are the values of p for which g becomes zero, make h become infinite and as such the zeros of $\tilde{S}_0$ emerge as the poles of $\hat{S}_0 = \mathscr{S}(\hat{A}_0, \hat{B}_0, \hat{C}_0, \hat{D}_0)$. Conversely the poles of $\tilde{S}_0$ make g infinite and h zero so that the poles of $\tilde{S}_0$ are the zeros of $\hat{S}_0$. In particular it is easy to show that for the functions $\phi_1(.)$ and $\phi_2(.)$, defined by eqn. (5.13), $\tilde{S}_0$ simply becomes the inverse system of $\hat{S}_0$. It follows then that the behaviour of approach to the zeros of $\hat{S}_0$ is related to the behaviour of departure from the poles of $\tilde{S}_0$ and hence from the poles of S_0.

Thus if we set $h = 1/l$ and we let $l \to \infty$ then p will tend to the FZ z_i of $\tilde{S}_0$ as

$$p = \left[\frac{\theta}{l}\right]^{1/j} + \hat{z}_i \tag{5.14}$$

where θ may be computed using the results of § 4. Now $\hat{z}_i$ is an FZ of $\hat{S}_0$ and therefore a pole $\tilde{\lambda}_i$ of $\tilde{S}_0$ so that a transformation of eqn. (5.14) back to the s-plane will readily provide a precise description of the manner of departure of s from λ_i as $k = 1/l$ tends to zero.

Note that a necessary condition for the derivation of the inverse system $\hat{S}_0$ under the transformation of eqn. (5.13) is that the D_0 operator of the direct system is itself invertible and therefore full rank. It follows then that the need for the transformation of the frequency variable s arises only for systems S_0 with D_0 singular when direct inversion is not possible ; the change of variable defines the new system $\tilde{S}_0$ which for $s_0 = \alpha/\gamma \neq z_i$ (where z_i is an FZ of S_0) will have a full rank $\tilde{D}_0$ operator and which therefore will be invertible. An alternative method for obtaining the behaviour of departure in a manner that circumvents the difficulty of invertibility is to substitute the gain transformation of eqn. (5.13) by

$$h = \frac{1}{g - a} \tag{5.15}$$

where a is now chosen to be a constant that does not coincide with any of the eigenvalues of D_0. Under this transformation the system $\hat{S}_0$ of eqn. (5.12) may be defined on S_0 directly since the term $\phi_2(D_0)$ now is always full rank ; the need for the preliminary transformation of the frequency variable is therefore obviated. Note however that the system $\hat{S}_0$ defined in this way is no longer the inverse of S_0 since by eqn. (5.15) we have that although the poles of g are still zeros of h the zeros of g are no longer poles of h.

The angles of departure may now be obtained from the angles of approach of $\hat{S}_0$ and these in turn, as discussed in § 4, may be obtained from the asymptotic directions of the system defined under the transformation of eqn. (4.9) for s_0 set equal to the FZ of $\hat{S}_0$. Remembering now that the FZ of $\hat{S}_0$, $\hat{z}_i$, are the poles of S_0, λ_i, we conclude that a convenient form of eqn. (4.9) may be chosen to be

$$p = \frac{1}{s - \lambda_i} \tag{5.16}$$

The asymptotic behaviour of the resulting system $\tilde{S}_0$ as $h = -1/l$ tends to zero for l tending to infinity is simply given by

$$p = [\tilde{\alpha}_i^{(j)} l]^{1/j} \tag{5.17}$$

where $\tilde{\alpha}_i^{(j)}$ denotes the non-zero eigenvalues of the jth PMP of $\tilde{S}_0$. A back transformation onto the s-plane and a substitution of $l = -1/h$ by eqn. (5.15) for $g = -1/k$ readily yield that the approximate behaviour of departure for $k \to 0$ is given by

$$s = \left[\frac{1}{\tilde{\alpha}^{(j)}}\right]^{1/j} k^{1/j} + \lambda_i \tag{5.18}$$

Hence the angles of departure of S_0 are equal and opposite in sign to the angles of the asymptotic directions of $\tilde{S}_0$. This observation of course, apart from unifying the ARLB to the behaviour of departure also dictates an easy method for the computation of the angles of departure. Furthermore, note that the methodology developed here, unlike that of § 5.1, is general and it applies uniformly to the use of distinct system poles, to that of repeated poles appearing in simple Jordan canonical blocks and to the case of repeated poles appearing in non-simple Jordan structures.

6. Numerical example

For the purposes of illustration we discuss now a numerical example. The model considered here does not correspond to a physical system and some of its aspects therefore may appear contrived. This model however was especially constructed in order to demonstrate with clarity some of the features of multivariable root locus behaviour discussed in this paper. Thus for example the system considered has a non-zero direct pass operator and a non-zero second PMP while the first PMP vanishes. The second order RL asymptotic patterns are chosen to appear in complex conjugate pairs with their pivots off the real axis. The system poles (with the exception of one) are multiple and appear in complex conjugate pairs, also two of the FZ appear in a complex conjugate pair. As a consequence both the departure from poles and the approach to FZ exhibit some interesting aspects of behaviour. The model is given in the state-space form described in eqn. (6.1).

$$\left.\begin{aligned}
A_0 &= \tfrac{1}{16}\begin{bmatrix} -32 & -80 & 16 & 0 & 0 & 0 & 0 \\ 16 & -64 & -16 & 0 & 0 & 0 & 0 \\ 0 & 0 & -48 & 0 & 0 & 0 & 0 \\ 0 & 0 & 0 & -32 & -80 & 0 & 0 \\ 0 & 0 & 0 & 16 & -64 & 0 & 0 \\ 1653 & 0 & 0 & 3424 & 0 & -32 & -80 \\ 76 & 0 & 0 & 928 & 0 & 16 & -64 \end{bmatrix} \\
B_0 &= \begin{bmatrix} 1 & 0 & 0 \\ -1 & 2 & -2 \\ 1 & 1 & 2 \\ 0 & 1 & 0 \\ 0 & 0 & 1 \\ 0 & 0 & 0 \\ 0 & 0 & 0 \end{bmatrix} \\
C_0 &= \tfrac{1}{8}\begin{bmatrix} 8 & -16 & 0 & 0 & 0 & 0 & 0 \\ -8 & -8 & 16 & -15 & -37 & 8 & 0 \\ 0 & 8 & 16 & -68 & 36 & 0 & 8 \end{bmatrix} \\
D_0 &= \begin{bmatrix} -8 & -2 & 10 \\ -4 & -1 & 5 \\ -8 & -2 & 10 \end{bmatrix}
\end{aligned}\right\} \quad (6.1\ a\text{–}d)$$

We examine the ARLB of $S_0 = \mathscr{S}(A_0, B_0, C_0, D_0)$ first and to this end we compute the spectral decomposition of the PMP $G_0 = D_0$

$$G_0 = \begin{bmatrix} 2 & -1 & -2 \\ 1 & -1 & 3 \\ 2 & -1 & -1 \end{bmatrix} \begin{bmatrix} 1 & 0 & 0 \\ 0 & 0 & 0 \\ 0 & 0 & 0 \end{bmatrix} \begin{bmatrix} -4 & -1 & 5 \\ -7 & -2 & 8 \\ -1 & 0 & 1 \end{bmatrix} \quad (6.2)$$

The parameter G_0 therefore is rank deficient so that S_0 does not have the full number of FZ and hence possesses IZ. To compute the latter we form the equivalent system $S_1=\mathscr{S}(A_1, B_1, C_1)$ of eqn. (2.45) using the parameters A_0, B_0, C_0 and D_0 together with the matrices U_0, Λ_0, V_0, M_0, N_0 defined by the spectral decomposition of D_0 above. Thus we obtain

$$\left.\begin{aligned} A_1 &= A_0-B_0G_0{}^{\dagger}\,C_0=A_0-B_0U_0\Lambda_0{}^{-1}\,V_0C_0 \\ &= \begin{bmatrix} 4\cdot0 & -33\cdot0 & -15\cdot0 & 81\cdot25 & -54\cdot25 & 2\cdot0 & -10\cdot0 \\ -11\cdot0 & 52\cdot0 & 31\cdot0 & -162\cdot5 & 108\cdot25 & -4\cdot0 & 20\cdot0 \\ 21\cdot0 & -98\cdot0 & -59\cdot0 & 284\cdot375 & -189\cdot875 & 7\cdot0 & -35\cdot0 \\ 3\cdot0 & -14\cdot0 & -8\cdot0 & 38\cdot625 & -32\cdot125 & 1\cdot0 & -5\cdot0 \\ 6\cdot0 & -28\cdot0 & 16\cdot0 & 82\cdot25 & -58\cdot25 & 2\cdot0 & -10\cdot0 \\ 103\cdot125 & 0\cdot0 & 0\cdot0 & 214\cdot0 & 0\cdot0 & -2\cdot0 & 5\cdot0 \\ 4\cdot75 & 0\cdot0 & 0\cdot0 & 58\cdot0 & 0\cdot0 & 1\cdot0 & 4\cdot0 \end{bmatrix} \\ B_1 &= B_0M_0=\begin{bmatrix} -1 & -2 \\ 1 & 10 \\ -4 & -1 \\ -1 & 3 \\ -1 & -1 \\ 0 & 0 \\ 0 & 0 \end{bmatrix} \\ C_1 &= N_0C_0=\begin{bmatrix} -5 & 24 & 12 & -64\cdot25 & 42\cdot25 & -2 & 8 \\ -1 & 3 & 2 & -8\cdot5 & 4\cdot5 & 0 & 1 \end{bmatrix} \end{aligned}\right\} \quad (6.3)$$

The first order type of ARLB of S_0 therefore now may be examined by forming the first PMP, $G_1=C_1B_1$. This however can be easily shown to be zero so that neither S_1 nor S_0 possess first order IZ. Therefore to study the higher order ARLB by Theorem 2.4, we introduce into eqn. (2.40) the substitution of $G_1{}^{\dagger}=0$, $N_1=M_1=I$, and for $i=1$ we obtain the state-space representation of the equivalent system S_2 from which the second PMP may be computed to be

$$G_2=C_2A_2B_2=C_1A_1B_1=\begin{bmatrix} -20 & 160 \\ -80 & 140 \end{bmatrix} \quad (6.4)$$

This parameter now is non-singular and has two non-zero eigenvalues at $\lambda_{1,2}{}^{(2)}=60\pm j80$. It follows that the nested projection of the MP is terminated here and that the highest order of IZ is 2. S_0 then has exactly 4 second order IZ and the corresponding asymptotes form two groups of Butterworth patterns with associated angles (with respect to the positive real axis)

$$\alpha_1{}^{(2)}=\tfrac{1}{2}\angle(-60-j80)=-63\cdot4^\circ \text{ or } =116\cdot6^\circ \quad (6.5\,a)$$

for the first pattern and

$$\alpha_2{}^{(2)}=\tfrac{1}{2}\angle(-60+j80)=63\cdot4^\circ \text{ or } =243\cdot4^\circ \quad (6.5\,b)$$

for the second.

In order to complete the description of the asymptotes one may compute the eigenvectors $\mathbf{u}_1{}^{(2)}$, $\mathbf{u}_2{}^{(2)}$ and dual eigenvectors $\mathbf{v}_1{}^{(2)}$, $\mathbf{v}_2{}^{(2)}$ of G_2 and hence calculate the location of the pivots of the first and second asymptotic patterns.

$$\eta_1^{(2)} = \frac{[\mathbf{v}_1^{(2)}]^t C_1 A_1^2 B_1 \mathbf{u}_1^{(2)}}{2\lambda_1^{(2)}} = 19{\cdot}24 - j11{\cdot}87 \tag{6.6 a}$$

and

$$\eta_2^{(2)} = \frac{[\mathbf{v}_2^{(2)}]^t C_1 A_1^2 B_1 \mathbf{u}_1^{(2)}}{2\lambda_2^{(2)}} = 19{\cdot}24 + j11{\cdot}87 \tag{6.6 b}$$

Confirmation of the ARLB of the system S_0 obtained by the application of the analytical results of §§ 2 and 3 is provided by means of the RL displayed in Fig. 9: This diagram was calculated by a PDP11/45 digital computer, by calculating pole positions as a function of feedback gain.

We now turn our attention to the FZ, and in order to calculate these we make use of the results discussed in § 4.1. In particular we select the parameters of the bilinear transformation of eqn. (4.9) to be $\alpha = 5$, $\beta = 1$, $\gamma = 1$, $\delta = 0$ and thus derive the equivalent system $\tilde{S}_0 = \mathscr{S}(\tilde{A}_0, \tilde{B}_0, \tilde{C}_0, \tilde{D}_0)$ so that $p = 1/(s-5)$

$$\left.\begin{aligned}
\tilde{A}_0 &= \begin{bmatrix} -0{\cdot}132 & 0{\cdot}074 & -0{\cdot}026 & 0 & 0 & 0 & 0 \\ -0{\cdot}015 & -0{\cdot}103 & 0{\cdot}011 & 0 & 0 & 0 & 0 \\ 0 & 0 & -0{\cdot}125 & 0 & 0 & 0 & 0 \\ 0 & 0 & 0 & -0{\cdot}132 & 0{\cdot}074 & 0 & 0 \\ 0 & 0 & 0 & -0{\cdot}015 & -0{\cdot}103 & 0 & 0 \\ -1{\cdot}764 & 0{\cdot}98 & -0{\cdot}343 & -3{\cdot}184 & 1{\cdot}769 & -0{\cdot}132 & 0{\cdot}074 \\ -0{\cdot}266 & 0{\cdot}147 & -0{\cdot}052 & -1{\cdot}207 & 0{\cdot}67 & -0{\cdot}015 & -0{\cdot}103 \end{bmatrix} \\
\tilde{B}_0 &= \begin{bmatrix} 0{\cdot}232 & -0{\cdot}121 & 0{\cdot}199 \\ -0{\cdot}099 & 0{\cdot}195 & -0{\cdot}228 \\ 0{\cdot}125 & 0{\cdot}125 & 0{\cdot}25 \\ 0 & 0{\cdot}132 & -0{\cdot}074 \\ 0 & 0{\cdot}015 & 0{\cdot}103 \\ 3{\cdot}086 & 1{\cdot}568 & 0{\cdot}876 \\ 0{\cdot}465 & 0{\cdot}963 & -0{\cdot}272 \end{bmatrix} \\
\tilde{C}_0 &= \begin{bmatrix} -0{\cdot}103 & 0{\cdot}279 & -0{\cdot}048 & 0 & 0 & 0 & 0 \\ -1{\cdot}616 & 1{\cdot}009 & -0{\cdot}578 & -2{\cdot}868 & 2{\cdot}107 & -0{\cdot}132 & 0{\cdot}074 \\ -0{\cdot}281 & 0{\cdot}045 & -0{\cdot}291 & -0{\cdot}148 & -0{\cdot}148 & -0{\cdot}015 & -0{\cdot}103 \end{bmatrix} \\
\tilde{D}_0 &= \begin{bmatrix} -7{\cdot}57 & -2{\cdot}511 & 10{\cdot}654 \\ -0{\cdot}796 & 0{\cdot}428 & 6{\cdot}067 \\ -7{\cdot}384 & -1{\cdot}651 & 11{\cdot}089 \end{bmatrix}
\end{aligned}\right\} \tag{6.7}$$

Next we obtain the FZ of $\tilde{S}_0$ by computing the eigenvalues of the matrix $\tilde{A}_0 - \tilde{B}_0 \tilde{D}_0^{-1} \tilde{C}_0$ and we separate those that are not located at the origin of the p-plane and which are given by

$$\tilde{z}_1 = -0{\cdot}124 + j0{\cdot}02, \quad \tilde{z}_2 = -0{\cdot}124 - j0{\cdot}02, \quad \tilde{z}_3 = 0{\cdot}01 \tag{6.8}$$

Hence the FZ of S_0 may be obtained by a back transformation onto the s-plane as

$$z_1 = -2{\cdot}833 - j1{\cdot}27, \quad z_2 = -2{\cdot}833 + j1{\cdot}27, \quad z_3 = -100 \tag{6.9}$$

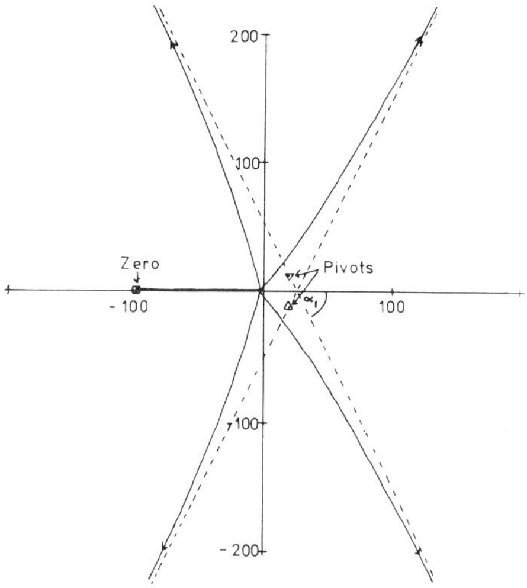

Figure 9.

The behaviour of approach to the FZ of S_0 according to the results of § 4.2 may be derived from the ARLB of $\tilde{S}_0$ where the equivalent system $\tilde{S}_0$ now must be computed under the transformation of eqn. (4.9) for $s_0=\alpha/\gamma=z_i$ $i=1, 2, 3$. Of the three FZ we chose to study z_1 only since the behaviour about z_2 may be obtained by a simple operation of complex conjugation and since the approach to z_3 which is real will be along the real axis and hence is of no intrinsic interest. The $\tilde{D}_0$ operator of $\tilde{S}_0$ when $\alpha=z_1$, $\beta=1$, $\gamma=1$, $\delta=0$, is, as expected, singular with rank defect 1. Following the procedure of § 2.5 therefore we compute the first PMP

$$\tilde{G}_1=\tilde{\lambda}_1^{(1)}=10{\cdot}656-j15{\cdot}102 \tag{6.10}$$

and invoking the result of Corollary 4.2.1 we derive the angle of approach as

$$\theta_1=\angle(-10{\cdot}656-j15.102)=54{\cdot}8^\circ \tag{6.11}$$

The RL diagram of Fig. 10 confirms both the location of and the approach to the FZ of S_0.

The last RL aspect we wish to illustrate through our numerical example is that of departure. The open-loop system, S_0, has only one simple pole at -3 while the remaining six poles appear with multiplicity 3 and are all located at $-3\pm j2$. It is the multiple poles that present the most challenging situation and hence it is those that will be considered here. Once again, of the two complex conjugate locations only one need be considered and $-3-j2$ is selected for this purpose. In order to obtain the behaviour of departure from this pole we apply the results of § 5.2 to the system S_0 of eqn. (6.1). Thus we introduce first the gain transformation

$$h=\frac{1}{g-2} \tag{6.12}$$

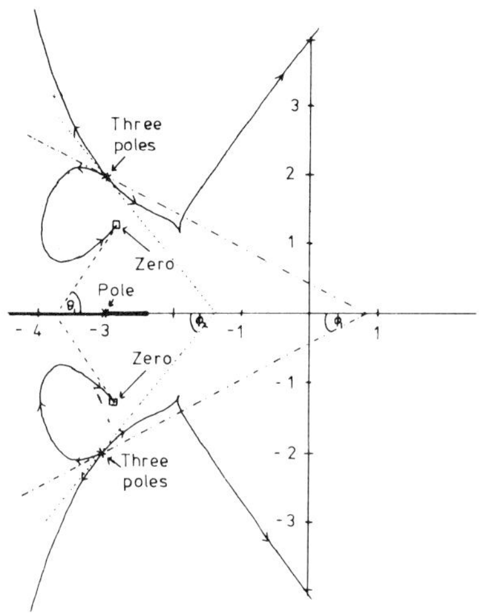

Figure 10.

and derive the transformed system $\hat{S}_0 = \mathscr{S}(\hat{A}_0, \hat{B}_0, \hat{C}_0, \hat{D}_0)$

$$\left.\begin{aligned}
\hat{A}_0 &= \begin{bmatrix} -4\cdot5 & 8\cdot0 & 9\cdot0 & -40\cdot63 & 27\cdot13 & -1\cdot0 & 5\cdot0 \\ 5\cdot5 & -33\cdot0 & -17\cdot0 & 87\cdot86 & -63\cdot38 & 3\cdot0 & -11\cdot0 \\ -10\cdot5 & 48\cdot5 & 28\cdot0 & -151\cdot63 & 97\cdot13 & -3\cdot0 & -18\cdot5 \\ -2\cdot0 & 6\cdot5 & 5\cdot0 & -23\cdot25 & 6\cdot25 & 0 & 2\cdot5 \\ -3\cdot0 & 14\cdot5 & 9 & -43\cdot88 & 25\cdot38 & -1\cdot0 & 5\cdot5 \\ 103\cdot31 & 0 & 0 & 214\cdot0 & 0 & -2\cdot0 & -5\cdot0 \\ 4\cdot75 & 0 & 0 & 58\cdot0 & 0 & -1\cdot0 & -4\cdot0 \end{bmatrix} \\
\hat{B}_0 &= \begin{bmatrix} 3\cdot50 & 1\cdot0 & -5\cdot00 \\ -7\cdot50 & -3\cdot0 & 11\cdot00 \\ 13\cdot50 & 3\cdot0 & -18\cdot50 \\ 2\cdot00 & 0 & -2\cdot50 \\ 4\cdot00 & 1\cdot0 & -5\cdot50 \\ 0 & 0 & 0 \\ 0 & 0 & 0 \end{bmatrix} \\
\hat{C}_0 &= \begin{bmatrix} -2\cdot50 & 13\cdot00 & 8\cdot00 & -40\cdot63 & 27\cdot13 & -1\cdot0 & 5\cdot00 \\ -2\cdot00 & 6\cdot50 & 5\cdot00 & -21\cdot25 & 11\cdot25 & 0 & 2\cdot50 \\ -3\cdot00 & 14\cdot50 & 9\cdot00 & -44\cdot88 & 29\cdot38 & -1\cdot0 & 5\cdot50 \end{bmatrix} \\
\hat{D}_0 &= \begin{bmatrix} 3\cdot50 & 1\cdot00 & -5\cdot00 \\ 2\cdot00 & 0 & -2\cdot50 \\ 4\cdot00 & 1\cdot00 & -5\cdot50 \end{bmatrix}
\end{aligned}\right\} \quad (6.13)$$

Subsequently we transform the frequency s according to the equation

$$p=\frac{1}{s-(-3-j2)} \tag{6.14}$$

and we obtain the resulting equivalent system $\tilde{S}_0=\mathscr{S}(\tilde{A}_0, \tilde{B}_0, \tilde{C}_0, \tilde{D}_0)$. The $\tilde{D}_0$ operator, as expected, is singular with rank defect 2, and hence applying the procedure of §§ 2.4 and 2.5 we obtain the first PMP

$$\tilde{G}_1=\begin{bmatrix} -0\cdot0074-j0\cdot06 & 0\cdot1107+j0\cdot0403 \\ -0\cdot0197-j0\cdot0498 & 0\cdot104 \ +j0\cdot0098 \end{bmatrix} \tag{6.15}$$

which is rank deficient and the second PMP

$$\tilde{G}_2=0\cdot00105+j0\cdot00535 \tag{6.16}$$

The non-zero eigenvalues if the PMPs of $\tilde{S}_0$ then are given as

$$\tilde{\alpha}^{(1)}=0\cdot1-j0\cdot05, \quad \tilde{\alpha}^{(2)}=0\cdot00105+j0\cdot00535 \tag{6.17}$$

By the results of § 5.2 therefore the angles of departure may be computed to be

$$\phi^{(1)}=\angle(-0\cdot1-j0\cdot05)=26\cdot56° \tag{6.18}$$

$$\begin{aligned} \phi^{(2)}&=\tfrac{1}{2}[\angle(-0\cdot00105+j0\cdot00595)+l\times 360°] \quad \text{for} \quad l=0,\ 1 \\ &=50\cdot6°+l\times180° \end{aligned} \tag{6.19}$$

The above behaviour is consistent with the observation that the pole at $-3-j2$ appears in two separate Jordan blocks in the canonical form of the matrix A_0, one of dimension 1 and the other of dimension 2 thus leading to both a first and a second order type of behaviour. Further confirmation of the predicted behaviour of departure is provided by the RL diagram of Fig. 10.

Appendix A

In order to prove that for $k\to\infty$ eqn. (2.12) reduces to (2.15) we observe that for s finite or $s^i=\lambda k$ with λ a constant and $i>1$ we may write

$$\left[I+\frac{k}{s}\Lambda_1+kV_1P_1(s)U_1\right]^{-1}\simeq\left[\frac{k}{s}I+k\Lambda_1^{-1}\,V_1P_1(s)U_1\right]^{-1}\Lambda_1^{-1} \tag{A 1}$$

Next, by making use of the identity $X[I+YX]^{-1}=[I+XY]^{-1}X$ we show that

$$U_1\left[\frac{k}{s}+k\Lambda_1^{-1}\,V_1P_1(s)U_1\right]^{-1}\Lambda_1^{-1}\,V_1=\frac{s}{k}\,[I+sG_1^{\dagger}\,P_1(s)]^{-1}G_1^{\dagger} \tag{A 2}$$

To complete the proof we now substitute the above into eqn. (2.12) in order to obtain

$$\begin{aligned} \Delta(s,k)&=|I+kN_1P_1(s)[I-s[I+sG_1^{\dagger}\,P_1(s)]^{-1}G_1^{\dagger}\,P_1(s)]M_1| \\ &=|I+kN_1P_1(s)[I+sG_1^{\dagger}\,P_1(s)]^{-1}M_1| \end{aligned} \tag{A 3}$$

Appendix B

The transfer function matrix $G_2(s)$ of eqn. (2.17) may be written as

$$G_2(s)=\frac{1}{s}\,N_1[sP_1(s)][I+G_1{}^\dagger(sP_1(s))]^{-1}M_1 \tag{B 1}$$

where $sP_1(s)$ may now be identified as the transfer function matrix of the system $\mathscr{S}(A_1, B_1, C_1A_1)$. Closure of the feedback loops around this system through the operator $G_1{}^\dagger$ results in the system $\mathscr{S}(A_1-B_1G_1{}^\dagger C_1A_1, B_1, C_1A_1)$ which after the appropriate projection indicated by N_1 and M_1 produces the triplet $\mathscr{S}(A_1-B_1G_1{}^\dagger C_1A_1, B_1M_1, N_1C_1A_1)$ which is the state-space representation of $sG_2(s)$. We now introduce the notation

$$T_1=I-B_1G_1{}^\dagger C_1,\quad A_2=T_1A_1,\quad B_2=B_1M_1,\quad C_2=N_1C_1 \tag{B 2}$$

and make use of the properties that

$$C_2A_1=C_2A_2 \quad\text{and}\quad C_2B_2=0 \tag{B 3}$$

from which it follows that the representation of $sG_2(s)$ becomes $\mathscr{S}(A_2, B_2, C_2, A_2)$ and that the triplet $S_2=\mathscr{S}(A_2, B_2, C_2)$ simply describes $G_2(s)$. To complete the proof of Lemma 2.2 we note that the first PMP of S_2 is C_2B_2 which by eqn. (B 3) is seen to be zero ; S_2 therefore does not possess any first order IZ.

Appendix C

The equivalence of the ARLB of S_1 and S_2 for all orders higher than 1 has already been discussed in § 2.3. Now S_2 has no first order IZ and yet it has the same number of poles ; it follows then that S_2 must have $m-d_1$ more FZ than S_1. The FZ of the two systems are given by the finite solutions of Kouvaritakis and MacFarlane (1976)

$$|sN^{(1)}M^{(1)}-N^{(1)}A_1M^{(1)}|=0 \quad\text{and}\quad |sN^{(2)}M^{(2)}-N^{(2)}A_1M^{(2)}|=0 \tag{C 1 a, b}$$

where $N^{(1)}$, $M^{(1)}$, $N^{(2)}$, $M^{(2)}$ are the maximum dimension full rank annihilators of the maps B_1, C_1, B_2 and C_2 defined by

$$N^{(1)}B_1=0,\quad C_1M^{(1)}=0,\quad N^{(2)}B_2=0 \quad\text{and}\quad C_2M^{(2)}=0 \tag{C 2 a–d}$$

Upon consideration of the above definition together with eqns. (2.8) and (2.19) it is easy to show that

$$N^{(2)}=\begin{bmatrix} N^{(1)} \\ \cdots\cdots \\ V_1C_1 \end{bmatrix} \quad\text{and}\quad M^{(2)}=[M^{(1)} \,\vdots\, B_1U_1] \tag{C 3}$$

which when substituted into eqn. (C 1 b) yields

$$\begin{vmatrix} sN^{(1)}M^{(1)}-N^{(1)}A_1M^{(1)} & -N^{(1)}A_2B_1U_1 \\ -V_1C_1A_2M^{(1)} & s\Lambda_1-V_1C_1A_2B_1U_1 \end{vmatrix}=0 \tag{C 4}$$

By eqn. (2.19) however we have that

$$N^{(1)}A_2M^{(1)}=N^{(1)}A_1M^{(1)},\quad V_1C_1A_2M^{(1)}=0 \quad\text{and}\quad V_1C_1A_2B_1U_1=0 \tag{C 5}$$

from which it follows that the FZ of S_2 given by eqn. (C 1) may be obtained as the solutions of

$$|sN^{(1)}M^{(1)} - N^{(1)}A_1M^{(1)}| = 0 \quad \text{and} \quad |s\Lambda_1| = 0 \tag{C 6}$$

the above two relations complete the proof of Theorem 2.3.

Appendix D

Consider the state-space equations of the system $\mathscr{S}(A_0, B_0, C_0, D_0)$ and take Laplace transforms in order to obtain

$$s\bar{x} = A_0\bar{\mathbf{x}} + B_0\bar{\mathbf{u}} \tag{D 1 a}$$

$$\bar{y} = C_0\bar{\mathbf{x}} + D_0\bar{\mathbf{u}} \tag{D 1 b}$$

Now introducing the bilinear transformation of eqn. (4.6) into the above we derive that

$$f_2(A_0)p\bar{\mathbf{x}} = f_1(A_0)\bar{\mathbf{x}} + B_0[-f'_2(p)\bar{\mathbf{u}}] \tag{D 2 a}$$

$$-f'_2(p)\bar{\mathbf{y}} = -C_0[f'_2(p)\bar{\mathbf{x}}] + D_0[-f'_2(p)\bar{\mathbf{u}}] \tag{D 2 b}$$

where the functions $f_1(.)$, $f_2(.)$ and $f'_2(.)$ are defined by eqns. (4.6) and (4.9). Both the input and the output vectors appear multiplied by the same scalar quantity $(-f'_2(p))$ so that from an input and output viewpoint one may re-define both $\bar{\mathbf{u}}$ and $\bar{\mathbf{y}}$ so as to include the scaling factor. Furthermore, solving eqn. (D 2 *a*) for $p\bar{\mathbf{x}}$ and substituting the result in (D 2 *b*) we obtain

$$p\bar{\mathbf{x}} = [f_2(A_0)]^{-1}[f_1(A_0)]\bar{\mathbf{x}} + [f_2(A_0)]^{-1}B_0\bar{\mathbf{u}} \tag{D 3 a}$$

$$\bar{\mathbf{y}} = \Delta C_0[f_2(A_0)]^{-1}\bar{\mathbf{x}} + [-\gamma C_0[f_2(A_0)]^{-1}B_0 + D_0]\bar{\mathbf{u}} \tag{D 3 b}$$

where $\Delta = \alpha\delta - \beta\gamma$ is the determinant of the bilinear transformation. A direct comparison of the above with the form of the transformed state-space equations of eqn. (D 1) obtains the state-space representation of $\tilde{G}_0(p)$ as described in eqns. (4.10 *a–d*).

Appendix E

For $s = z_i$, where z_i is a FZ of the system $S_1 = \mathscr{S}(A_1, B_1, C_1)$ the direct pass operator of the transformed system $\tilde{S}_0$ of eqn. (4.5) satisfies the conditions

$$\tilde{D}_0\tilde{\mathbf{M}}_0 = G_1(z_i)\tilde{\mathbf{M}}_0 = 0 \quad \text{and} \quad \tilde{\mathbf{N}}_0\tilde{D}_0 = \tilde{\mathbf{N}}_0G_1(z_i) = 0 \tag{E 1 a, b}$$

where $\tilde{\mathbf{M}}_0$ and $\tilde{\mathbf{N}}_0$ are two vectors spanning the right and left null-space of $G(s)$ respectively.

For C_1B_1 full rank it is easy to show from eqns. (E 1 *a*, *b*) that

$$\tilde{\mathbf{M}}_0 = -[C_1B_1]^{-1}C_1A_1M^{(1)}\mathbf{u}_{z_i} \quad \text{and} \quad \tilde{\mathbf{N}}_0 = -\gamma\mathbf{v}_{z_i}{}^t N^{(1)}A_1B_1[C_1B_1]^{-1} \tag{E 2 a, b}$$

where $\mathbf{u}_{z_i}$ and $\mathbf{v}_{z_i}$ are some appropriate vectors of constants and γ is a scalar such that

$$\tilde{\mathbf{N}}_0\tilde{\mathbf{M}}_0 = \gamma\mathbf{v}_{z_i}{}^t N^{(1)}A_1B_1[C_1B_1]^{-2}C_1A_1M^{(1)}\mathbf{u}_{z_i} = 1 \tag{E 3}$$

Now by definition the parameter $\tilde{\lambda}_j^{(1)}$ is given as

$$\tilde{\lambda}_j^{(1)} = \tilde{\mathbf{N}}_0 \tilde{C}_0 \tilde{B}_0 \tilde{\mathbf{M}}_0 = \tilde{\mathbf{N}}_0 C_1 (z_i I - A)^{-1} (z_i I - A)^{-1} B_1 \tilde{\mathbf{M}}_0 \quad \text{(E 4)}$$

which by eqns. (E 1), (E 2) and (E 3) becomes

$$\tilde{\lambda}_j^{(1)} = \gamma = 1/(\mathbf{v}_{z_i}{}^t N^{(1)} A_1 B_1 [C_1 B_1]^{-2} C_1 A_1 M^{(1)} \mathbf{u}_{z_i}) \quad \text{(E 5)}$$

Equation (4.19) now follows immediately.

References

Desoer, C. A., and Schulman, J. D., 1974, *I.E.E.E. Trans. Circuit Syst.*, **21,** 3.
Edmunds, J. M., 1978, *Int. J. Control*, to appear.
Evans, W. R., 1948, *Am. Inst. elect. Electron. Engrs*, **67,** 547.
Gantmacher, F. R., 1959, *Theory of Matrices*, Vol. 1 (New York : Chelsea).
Kouvaritakis, B., 1974, Ph.D. Thesis, University of Manchester, U.K. ; 1978 a, *Int. J. Control*, **27,** 705 ; 1978 b, *Ibid.* (to appear) ; 1978 c, *Ibid.*, **28,** 33.
Kouvaritakis, B., and MacFarlane, A. G. J., 1976, *Int. J. Control*, **23,** 140.
Kouvaritakis, B., and Shaked, U., 1976, *Int. J. Control*, **23,** 297.
MacFarlane, A. G. J., and Karcanias, N., 1976, *Int. J. Control*, **24,** 33.
MacFarlane, A. G. J., Kouvaritakis, B., and Edmunds, J. M., 1977, N.E.C. Conference, Chicago, U.S.A.
MacFarlane, A. G. J., and Postlethwaite, I., 1977 a, *Int. J. Control*, **25,** 81 ; 1977 b, *Ibid.*, **26,** 265.
Owens, D. H., 1977, *Int. J. Control*, **25,** 819.
Patel, R. V., 1976, *Int. J. Control*, **24,** 145.
Shaked, U., 1976, *Int. J. Control*, **23,** 445.
Phillips, E. G., 1940, *Functions of a Complex Variable* (Edinburgh : Oliver & Boyd).
Pringle, R. M., and Rayner, A. A., 1971, *Generalised Inverse Matrices* (London : Griffin).

The optimal root loci of linear multivariable systems

B. KOUVARITAKIS

Optimally regulated linear systems possess a number of properties attractive to the designer of feedback control loops. On account of these properties both the optimal controller and the compensated system have formed the subject of extensive research in recent years. One aspect of the behaviour of optimal systems which has received little attention, however, concerns the movement of the closed-loop poles of optimal systems when the weight on the control effort in the index of performance is relaxed to arbitrarily low levels. Relevant results on this topic have been derived for scalar systems only while extensions of these to the multivariable case have been presented in a qualitative manner only with the exception of the lowest order of behaviour for which some explicit results exist. Here a new root-locus approach is used in order to study all orders of behaviour. The results derived relate the asymptotic behaviour of the poles of the optimal system to that of the original system and for this reason more light is cast onto the function of the optimal controller. Explicit results are given for all orders up to five although the symmetric nature of the pattern of behaviour established enables by conjecture the extension of these results to all orders. A numerical example is included at the end of this paper for the purposes of illustration.

1. Introduction

The concept of root loci first introduced by Evans (1948) has long since been established as a valuable tool for the analysis and design of feedback systems. However, until recently the root-locus technique has been restricted to systems with one input and one output only. Considerable interest concerning possible extensions to multivariable systems has developed in the past five years and a number of useful results have already appeared in the literature.

One particular topic relating to the movement of closed-loop poles as some gain parameter is varied that has received some research effort is the area of the root loci of linear optimal regulators. These are not root loci in the conventional sense, since the gain parameter that is varied relates to the weight assigned on the control effort in the index of performance ; different weights however require a different structure for the feedback connections if the condition of optimality is to be preserved. On the other hand, root loci in the normal sense describe the motion of closed-loop poles for a fixed feedback structure, as the scalar loop gain of the system alone is allowed to vary. For this reason we shall make the distinction between the two by referring to the latter simply as the root loci (RL) and to the loci of the optimal regulator as the optimal root loci (ORL).

The present paper is intended to study the asymptotic behaviour of the ORL as the demand on the control power is arbitrarily relaxed. This subject has already received some attention and a number of useful results have been obtained : Kwakernaak and Sivan (1972), Wonham (1974) and Kwakernaak (1976). It is felt however that the particular approaches used thus far do little to relate the ORL to the elementary state-space matrices of the system and the aymptotic behaviour of the RL of the non-optimal system itself and for this reason a lot of insight is lost. Furthermore, the description of the asymptotic behaviour obtained thus far is very complex and non-explicit for all orders of behaviour other than the lowest order. Here, following a suggestion of MacFarlane (1970, 1977), we apply the recently developed multivariable root locus techniques introduced by Kouvaritakis and Shaked (1976) to the optimally regulated system. The approach derived offers a new angle to the study of the ORL and fosters the extension of the explicit description of the asymptotic behaviour to higher orders of behaviour.

The basic philosophy of our approach is outlined in § 2. Here a brief review of the multivariable root-locus technique is given and the relation between the ORL and the RL of a suitably augmented system is discussed.

The asymptotic behaviour of the ORL that remain finite is treated in § 3. It is shown that for systems with the same number of inputs as outputs the finite limits of the ORL are closely related to the zeros of the original system ; they are the same in number and are equal for minimum-phase systems. The case of non-minimum-phase systems is also discussed.

The ORL that vanish off to infinity are also for convenience thought of as being attracted by another set of zeros which are called the optimal infinite zeros (OIZ). It is shown in § 4 that, depending on the rate of divergence of the ORL and the spatial symmetry of the asymptotes, the OIZ may be grouped into sets of OIZ of different orders. Each set of OIZ of the same order appears in Butterworth patterns which are associated with asymptotes displaced symmetrically in the left-half of the complex plane ; it is shown that the angular displacement of the asymptotes is uniquely determined by the order of OIZ. Certain important properties of the OIZ then emerge. First we see that no OIZ of odd orders exist ; this is proved to be essential to the symmetrical arrangement of the asymptotes. Then it is found that the even-order OIZ, say order $2q$, are closely related to the IZ of the non-optimal system of order q. Thus it is found that just as the IZ of order q are determined by the qth projected Markov parameter (PMP) of the system the OIZ of order $2q$ are determined by a quadratic function of this PMP. The implications of this result are interesting ; first we show that there are as many $2q$th-order OIZ as there are IZ of order q and next that the spatial arrangement of these OIZ is dictated by the spatial arrangement of the roots of $x^{2q} = \pm 1$, the sign of the unity being determined by whether q is odd or even. These results are derived for all orders of OIZ less than six, but the symmetry of the parameters involved enables the extension of the results to all orders.

The results obtained are illustrated by means of a simple numerical example in § 5.

2. Root loci and optimal root loci

2.1. *The statement of the problem*

Consider the linear time-invariant and continuous system described by the set of state-space equations

$$\left.\begin{aligned} \frac{d\mathbf{x}}{dt} &= A\mathbf{x} + B\mathbf{u} \\ \mathbf{y} &= C\mathbf{x} \end{aligned}\right\} \tag{2.1}$$

where $\mathbf{x}$, $\mathbf{u}$ and $\mathbf{y}$ are the n, l and m-dimensional state input and output vectors, A, B, C are matrices of real constants of appropriate dimension and the pair (A, C) is assumed to be observable. Note that the majority of the results derived refer to square systems, that is systems for which $l=m$; for this reason the symbols l and m will be used interchangably. Then the optimal-control action required for the minimization of the performance index

$$V(\infty)=\int_0^\infty (y^tQy+\mathbf{u}^tR\mathbf{u})\,dt \tag{2.2}$$

is dictated by the equation

$$\mathbf{u}=-R^{-1}B^tP\mathbf{x} \tag{2.3}$$

where P, for Q and R positive definite, denotes the unique positive definite solution of the steady-state matrix Riccati equation

$$-PA-A^tP+PBR^{-1}B^tP=C^tQC \tag{2.4}$$

If now we define K to be the feedback operator given by

$$K=R^{-1}B^tP \tag{2.5}$$

then upon substitution of eqns. (2.3) and (2.5) into (2.1) we derive

$$\left.\begin{aligned}\frac{d\mathbf{x}}{dt}&=(A-BK)\mathbf{x}\\ \mathbf{y}&=C\mathbf{x}\end{aligned}\right\} \tag{2.6}$$

where the expression

$$A_c=A-BK \tag{2.7}$$

may now be identified as the closed-loop state matrix of the optimal feedback system.

It should be apparent by now that, since K is a function of P which in turn depends on both the weighting matrices Q and R, the optimality of the feedback system described by eqn. (2.6) holds good only for the specific pair Q and R used for the computation of K. Thus in order to preserve the optimality of the above system one must update the solution for P and hence the solution for K. An immediate question that arises out of these considerations concerns the location of the closed-loop poles, which under the circumstances will be referred to as the 'optimal poles', for varying P and Q. Although a considerable amount of research has been successfully undertaken with respect to computing the optimal poles for specific weighting matrices (MacFarlane 1970 a, b and Walter 1970) little is known about the loci traced by the optimal poles for varying Q and/or R. An exception to this is to be found in the case of the special choice for Q and R given by

$$Q=qI_m \tag{2.8 a}$$

$$R=rI_1 \tag{2.8 b}$$

where I_1 denotes the $l\times l$ identity matrix and q and r are two positive constants. The problem of identifying the movement of the optimal poles as r is varied with respect to q has attracted a lot of research effort in recent years. In particular, extensive attention was given to the problem of determining the loci that the optimal poles traverse as the weight on the input is arbitrarily decreased with respect to the weight on the outputs ; that is, as the ratio r/q is allowed to attain arbitrarily small values. The loci thus defined will be referred to as the 'optimal root loci' (ORL) where it should be remembered that these loci are associated with a varying controller structure and, as such, are not to be confused with the root loci obtained for a fixed optimal controller K and varying scalar gain k, a study of which is given by Shaked and Kouvaritakis (1976).

The investigation of the asymptotic behaviour of the optimal root loci (ORL) was first undertaken by Kwakernaak and Sivan (1972) who obtained results for scalar systems and subsequently by Wonham (1974) and Kwakernaak (1976) who extended the results to the multivariable case. It is felt that the different approaches employed by the above authors in the derivation of the results relating to the asymptotic behaviour of the ORL are rather involved and hence do not cast enough light onto the relation of the ORL and the root locus structure of the system $\mathscr{S}(A, B, C)$. It is the intention of the present paper to give an alternative treatment of the subject based on the root-locus techniques developed by Kouvaritakis and Shaked (1976) and thus to tie up the asymptotic structure of the ORL to the zero and infinite zero structure of the system of eqn. (2.1).

2.2. *Multivariable root locus : the finite and infinite zeros*

The discussion included in this subsection is a *brief* review of some of the important features of the asymptotic behaviour of the root loci of multivariable systems and hence for further detail the interested reader is referred to Kouvaritakis and MacFarlane (1967 a, b) and Kouvaritakis and Shaked (1976). Consider the system of eqn. (2.1) embedded in the feedback configuration shown below

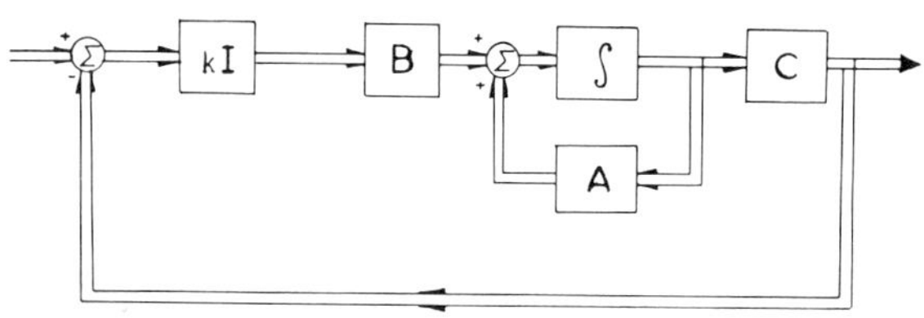

Figure 1.

where **r** denotes a reference input and k a scalar gain. We define as root loci (RL) the loci along which the poles of the closed-loop system move as k is varied ; then the asymptotic behaviour of the RL refers to the behaviour of the RL as k tends to infinity. Here we study this behaviour by using the 'CB approach', an approach which is applicable to the generic situation.

Thus the CB approach is valid for all cases except when the elements of the Markov parameters (MP) defined by

$$P_i \triangleq CA^{i-1}B, \quad i=1, 2, \ldots \tag{2.9}$$

satisfy certain special conditions that result in some non-simple Jordan forms in a manner to be made precise later on. An alternative approach, the BC approach, which deals with these special cases will not be discussed here.

We begin our review by examining the behaviour of those closed-loop poles that remain finite with increasing gain. Thus let k tend to infinity; then a number of characteristic frequencies, say n_z, where $n_z \leqslant n-m$, will tend one each to the n_z finite solutions of the equation

$$|sNM - NAM| = 0 \tag{2.10}$$

where N and M are full-rank matrix representations of the left null-space of B and the kernel of C, such that $NB=0$ and $CM=0$. The set of solutions to eqn. (2.3) is the set of the finite zeros of $\mathscr{S}(A, B, C)$.

The remaining poles of $\mathscr{S}(A, B, C)$ migrate off to infinity along asymptotes that belong to patterns characterized by spatial symmetry. For the convenience of visualization a set of zeros which lie on the point at infinity, called the *infinite zeros* (IZ), can be associated with each pattern of asymptotes (Kouvaritakis and Shaked 1976). Attached to each such set of IZ there exists an order which corresponds to the number of asymptotes in a pattern and the angular displacement between them.

The introduction of IZ offers to the designer the conceptional advantage that he may now consider that for increasing feedback gains *all* the poles are uniformly attracted by zeros (some finite and some infinite). The study of the properties of the IZ therefore seems to be an undertaking of equal importance to the study of the properties of the finite zeros.

In order to investigate the behaviour of the root loci in this manner we must study the location of the IZ and we do this by considering some of the eigen-properties of the Markov parameters (MP). Define as G_1 the first MP such that

$$G_1 \triangleq CB \tag{2.11}$$

Now assume that G_1 is rank-deficient by d_1 degrees and has z_1 zero eigenvalues. Then for $d_1=z_1$ we may write the spectral decomposition of G_1 as

$$G_1 = [U_1 \vdots M_1] \begin{bmatrix} \Lambda_1 & 0 \\ 0 & 0_{d_1} \end{bmatrix} \begin{bmatrix} V_1 \\ N_1 \end{bmatrix} \tag{2.12}$$

where

$$\begin{bmatrix} V_1 \\ N_1 \end{bmatrix} [U_1 \vdots M_1] = I,$$

and where $\operatorname{diag}\{\Lambda_1, 0_{d_1}\}$ is the Jordan form of G_1 and Λ_1 is full rank. The condition $z_1=d_1$ ensures that all the zero eigenvalues of G_1 appear in a canonical block with simple structure. For this reason G_1 will be said to have *simple null structure* (SNS).

Consider next the second MP, CAB, and define G_2 to be the projection of CAB onto the null space of G_1 so that

$$G_2 \triangleq N_1\, CAB\, M_1 \tag{2.13}$$

We once again assume G_2 to have SNS and write

$$G_2 = [U_2 \vdots M_2] \begin{bmatrix} \Lambda_2 & 0 \\ 0 & 0_{d_2} \end{bmatrix} \begin{bmatrix} V_2 \\ N_2 \end{bmatrix} \tag{2.14}$$

where d_2 denotes the rank-deficiency of G_2. In a similar manner G_3 may then be defined as the third MP seen through the null spaces of G_1 and G_2:

$$G_3 \triangleq N_2 N_1 C A^2 B M_1 M_2 \tag{2.15}$$

The process of defining the parameters G_1 is terminated at $i=\nu$, where $|G_\nu| \neq 0$. For convenience the parameters $G_1, G_2, \ldots, G_\nu$ will be called the *projected Markov parameters* (PMP). Although the PMP have been defined for the case of SNS, such a definition can be extended to the case of NSNS† simply by ensuring that the projecting operators M_i and N_i contain as column vectors and row vectors the eigenvectors and reciprocal eigenvectors that correspond to the zero eigenvalues of G_i for all $i=1, 2, \ldots, \nu$. The result concerning the asymptotic behaviour of the closed-loop poles that vanish at infinity can be stated as follows.

Theorem 2.1

If all the PMP of a system have SNS and $d_{i+1} < d_i$ for all $i=1, 2, \ldots, \nu-1$, then the RL that tend to the ith-order IZ as k goes to infinity travel along asymptotes whose angles are

$$\alpha_{j,l}{}^{(i)} = \frac{1}{i}[\angle \lambda_j{}^{(i)} + (2l+1)\pi] \quad \text{for } l=1, 2, \ldots, i$$
$$j=1, 2, \ldots, d_{i-1}-d_i$$
$$i=1, 2, \ldots, \nu$$

where $d_0 = m$, $d_\nu = 0$ and $\lambda_j{}^{(i)}$ is the jth non-zero eigenvalue of the ith PMP.

As a consequence of the above theorem we have the following important result.

Corollary 2.1.1

A system $\mathscr{S}(A, B, C)$ with $\nu-1$ rank-deficient PMP, each by d_i degrees for $i=1, 2, \ldots, \nu-1$ (assume $d_\nu = 0$), has $l + \sum_{i=1}^{\nu-1} d_i$ asymptotes and $n - l - \sum_{i-1}^{\nu-1} d_i$ finite zeros.

The characterization of the asymptotes or equivalently the characterization of the IZ becomes complete only after the introduction of the concept of the pivots. These are the points where all asymptotes associated with any one non-zero eigenvalue of the PMP meet. The computation of the pivots

† Non-simple Null Structure.

is an aspect beyond the scope of this report. Systems with one PMP, G_1, have IZ of order 1 only. In particular, systems like this will have $n-m$ of their closed-loop poles tend to finite zeros given by the eigenvalues of NAM for $NM=I$ and the remaining m tend to first-order IZ along asymptotes that with the negative real axis make angles equal to the phase of the eigenvalues of G_1. If our system now had two PMP, that is to say, if G_1 were rank-deficient, say by d_1 degrees, but G_2 were non-singular, then IZ of order 2 will be introduced, replacing some of the finite and first-order infinite zeros. The number of finite zeros will be now reduced by d_1 and these zeros will be given by the $n-m-d_1$ finite solutions of eqn. (2.3). There will exist $m-d_1$ *first*-order IZ determined by the non-zero eigenvalues of G_1 and $2d_1$ *second*-order IZ determined by the eigenvalues of G_2. In general, if G_i is rank-deficient by d_i degrees for all $i=1, 2, \ldots, \nu-1$ and G_ν is full rank, then there will exist $n-m-\sum_{i=1}^{\nu-1} d_i$ finite zeros given by eqn. (2.3), $m-d_1$ first-order IZ determined by the non-zero eigenvalues of G_1, $2(d_1-d_2)$ second-order IZ determined by the non-zero eigenvalues of G_2 and so on up to $\nu d_{\nu-1}$ νth-order IZ determined by the eigenvalues of G_ν.

3. The finite limits of the optimal root loci

3.1. *The return-difference matrix and the optimal root loci*

The discussion in this subsection stems from certain fundamental results obtained by MacFarlane (1970 a, b, 1977 c). One first such result we propose to use concerns the relationship between the optimal return-difference matrix, $F(s)$, and the system transfer-function matrix, $G(s)$. Thus, defining $F(s)$ to be the return-difference matrix of $\mathscr{S}(A, B, K)$:

$$F(s) \triangleq I + K(sI-A)^{-1}B \tag{3.1}$$

and $G(s)$ the transfer-function matrix of $\mathscr{S}(A, B, C)$:

$$G(s) \triangleq C(sI-A)^{-1}B \tag{3.2}$$

After suitable manipulation of the matrix Riccati equation, MacFarlane (1970) was able to show that

$$F^{t}(-s)RF(s) = R + G^{t}(-s)QG(s) \tag{3.3}$$

For the special choice of Q and R indicated by eqn. (2.8) the above, in determinental form, reduces to

$$|F^{t}(-s)F(s)| = \left| I + \frac{q}{r}\, G^{t}(-s)G(s) \right| \tag{3.4}$$

Subsequent to this, it is readily suggested that the ORL together with their mirror images about the imaginary axis could be obtained by the RL of the system described by the transfer-function matrix $G^{t}(-s)G(s)$ or the equivalent state-space description

$$\left.\begin{aligned} \hat{A}=\begin{bmatrix} A & BB^{\mathrm{t}} \\ 0 & -A^{\mathrm{t}} \end{bmatrix}, \quad \hat{B}=\begin{bmatrix} 0 \\ -C^{\mathrm{t}} \end{bmatrix} \\ \hat{C}=[C \quad 0] \end{aligned}\right\} \tag{3.5}$$

The scalar gain k for the RL of $\mathscr{S}(\hat{A}, \hat{B}, \hat{C})$ must be suitably defined to be

$$k=\frac{q}{r} \tag{3.6}$$

hence we may study the asymptotic behaviour of the ORL as the weight on the control effort is arbitrarily relaxed $(r/q \rightarrow 0)$ by investigating the asymptotic behaviour of the RL of $\mathscr{S}(\hat{A}, \hat{B}, \hat{C})$ for $k \rightarrow \infty$. It is the undertaking of the present paper to examine the asymptotic behaviour of the RL of $\mathscr{S}(\hat{A}, \hat{B}, \hat{C})$.

It is important to realize that the generality of the results derived in this report is not restricted by the special choice for weighting matrices R and Q of eqn. (2.8). To see this consider a general R and Q given as

$$R=rR_0 \tag{3.7 a}$$

$$Q=qQ_0 \tag{3.7 b}$$

where r and q are some positive scalars and R_0 and Q_0 denote two positive-definite matrices and as such enable R and Q to be written as

$$R=rR_0^{1/2}\,R_0^{1/2} \tag{3.8 a}$$

$$Q=qQ_0^{1/2}\,Q_0^{1/2} \tag{3.8 b}$$

Then, expressing eqn. (3.3) in determinantal form we obtain

$$\left.\begin{aligned} |F^{\mathrm{t}}(-s)RF(s)| &= |rR_0^{1/2}|\;|F^{\mathrm{t}}(-s)|\;|R_0^{1/2}| \\ &= |rR_0+qG(s)Q_0^{1/2}\,Q_0^{1/2}\,G^{\mathrm{t}}(-s)| \\ \text{or} \qquad |F^{\mathrm{t}}(-s)F(s)| &= \left|I+\frac{q}{r}\,R_0^{-1/2}\,G(s)Q_0^{1/2}\,Q_0^{1/2}\,G^{\mathrm{t}}(-s)Q_0^{1/2}\,R_0^{-1/2}\right| \\ &= \left|I+\frac{q}{r}\,G_0(s)G_0^{\mathrm{t}}(-s)\right| \end{aligned}\right\} \tag{3.9}$$

where

$$G_0(s)=R_0^{-1/2}\,C(sI-A)^{-1}BQ_0^{1/2}$$

Thus if we define a new set of state-space matrices

$$A_0=A, \quad B_0=BQ_0^{1/2} \quad \text{and} \quad C_0=R_0^{-1/2}\,C \tag{3.10}$$

we see that the ORL of $\mathscr{S}(A, B, C)$ for the general weighting matrices R and Q given above may simply be obtained as the ORL of the system $\mathscr{S}(A_0, B_0, C_0)$ for the special choice of weighting matrices given by eqn. (2.8).

3.2. The optimal finite zeros

A consequence of the observation made in the previous subsection is that the zeros of the ORL, that is the finite limits to which the optimal poles tend as $r/q \to 0$, together with their mirror images about the imaginary axis, will be given by the set of finite zeros, $Z(\hat{A}, \hat{B}, \hat{C})$, of the system $\mathscr{S}(\hat{A}, \hat{B}, \hat{C})$. Once again it is important to make a distinction between the optimal finite zeros (OFZ) as defined above and the finite zeros of an optimal system. The latter are the limits to which poles of $\mathscr{S}(A, B, K)$ for a fixed K tend as the scalar gain around the system is increased, and may be computed as suggested by Shaked and Kouvaritakis (1977). If we now compute these for a series of optimal controllers K obtained for increasing values of the ratio q/r, then we obtain a set of loci which, for $q/r \to \infty$, terminate at the OFZ.

In order to obtain the OFZ, let us then calculate the zeros of $\mathscr{S}(\hat{A}, \hat{B}, \hat{C})$. We do this by applying the NAM algorithm suggested by Kouvaritakis and MacFarlane (1976 a, b). Thus define N, M, $\hat{N}$ and $\hat{M}$ as the full-rank annihilators of the matrices B, C, $\hat{B}$ and $\hat{C}$, respectively so that

$$NB = 0_{n-l,\,l} \qquad CM = 0_{l,\,n-l} \tag{3.11 a}$$

$$\hat{N}\hat{B} = 0_{2n-l,\,l} \qquad \hat{C}\hat{M} = 0_{l,\,2n-l} \tag{3.11 b}$$

Note that we assume here that the original system $\mathscr{S}(A, B, C)$ is square, that is, that it has the same number of inputs as outputs. Then a set of obvious solutions for $\hat{N}$ and $\hat{M}$ in terms of the annihilators of B and C is given as

$$\hat{N} = \begin{bmatrix} I_n & 0_{n,n} \\ 0_{n-l,n} & M^{\mathrm{t}} \end{bmatrix} \tag{3.12 a}$$

$$\hat{M} = \begin{bmatrix} M & 0_{n,n} \\ 0_{n,n-l} & I_n \end{bmatrix} \tag{3.12 b}$$

so that the zero polynomial matrix for $\mathscr{S}(\hat{A}, \hat{B}, \hat{C})$ becomes

$$\hat{Z}_0(s) = s\begin{bmatrix} M & 0_{n,n-l} \\ 0_{n-l,n} & M^{\mathrm{t}} \end{bmatrix} - \begin{bmatrix} AM & BB^{\mathrm{t}} \\ 0 & -(AM)^{\mathrm{t}} \end{bmatrix} \tag{3.13}$$

which, after the application of the full-rank square transformations

$$\left[\begin{array}{c:c} N & 0_{n-l,n-l} \\ \hdashline B^{\dagger} & 0_{l,n-l} \\ \hdashline 0_{n-l,n} & I_{n-l} \end{array}\right] \quad \text{and} \quad \left[\begin{array}{c:c:c} I_{n-l} & 0_{n-l,n-l} & 0_{n-l,n} \\ \hdashline 0_{n,n-l} & N^{\mathrm{t}} & (B^{\dagger})^{\mathrm{t}} \end{array}\right]$$

from the left and right, respectively becomes

$$\hat{Z}(s)=\begin{bmatrix} sNM-NAM & 0 & 0 \\ sB^{\dagger}M-B^{\dagger}AM & 0 & -I \\ 0 & (SNM+NAM)^{t} & (sB^{\dagger}M+B^{\dagger}AM)^{t} \end{bmatrix} \tag{3.14}$$

where $B^{\dagger}$ denotes a left inverse of B.

The zeros of $\mathscr{S}(\hat{A}, \hat{B}, \hat{C})$ however are simply given by the values of s for which the zero polynomial matrix loses rank, which in turn, because of the special form of $\hat{Z}(s)$ above, can be directly obtained as the values of s for which

$$|sNM-NAM|=0 \tag{3.15 a}$$

$$|(sNM+NAM)^{t}|=0 \tag{3.15 b}$$

The polynomial matrices involved in eqn. (3.15) may be readily identified as the zero polynomial matrices of $\mathscr{S}(A, B, C)$ and $\mathscr{S}(-A^{t}, C^{t}, B^{t})$ which leads us to the following important result.

Theorem 3.1

The optimal finite zeros of a square system $\mathscr{S}(A, B, C)$ together with their mirror images about the imaginary axis are given by the finite zeros of $\mathscr{S}(A, B, C)$ together with their mirror images about the imaginary axis.

It was mentioned earlier that the OFZ may be thought of as the limits of the FZ of $\mathscr{S}(A, B, K)$ for an optimal K computed as $r/q \to 0$. It is a well-known property however that optimal systems are minimum-phase systems (Kwakernaak and Sivan 1972 and Kwakernaak 1976) so that upon reflection of the statement of the theorem above we may readily derive the following two simple results.

Corollary 3.1.1

The optimal finite zeros of a square minimum-phase system $\mathscr{S}(A, B, C)$ are equal to the zeros of $\mathscr{S}(A, B, C)$.

Corollary 3.1.2

The optimal finite zeros of a square system $\mathscr{S}(A, B, C)$ are given by the union of the set of the left-hand plane zeros of $\mathscr{S}(A, B, C)$ together with the set of the mirror images of the right-hand zeros of $\mathscr{S}(A, B, C)$ about the imaginary axis.

These results are by no means claimed to be new or even not well known (Kwakernaak and Sivan 1972 and Shaked and Kouvaritakis 1977) yet it is felt that the approach used in their derivation here is both original and simple.

We bring the discussion on the OFZ to a close by making some remarks concerning their number. Let the MP, P_i of $\mathscr{S}(A, B, C)$ be as defined by eqn. (2.9), let the PMP, G_1, G_2, G_3, ... be as defined by eqns. (2.11), (2.14) and (2.15) and denote the rank deficiency of the PMP, G_i, by d_i for $i = 1, 2, \ldots, \nu - 1$, where G_ν is assumed to be the first full-rank PMP. Then by Corollary (2.1.1) we find that the system $\mathscr{S}(A, B, C)$ has $n - l - \sum_{i=1}^{\nu-1} d_i$ FZ which leads us to the important result that :

Theorem 3.2

A square system $\mathscr{S}(A, B, C)$ with $\nu - 1$ rank-deficient PMP has exactly $n - l - \sum_{i=1}^{\nu-1} d_i$ OFZ, where d_i denotes the degree of rank deficiency of the ith PMP.

Corollaries to this, which will be shown to be useful for the development of the material in this work, can be derived by simple deduction.

Corollary 3.2.1

The augmented system $\mathscr{S}(\hat{A}, \hat{B}, \hat{C})$ of eqn. (3.5) has $2l + 2 \sum_{i=1}^{\nu-1} d_i$ IZ.

Corollary 3.2.2

The optimal poles of a square system $\mathscr{S}(A, B, C)$, as the weight on the control inputs is decreased arbitrarily, tend to $n - l - \sum_{i=1}^{\nu-1} d_i$ finite limits and $l + \sum_{i=1}^{\nu-1} d_i$ limits at infinity.

We define these infinite limits as the optimal infinite zero (OIZ) and proceed to examine them. Note that for a fixed K the optimal system $\mathscr{S}(A, B, K)$ has exactly $n - l$ first-order IZ and no IZ of higher order (Shaked and Kouvaritakis 1977) yet these IZ are not related in any way to the OIZ ; for this reason the distinction between the IZ of an optimal system and the OIZ must be borne in mind.

4. The asymptotes to the optimal root loci

4.1. *The first- and second-order optimal infinite zeros*

Following MacFarlane's suggestion, discussed in § 3.1, we may now initiate the study of the asymptotic behaviour of the ORL by examining the asymptotic behaviour of the RL of the augmented system $\mathscr{S}(\hat{A}, \hat{B}, \hat{C})$. In order to do this however we must first examine the MP of $\mathscr{S}(\hat{A}, \hat{B}, \hat{C})$, $\hat{P}_i$ and from these form the PMP, $\hat{G}_i$. It turns out that what would have been the first PMP is zero, a result which is also true for the third, fifth and in general all the odd-numbered PMP. This observation of course plays a crucial role in the symmetry of the asymptotes of the ORL, as will be seen later.

From eqn. (3.5) we may compute the MP in terms of the elementary matrices of the original system as

$$\hat{P}_1 = \hat{C}\hat{B} = 0 \tag{4.1 a}$$

$$\hat{P}_2 = \hat{C}\hat{A}\hat{B} = -CB(CB)^t = -P_1 P_1^t \tag{4.1 b}$$

$$\hat{P}_3 = \hat{C}\hat{A}^2\hat{B} = -CAB(CB)^t + CB(CAB)^t = -P_2 P_1^t + P_1 P_2^t \tag{4.1 c}$$

$$\hat{P}_4 = \hat{C}\hat{A}^3\hat{B} = CA^2B(CB)^t + CAB(CAB)^t - CB(CA^2B)^t$$
$$= -P_3P_1^t + P_2P_2^t - P_1P_3^t \quad (4.1\,d)$$

$$\hat{P}_5 = \hat{C}\hat{A}^4\hat{B} = -CA^3B(CB)^t + CA^2B(CAB)^t - CAB(CA^2B)^t + CB(CA^3B)^t$$
$$= -P_4P_1^t + P_3P_2^t - P_2P_3^t + P_1P_4^t \quad (4.1\,e)$$

$$\vdots$$

$$\hat{P}_i = \hat{C}\hat{A}^{i-1}\hat{B} = -CA^{i-2}B(CB)^t + CA^{i-3}B(CAB)^t - CA^{i-4}B(CA^2B)^t + \dots + (-1)^{i-1}CB(CA^{i-2}B)^t$$
$$= -P_{i-1}P_1^t + P_{i-2}P_2^t - P_{i-3}P_3^t - \dots + (-1)^{i-1}P_1P_{i-1}^t \quad (4.1\,f)$$

Thus eqn. (4.1 a) verifies our earlier statement that the first PMP of $\mathscr{S}(\hat{A}, \hat{B}, \hat{C})$ is zero :

$$\hat{G}_1 = \hat{P}_1 = 0 \quad (4.2)$$

The implication of this is two-fold, as expressed in the propositions below.

Lemma 4.1.1

There are no first-order OIZ.

Lemma 4.1.2

The null space of the first PMP of $\mathscr{S}(\hat{A}, \hat{B}, \hat{C})$ is the whole l-dimensional space.

We proceed now to examine the second-order infinite zero behaviour by forming the second PMP which by Lemma 4.1.2 may be obtained directly as

$$\hat{G}_2 = \hat{P}_2 = -CB(CB)^t$$
$$= -G_1G_1^t \quad (4.3)$$

Thus we have that

Lemma 4.1.3

The second PMP of $\mathscr{S}(\hat{A}, \hat{B}, \hat{C})$ is a quadratic of the first PMP of $\mathscr{S}(A, B, C)$ with the sign inverted. Clearly then, by Theorem 2.1, we have that :

Lemma 4.1.4

For a non-trivial product CB, the system $\mathscr{S}(\hat{A}, \hat{B}, \hat{C})$ has second-order IZ ; the corresponding asymptotes make angles $\hat{\alpha}_{j,l}^{(2)}$ dictated by the non-zero eigenvalues of $-G_1G_1^t$, $\hat{\lambda}_j^{(2)}$, as shown below :

$$\hat{\alpha}_{j,l}^{(2)} = \tfrac{1}{2}[\angle \hat{\lambda}_j^{(2)} + (2l+1)\pi] \quad \text{for } l = 0, 1 \quad (4.4)$$

Although nothing can be said, in general, about the relationship between the eigenvalues of $\hat{G}_2$ in terms of the eigenvalues of the first PMP of $\mathscr{S}(A, B, C)$, G_1 (unless G_1 commutes with G_1^t) there are several properties of $\hat{G}_2$ relevant to the asymptotic behaviour that hold true for all CB ; we summarize two of these in the following lemma.

Lemma 4.1.5

The second PMP of $\mathscr{S}(\hat{A}, \hat{B}, \hat{C})$ is a negative semi-definite matrix and loses rank by d_1 degrees, where d_1 denotes the rank-deficiency of the first PMP of $\mathscr{S}(A, B, C)$.

The proof of this lemma is easy to derive if one considers the spectral decomposition of G_1 given in eqn. (2.12). Then we have that

$$V_1\hat{G}_2V_1^{\mathrm{t}} = -\Lambda_1 V_1 V_1^{\mathrm{t}}\Lambda_1 = \text{full rank} \tag{4.5}$$

$$N_1\hat{G}_2 = 0_{d_1} \tag{4.6}$$

Thus $\hat{G}_2$ has at least rank $l-d_1$ and is rank-deficient by at least d_1 degrees, which implies that $\hat{G}_2$ is rank-deficient by exactly d_1 degrees. Further, $\hat{G}_2$ is symmetric, and since $G_1G_1^{\mathrm{t}}$ does not have negative eigenvalues $\hat{G}_2$ is negative semi-definite.

We may now combine Lemmas 4.1, 1–5 in order to formulate the following important theorem.

Theorem 4.1

The lowest order of OIZ for a system $\mathscr{S}(A, B, C)$ is 2, since there exist no first-order OIZ. There exist exactly $l-d_1$ second-order OIZ and all the associated second-order asymptotes lie on the left-half of the real axis.

This follows immediately from the lemmas. Thus all the poles of $\mathscr{S}(\hat{A}, \hat{B}, \hat{C})$ that vanish off to infinity according to a second-order rule obey equations of the kind

$$s^2 = -k\hat{\lambda}_j^{(2)}, \quad j = 1, 2, \ldots, l-d_1 \tag{4.7}$$

The $\hat{\lambda}_j^{(2)}$'s however are the non-zero eigenvalues of $\hat{G}_2$ and hence they are real and negative so that eqn. (4.7) yields $2(l-d_1)$ solutions for s :

$$s = \pm\sqrt{(-\hat{\lambda}_j^{(2)}k)}, \quad j = 1, 2, \ldots, l-d_1 \tag{4.8}$$

Clearly the negative roots belong to the ORL and the positive roots are simply the mirror images of the ORL about the imaginary axis. Thus there are exactly $l-d_1$ asymptotes all parallel to the negative real axis. Note that simple calculation of the point of radiation of these asymptotes, that is the pivots to the second-order IZ of $\mathscr{S}(\hat{A}, \hat{B}, \hat{C})$, will readily yield that all the second-order OIZ will lie on the real axis. Due to the symmetry of asymptotes of all orders about the origin the location of the pivots to the asymptotes associated with any order IZ will be the origin. It should be mentioned here that inherent in the statement of Theorem 4.1 is the assumption that $\hat{G}_2$ has simple null-structure. Such an assumption is fully justified if one considers the fact that G_2 is symmetric ; a fact that precludes the possibility of non-simple null-structures. This observation holds true for all PMP defined later on in this paper and for this reason the subject of simple or non-simple null-structures will not be touched upon again.

We may summarize the situation now as follows. The ORL have no *first-order* IZ but have exactly as many *second-order* IZ as the original system has *first-order* IZ. Indeed, if one excepts the order of divergence these second-order OIZ in a sense possess first-order IZ properties in as far as they all lie in the left-half of the real axis and appear in Butterworth patterns of ones, that is to say that to each non-zero eigenvalue there corresponds a symmetrical pattern of two asymptotes of which only one is stable.

The qualitative nature of the results developed here extends to all orders of OIZ. Thus it will be seen that only even orders of OIZ exist, but they all come in Butterworth patterns and the angular displacement between asymptotes belonging to the same pattern is a fraction of π as dictated by the order of the OIZ.

4.2. *The third-order optimal infinite zeros*

We proceed next to determine the third-order OIZ and to do so we must first compute the third PMP, $\hat{G}_3$. For this we need to identify the spectral decomposition of the previous PMP expressed by

$$\hat{G}_2=[\hat{U}_2 \vdots \hat{M}_2]\begin{bmatrix} \hat{\Lambda}_2 & 0 \\ 0 & 0_{d_1} \end{bmatrix}\begin{bmatrix} \hat{V}_2 \\ \hat{N}_2 \end{bmatrix} \tag{4.9}$$

where $\hat{\Lambda}_2$ denotes the Jordan canonical block corresponding to the non-zero eigenvalues of $\hat{G}_2$. First we note that due to the symmetry of $\hat{G}_2$ we may always choose $\hat{N}_2$ and $\hat{M}_2$ such that

$$[\hat{U}_2 \vdots \hat{M}_2]=[\hat{V}_2{}^{\mathrm{t}} \vdots \hat{N}_2{}^{\mathrm{t}}] \tag{4.10}$$

Next, by direct comparison of eqns. (4.6) and (4.9) we have

$$\hat{N}_2=R_1N_1 \tag{4.11}$$

where R_1 is a $d_1 \times d_1$ full-rank matrix, which by eqn. (4.10) yields

$$R_1N_1N_1{}^{\mathrm{t}}\, R_1{}^{\mathrm{t}}=I \tag{4.12 a}$$

or

$$R_1{}^{\mathrm{t}}\, R_1=(N_1N_1{}^{\mathrm{t}})^{-1} \tag{4.12 b}$$

A reversal of the arguments above readily shows that any matrix R_1 chosen to satisfy eqn. (4.12 *b*) leads to an annihilator which is a suitable candidate for the decomposition of $\hat{G}_2$.

Projecting the third MP of $\mathscr{S}(\hat{A}, \hat{B}, \hat{C})$, $\hat{P}_3$ onto the right and left null-spaces of $\hat{G}_2$ we derive the third PMP as

$$\begin{aligned}\hat{G}_3&=\hat{N}_2\hat{P}_3\hat{M}_2\\ &=R_1N_1(-P_2G_1{}^{\mathrm{t}}+G_1P_2{}^{\mathrm{t}})N_1{}^{\mathrm{t}}\, R_1{}^{\mathrm{t}}\\ &=0\end{aligned} \tag{4.13}$$

Hence, as mentioned earlier, the third PMP of $\mathscr{S}(\hat{A}, \hat{B}, \hat{C})$ is zero, so that the ORL will not have any third-order IZ. Normally one would proceed now to the investigation of higher-order OIZ by projecting the fourth MP of $\mathscr{S}(\hat{A}, \hat{B}, \hat{C})$ onto the null-spaces of $\hat{G}_2$ and studying the eigen-structure of the result. The vanishing of the third PMP however violates the condition of Theorem 2.1 concerning the rank of successive PMP; namely, that $d_{i+1}>d_i$ for all i. Thus some extension of the RL approach to the case of a vanishing

PMP is needed. We derive one such extension by considering a system $\mathscr{S}(A, B, C)$ for which $G_i = 0$. Clearly the presence of a zero PMP does not affect the IZ of order $i-1$ or less; for this reason we concentrate on obtaining a solution conforming to the behaviour $s^j = \alpha^{(j)}k$, where $j \geqslant i$. We do this by considering the expansion

$$|I+kG(s)| = \left| I + \frac{k}{s}CB + \frac{k}{s^2}CAB + \dots \right| = 0 \qquad (4.14)$$

after suitable projection according to the spectral decompositions of all the PMP up to the $(i-2)$th. The behaviour of the IZ of higher order then may be obtained from

$$\left| I + \frac{k}{s^{i-1}} G_{i-1} + \frac{k}{s^i} T_{i-1}{}^{(i)} + \frac{k}{s^{i+1}} T_{i-1}{}^{(i-1)} + \dots \right| = 0 \qquad (4.15)$$

where $T_{i-1}{}^{(p)}$ denotes the projection of the pth MP onto the null-space of all the PMP up to the $(i-2)$th. Then diagonalizing G_{i-1} we get

$$\left| I + \frac{k}{s^{i-1}} \left[\begin{array}{c|c} \Lambda_{i-1} & 0 \\ \hline 0 & 0_{d_{i-1}} \end{array}\right] + \frac{k}{s^i} \left[\begin{array}{c|c} V_{i-1}T_{i-1}{}^{(i)}U_{i-1} & V_{i-1}T_{i-1}{}^{(i)}M_{i-1} \\ \hline N_{i-1}T_{i-1}{}^{(i)}U_{i-1} & G_i \end{array}\right] + \frac{k}{s^{i+1}} \left[\begin{array}{c|c} V_{i-1}T_{i-1}{}^{(i+1)}U_{i-1} & V_{i-1}T_{i-1}{}^{(i+1)}M_{i-1} \\ \hline N_{i-1}T_{i-1}{}^{(i+1)}U_{i-1} & N_{i-1}T_{i-1}{}^{(i+1)}M_{i-1} \end{array}\right] + \dots \right| = 0 \qquad (4.16)$$

Assuming G_i now to be zero we may reduce the above to

$$\left| \begin{array}{c|c} I + \dfrac{k}{s^{i-1}}\Lambda_{i-1} & V_{i-1}\left[\dfrac{k}{s^i}T_{i-1}{}^{(i)} + \dfrac{k}{s^{i+1}}T_{i-1}{}^{(i+1)} + \dots\right]M_{i-1} \\ \hline N_{i-1}\left[\dfrac{k}{s^i}T_{i-1}{}^{(i)} \quad \dfrac{k}{s^{i+1}}T_{i-1}{}^{(i+1)} + \dots\right]U_{i-1} & I + N_{i-1}\left[\dfrac{k}{s^{i+1}}T_{i-1}{}^{(i+1)} + \dfrac{k}{s^{i+2}}T_{i-1}{}^{(i+2)} + \dots\right]M_{i-1} \end{array} \right| = 0 \qquad (4.17)$$

which upon the application of the Schur formula (Gantmacher 1959) yields the RL we wish to study as the solutions of

$$\left| I + \frac{k}{s^{i+1}} [N_{i-1} T_{i-1}{}^{(i+1)} M_{i-1} - N_{i-1} T_{i-1}{}^{(i)} U_{i-1} \Lambda_{i-1}{}^{-1} V_{i-1} T_{i-1}{}^{(i)} M_{i-1} + \frac{k}{s^{i+2}} [N_{i-1} T_{i-1}{}^{(i+2)} M_{i-1} - N_{i-1} T_{i-1}{}^{(i)} U_{i-1} \Lambda_{i-1}{}^{-1} V_{i-1} T_{i-1}{}^{(i+1)} M_{i-1} - N_{i-1} T_{i-1}{}^{(i+1)} U_{i-1} \Lambda_{i-1}{}^{-1} V_{i-1} T_{i-1}{}^{(i)} M_{i-1}] + \ldots \right| = 0 \tag{4.18}$$

We see then that in the case of the parameter $G_i = N_{i-1} T_{i-1}{}^{(i)} M_{i-1}$ becoming zero it does not suffice to look at $N_{i-1} T_{i-1}{}^{(i+1)} M_{i-1}$ in order to determine the $(i+1)$th-order zeros, but rather one must correct this by subtracting the effect of off-diagonal blocks. For this reason we define as the $(i+1)$th PMP the matrix coefficient of k/s^{i+1} above while the remaining coefficients in eqn. (4.18) will be defined as the new T parameters. Thus we have

$$G_{i+1} \triangleq N_{i-1} T_{i-1}{}^{(i+1)} M_{i-1} - N_{i-1} T_{i-1}{}^{(i)} U_{i-1} \Lambda_{i-1}{}^{-1} V_{i-1} T_{i-1}{}^{(i)} M_{i-1} \tag{4.19}$$

$$\bar{T}_{i+1}{}^{(i+2)} = N_{i-1} T_{i-1}{}^{(i+2)} M_{i-1} - N_{i-1} T_{i-1}{}^{(i)} U_{i-1} \Lambda_{i-1}{}^{-1} V_{i-1} T_{i-1}{}^{(i+1)} M_{i-1} - N_{i-1} T_{i-1}{}^{(i+1)} U_{i-1} \Lambda_{i-1}{}^{-1} V_{i-1} T_{i-1}{}^{(i)} M_{i-1} \tag{4.20}$$

etc. Then eqn. (4.18) reduces to

$$\left| I + \frac{k}{s^{i+1}} G_{i+1} + \frac{k}{s^{i+2}} \bar{T}_{i+1}{}^{(i+2)} + \frac{k}{s^{i+3}} \bar{T}_{i+1}{}^{(i+3)} + \ldots \right| = 0 \tag{4.21}$$

which of course may be seen to have the same format as (4.15); hence the RL analysis may proceed as before.

The formulae derived are rather cumbersome, due to their complexity; however, for the case of ORL the system under consideration is the augmented system $\mathscr{S}(\hat{A}, \hat{B}, \hat{C})$, the MP of which have a special symmetry; this symmetry enables the reduction of all the above equations to a very simple form. We proceed to illustrate this by investigating the fourth-order asymptotic behaviour of $\mathscr{S}(\hat{A}, \hat{B}, \hat{C})$.

4.3. *The fourth-order optimal infinite zeros*

The extension of the RL analysis to the case of vanishing PMP now enables us to discuss the behaviour of the fourth-order ORL. In order to do this we compute first the parameters of eqn. (4.15) setting $i=3$, for which we obtain

$$\hat{T}_2{}^{(3)} = \hat{P}_3, \quad \hat{T}_2{}^{(4)} = \hat{P}_4, \ldots \tag{4.22}$$

and $\hat{G}_2$ of course is as defined in eqn. (4.3). Substituting the above in eqn. (4.15) we derive that the fourth and higher-order behaviour is described by

$$\left| I+\frac{k}{s^2}\hat{G}_2+\frac{k}{s^3}\hat{P}_3+\frac{k}{s^4}\hat{P}_4+\dots \right| =0 \tag{4.23}$$

Since $\hat{G}_3$ is zero we may now apply eqn. (4.19) in order to compute the fourth PMP :

$$\begin{aligned}\hat{G}_4&=\hat{N}_2\hat{P}_4\hat{M}_2-\hat{N}_2\hat{P}_3\hat{U}_2\hat{\Lambda}_2^{-1}\hat{V}_2\hat{P}_3\hat{M}_2\\ &=\hat{N}_2CAB(CAB)^{\mathrm{t}}\hat{N}_2^{\mathrm{t}}+\hat{N}_2CAB(CB)^{\mathrm{t}}\hat{U}_2\hat{\Lambda}_2^{-1}\hat{V}_2CB(CAB)^{\mathrm{t}}\hat{N}_2^{\mathrm{t}}\end{aligned} \tag{4.24}$$

where use is made of the MP defining eqns. (4.1 *c*) and (4.1 *d*) and of eqn. (4.11) which relates the annihilator $\hat{N}_2$ to the annihilator N_1. Thus we may write $\hat{G}_4$ as

$$\hat{G}_4=\hat{N}_2CABX_4(CAB)^{\mathrm{t}}\hat{N}_2^{\mathrm{t}} \tag{4.25}$$

where

$$X_4=I+(CB)^{\mathrm{t}}\hat{U}_2\hat{\Lambda}_2^{-1}\hat{V}_2CB \tag{4.26}$$

Upon pre- and post-multiplication by CB and $(CB)^{\mathrm{t}}$ respectively first and M_1 and M_1^{t} respectively next, we conclude that X_4 loses rank by at least $m-d_1$ degrees and has rank d_1, at least, so that X_4 has rank d_1 exactly. In particular we have that

$$\begin{aligned}CBX_4(CB)^{\mathrm{t}}&=CB(CB)^{\mathrm{t}}+CB(CB)^{\mathrm{t}}\hat{U}_2\hat{\Lambda}_2^{-1}\hat{V}_2CB(CB)^{\mathrm{t}}\\ &=-\hat{G}_2+\hat{G}_2\\ &=0\end{aligned} \tag{4.27}$$

and

$$M_1^{\mathrm{t}}X_4M_1=M_1^{\mathrm{t}}M_1 \tag{4.28}$$

Now eqn. (4.27) implies that X_4 has the form

$$X_4=M_1L_1L_1^{\mathrm{t}}M_1^{\mathrm{t}} \tag{4.29}$$

where L_1 is some $d_1\times d_1$ full-rank matrix of constants, which after substitution into eqn. (4.28) yields

$$(L_1L_1^{\mathrm{t}})=(M_1^{\mathrm{t}}M_1)^{-1} \tag{4.30}$$

Substituting eqns. (4.29) and (4.30) into eqn. (4.25) we may now write

$$\hat{G}_4=\hat{N}_2CABM_1(M_1^{\mathrm{t}}M_1)^{-1}M_1^{\mathrm{t}}(CAB)^{\mathrm{t}}\hat{N}_2^{\mathrm{t}}$$

Making use of eqns. (4.11) and (4.12) we may express R_1 as

$$R_1=R_{N_1}=(N_1N_1^{\mathrm{t}})^{-1/2} \tag{4.31}$$

and

$$(M_1^{\mathrm{t}}M_1)^{-1}=R_{M_1}R_{M_1}^{\mathrm{t}}$$

where

$$R_{M_1}=(M_1^{\mathrm{t}}M_1)^{-1/2}$$

we obtain the following important result :

$$G_4 = R_{N_1}N_1CABM_1R_{M_1}R_{M_1}{}^{\mathrm{t}}\, M_1{}^{\mathrm{t}}(CAB)^{\mathrm{t}}N_1{}^{\mathrm{t}}\, R_{N_1}{}^{\mathrm{t}}$$
$$= (R_{N_1}G_2R_{M_1})(R_{N_1}G_2R_{M_1})^{\mathrm{t}} \tag{4.32}$$

Thus we have

Lemma 4.2.1

The fourth PMP of $\mathscr{S}(\hat{A}, \hat{B}, \hat{C})$ is a quadratic of the second PMP of $\mathscr{S}(\hat{A}, \hat{B}, \hat{C})$ and

Lemma 4.2.2

The fourth PMP of $\mathscr{S}(\hat{A}, \hat{B}, \hat{C})$ is a positive semi-definite matrix and loses rank by d_2 degrees, where d_2 denotes the rank-deficiency of the second PMP of $\mathscr{S}(A, B, C)$.

Furthermore, by eqn. (4.21) for $i=3$ we have that

Lemma 4.2.3

For a non-trivial PMP, G_1, of the system $\mathscr{S}(A, B, C)$ the system $\mathscr{S}(\hat{A}, \hat{B}, \hat{C})$ has fourth-order IZ ; the corresponding asymptotes make with the positive real axis angles $\hat{\alpha}_{j,l}{}^{(4)}$ dictated by the non-zero eigenvalues of $\hat{G}_4$, $\hat{\lambda}_j{}^{(4)}$, as shown below :

$$\hat{\alpha}_{j,l}{}^{(4)} = \tfrac{1}{4}[\angle \hat{\lambda}_j{}^{(4)} + (2l+1)\pi] \quad \text{for } l=0, 1, 2, 3 \tag{4.33}$$

Although once again in general we may not relate the eigenvalues of $\hat{G}_4$, $\hat{\lambda}_j{}^{(4)}$ to the eigenvalues of G_2, following Lemmas 4.2.1–3 we may state the important result given below.

Theorem 4.2

The next higher order of OIZ to 2, for a system $\mathscr{S}(A, B, C)$, is 4 since there exist no third-order OIZ. There exist exactly, $2(d_1-d_2)$ fourth-order OIZ and all these OIZ are associated with Butterworth patterns of two asymptotes each, one asymptote making an angle of $+\pi/4$ with the negative real axis and the other making an angle of $-\pi/4$.

This follows immediately from the lemmas. Thus all the poles of $\mathscr{S}(\hat{A}, \hat{B}, \hat{C})$ that vanish off to infinity according to a fourth-order rule, obey equations of the kind

$$s^4 = -\hat{\lambda}_j{}^{(4)}k, \quad j=1, 2, \ldots, d_1-d_2 \tag{4.34}$$

The $\hat{\lambda}_j{}^{(4)}$'s however are the non-zero eigenvalues of $\hat{G}_4$ and hence they are real and positive so that the above equation yields $4(d_1-d_2)$ solutions for s :

$$s = {}^4\sqrt{(-\lambda_j{}^{(4)}k)}, \quad j=1, 2, \ldots, d_1-d_2 \tag{4.35}$$

Clearly the quantity under the root sign is negative, so that all the solutions will vanish along asymptotes whose directions are symmetrically displaced in the same manner as the four fourth roots of $+1$. Thus each Butterworth pattern contains four asymptotes, only two of which however lie in the left-half of the complex plane, the remaining two being mere mirror images of the above two about the imaginary axis.

We have so far examined all orders of behaviour up to the fourth order. The pattern of OIZ emerges quite clearly now :

(i) There are no first or third-order OIZ.

(ii) There are second-order OIZ given by the eigenvalues of a quadratic of the first PMP of $\mathscr{S}(A, B, C)$ and are exactly the same in number as the number of first-order IZ of $\mathscr{S}(A, B, C)$.

(iii) The second-order OIZ are on the negative real axis.

(iv) There are fourth-order OIZ given by the eigenvalues of a quadratic of the second PMP of $\mathscr{S}(A, B, C)$ and are exactly the same in number as the second-order IZ of $\mathscr{S}(A, B, C)$.

(v) The fourth-order OIZ make angles of $\pm\pi/4$ with the negative real axis.

Naturally, as expected this pattern of behaviour extends to the fifth, sixth etc., OIZ. The proof for the general ith OIZ is difficult to derive in view of the relative complexity of the parameters involved. However, these parameters preserve the same symmetry for all orders, so that the nature of the results derived thus far may be conjectured to extend to higher orders of OIZ. In this paper we just give the proof for $\hat{G}_5=0$ and conclude the work by drawing up some general remarks.

4.4. *The fifth and higher-order optimal infinite zeros*

We may now apply the analysis developed in § 4.2 in order to prove that the fifth PMP of the augmented system $\mathscr{S}(\hat{A}, \hat{B}, \hat{C})$ is zero. Thus from eqns. (4.19–21), after suitable projection of the parameter $\bar{T}_{i+1}{}^{(i+2)}$ for $i=3$ onto the null-space of the fourth PMP, $\hat{G}_4$, we have

$$\hat{G}_5=\hat{N}_4\bar{T}_4{}^{(5)}\hat{N}_4{}^{\mathrm{t}} \tag{4.36}$$

where $\hat{N}_4$ and $\hat{N}_4{}^{\mathrm{t}}$ are defined by the spectral decomposition of $\hat{G}_4$

$$\hat{G}_4=[\hat{U}_4 \vdots \hat{N}_4{}^{\mathrm{t}}]\begin{bmatrix} \hat{\Lambda}_4 & 0 \\ 0 & 0 \end{bmatrix}\begin{bmatrix} \hat{V}_4 \\ \hat{N}_4 \end{bmatrix} \tag{4.37}$$

and $\bar{T}_4{}^{(5)}$ is given by eqn. (4.20) for $i=3$ as

$$\begin{aligned}\bar{T}_4{}^{(5)}=\hat{N}_2\hat{T}_2{}^{(5)}\hat{N}_2{}^{\mathrm{t}}-\hat{N}_2\hat{T}_2{}^{(3)}\hat{U}_2\hat{\Lambda}_2{}^{-1}\hat{V}_2\hat{T}_2{}^{(4)}\hat{N}_2{}^{\mathrm{t}}\\ -\hat{N}_2\hat{T}_2{}^{(4)}\hat{U}_2\hat{\Lambda}_2{}^{-1}\hat{V}_2\hat{T}_2{}^{(3)}\hat{N}_2{}^{\mathrm{t}}\end{aligned} \tag{4.38}$$

We may now substitute for $\hat{T}_2{}^{(3)}$, $\hat{T}_2{}^{(4)}$ and $\hat{T}_2{}^{(5)}$ according to eqns. (4.2) and (4.1 *c–e*) in order to obtain

$$\hat{N}_2\hat{T}_2{}^{(5)}\hat{N}_2{}^{\mathrm{t}}=\hat{N}_2CA^2B(CAB)^{\mathrm{t}}\hat{N}_2-\hat{N}_2CAB(CA^2B)^{\mathrm{t}}\hat{N}_2{}^{\mathrm{t}} \tag{4.39 a}$$

$$\hat{N}_2\hat{T}_2{}^{(3)}=-\hat{N}_2CAB(CB)^{\mathrm{t}} \tag{4.39 b}$$

$$\hat{T}_2{}^{(3)}\hat{N}_2{}^{\mathrm{t}}=CB(CAB)^{\mathrm{t}}\hat{N}_2{}^{\mathrm{t}} \tag{4.39 c}$$

$$\hat{N}_2\hat{T}_2{}^{(4)}=-\hat{N}_2CA^2B(CB)^{\mathrm{t}}+\hat{N}_2CAB(CAB)^{\mathrm{t}} \tag{4.39 d}$$

$$\hat{T}_2{}^{(4)}\hat{N}_2{}^{\mathrm{t}}=CAB(CAB)^{\mathrm{t}}\hat{N}_2{}^{\mathrm{t}}-CB(CA^2B)^{\mathrm{t}}\hat{N}_2{}^{\mathrm{t}} \tag{4.39 e}$$

Upon substitution of the above and eqn. (4.38) into eqn. (4.36) it is easy to verify that

$$\hat{G}_5 = 0 \tag{4.40}$$

In order to do this we make use of the fact that $\hat{N}_4$, the annihilator of $\hat{G}_4$, also annihilates $R_{N_1} G_2 R_{M_1}$, so that

$$\hat{N}_4 R_{N_1} G_2 R_{M_1} = \hat{N}_4 R_1 N_1 CABM_1 R_{M_1} = 0 \tag{4.41}$$

which when compared with eqn. (2.14) readily yields that

$$\hat{N}_4 R_{N_1} = R_2 N_2 \tag{4.42}$$

Thus, for instance, in order to show that all the terms of $\hat{G}_5$ containing the second power of A, A^2 vanish we write

$$\begin{aligned}
&\hat{N}_4 \hat{N}_2 [CA^2B(CAB)^t + CA^2B(CB)^t \hat{U}_2 \hat{\Lambda}_2^{-1} \hat{V}_2 CB(CAB)^t \\
&\qquad - CAB(CA^2B)^t - CAB(CB)^t \\
&\qquad\qquad \times \hat{U}_2 \hat{\Lambda}_2^{-1} \hat{V}_2 CB(CA^2B)^t] \hat{N}_2^t \hat{N}_4 \\
&= R_2 N_2 N_1 [CA^2B[I + (CB)^t \hat{U}_2 \hat{\Lambda}_2^{-1} \hat{V}_2 CB](CAB)^t] \\
&\quad \times N_1^t N_2^t P_2^t - R_2 N_2 N_1 [CAB[I + (CB)^t \hat{U}_2 \hat{\Lambda}_2^{-1} \hat{V}_2 CB] \\
&\qquad\qquad \times (CA^2B)^t] N_1^t N_2^t R_2^t \\
&= 0
\end{aligned} \tag{4.43}$$

since by eqns. (4.26) and (4.29) we have that

$$I + (CB)^t \hat{U}_2 \hat{\Lambda}_2^{-1} \hat{V}_2 CB = M_1 L_1 L_1^t M_1^t \tag{4.44}$$

and since

$$R_2 N_2 N_1 CABM_1 = 0$$

or

$$M_1^t (CAB)^t N_1^t N_2^t R_2 = 0 \tag{4.45}$$

working along similar lines to the above we may complete the proof of condition (4.40).

Thus, as expected, we have that $\mathscr{S}(\hat{A}, \hat{B}, \hat{C})$ has no IZ of order 5 so that

Theorem 4.3

There are no fifth-order OIZ.

This result is in agreement with the pattern of asymptotic behaviour established earlier ; namely, that there are no odd orders of OIZ. The proof for a general odd power would be extremely cumbersome, due to the complexity of the expressions of the various parameters involved. However, due to the symmetry involved in these expressions we may make the following conjecture.

Conjecture 4.1

The ORL exhibit no odd orders of asymptotic behaviour.

Rather than attempt to establish an expression for the next PMP of $\mathscr{S}(\hat{A}, \hat{B}, \hat{C})$, an exercise which would prove to be extremely tedious, we emphasize once again the symmetrical nature of the relevant equations so that we may stipulate the following conjecture.

Conjecture 4.2

The ORL exhibit even-order asymptotic behaviour only, wnere there are exactly as many $2q$th-order OIZ as there are qth-order IZ for the non-optimal system $\mathscr{S}(A, B, C)$; the asymptotic behaviour of the ORL tending to these OIZ is described by the equation

$$s^{2q} = -\hat{\lambda}_j^{(2q)} k, \quad j = 1, 2, \ldots, d_{q-1} - d_q \tag{4.46}$$

where $\hat{\lambda}_j^{(2q)}$ denotes the non-zero eigenvalues of the $2q$th PMP of $\mathscr{S}(\hat{A}, \hat{B}, \hat{C})$, $\hat{G}_{2q}$.

Concerning the spatial arrangement of all the asymptotes of order $2q$ we may state the following conjecture.

Conjecture 4.3

The $2q$th PMP of $\mathscr{S}(\hat{A}, \hat{B}, \hat{C})$ is a quadratic function of the qth PMP of $\mathscr{S}(A, B, C)$ with the sign inverted for all odd values of q. For this reason and by Conjecture 4.2 all asymptotes of order $2q$ appear in Butterworth patterns of q asymptotes, each of whose spatial arrangement is given by the q left-hand plane solutions of

$$x^{2q} = -1, \quad q = \text{odd} \tag{4.47 a}$$

$$x^{2q} = +1, \quad q = \text{even} \tag{4.47 b}$$

A diagrammatic representation of the asymptotic behaviour described by the above conjuncture is given below :

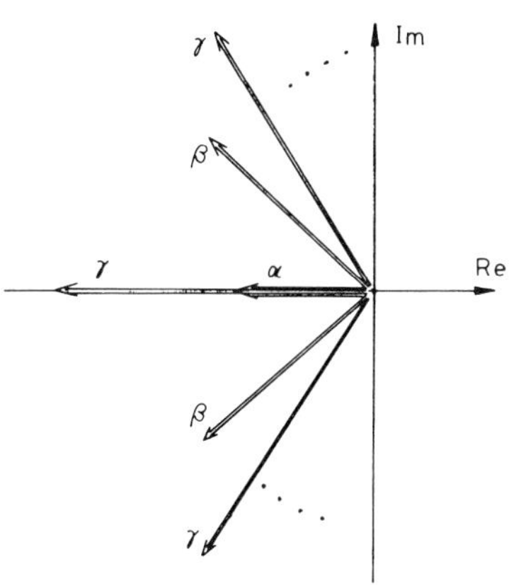

Figure 2.

The arrows in Fig. 2 indicate the directions of the asymptotes of the ORL and the letters, α, β and γ denote the first, second and third-order of behaviour, respectively.

We conclude this paper by discussing a simple numerical example.

5. Numerical example

Consider the system described by the state-space matrices

$$A=\begin{bmatrix} -1 & 0 & 0 & 0 & 0 & 0 & 0 \\ 0 & -2 & 0 & 0 & 0 & 0 & 0 \\ 0 & 0 & -2 & 0 & 0 & 0 & 0 \\ 0 & 0 & 1 & -3 & 0 & 0 & 0 \\ 0 & 0 & 0 & 0 & -4 & 0 & 0 \\ 0 & 0 & 0 & 0 & 0 & -5 & 0 \\ 0 & 0 & 0 & 0 & -3 & 0 & -6 \end{bmatrix} \tag{5.1 a}$$

$$B=\begin{bmatrix} 1 & 0 & 0 \\ 1 & 0 & 0 \\ 0 & 1 & 0 \\ 0 & -1 & 0 \\ 0 & 0 & 1 \\ 0 & 0 & 1 \\ 0 & 0 & -2 \end{bmatrix} \tag{5.1 b}$$

$$C=\begin{bmatrix} 1 & 1 & 0 & 0 & 0 & 0 & 0 \\ 0 & 0 & 1 & 1 & 0 & 0 & 0 \\ 0 & 0 & 0 & 0 & 1 & 1 & 1 \end{bmatrix} \tag{5.1 c}$$

We first compute the finite zeros of $\mathscr{S}(A, B, C)$ using the NAM algorithm. Note that CB is rank deficient by two degrees, so that the maximum number of FZ expected is $n-l-2=7-3-2=2$.

Employing standard procedures (Kouvaritakis and MacFarlane 1976) we compute the annihilators N and M to be

$$N=\begin{bmatrix} 1 & -1 & 0 & 0 & 0 & 0 & 0 \\ 0 & 0 & 0 & 0 & 1 & -1 & 0 \\ 0 & 0 & 1 & 1 & 0 & 0 & 0 \\ 0 & 0 & 0 & 0 & 1 & 1 & 1 \end{bmatrix} \tag{5.2 a}$$

$$M=\begin{bmatrix} 1 & 0 & 0 & 0 \\ -1 & 0 & 0 & 0 \\ 0 & 0 & 1 & 0 \\ 0 & 0 & -1 & 0 \\ 0 & 1 & 0 & 1 \\ 0 & -1 & 0 & 1 \\ 0 & 0 & 0 & -2 \end{bmatrix} \tag{5.2 b}$$

from which the zero-polynomial matrix $Z(s)$ turns out to be

$$Z(s)=sNM-NAM$$

$$=s\begin{bmatrix} 2 & 0 & 0 & 0 \\ 0 & 2 & 0 & 0 \\ 0 & 0 & 0 & 0 \\ 0 & 0 & 0 & 0 \end{bmatrix}-\begin{bmatrix} -3 & 0 & 0 & 0 \\ 0 & -9 & 0 & 1 \\ 0 & 0 & 2 & 0 \\ 0 & -2 & 0 & 0 \end{bmatrix} \tag{5.3}$$

It is easy now to show either by direct calculation or by the procedure outlined by Kouvaritakis and MacFarlane that $Z(s)$ above loses rank for $s=-\frac{3}{2}$ only. Thus the system $\mathscr{S}(A, B, C)$ has one finite zero only at

$$z=-\tfrac{3}{2} \tag{5.4}$$

and as such it will therefore have six IZ associated with six asymptotes. We may now study the asymptotic behaviour of the RL of $\mathscr{S}(A, B, C)$ by computing the PMP. Upon forming the first parameter we obtain

$$G_1=CB=\begin{bmatrix} 2 & 0 & 0 \\ 0 & 0 & 0 \\ 0 & 0 & 0 \end{bmatrix} \tag{5.5}$$

from which, clearly, we see that the rank defect of G_1 is $d_1=2$ and that annihilators N_1 and M_1 become

$$N_1=\begin{bmatrix} 0 & 1 & 0 \\ 0 & 0 & 1 \end{bmatrix} \tag{5.6 a}$$

$$M_1=\begin{bmatrix} 0 & 0 \\ 1 & 0 \\ 0 & 1 \end{bmatrix} \tag{5.6 b}$$

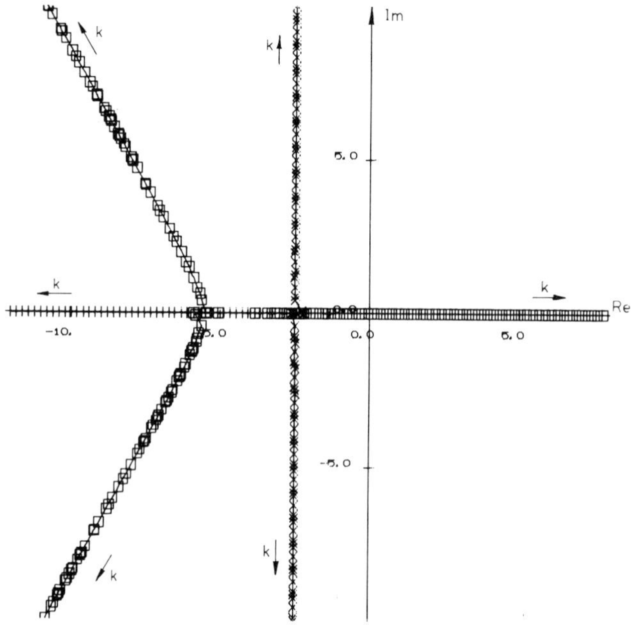

Figure 3 (a). The complete root locus for $\mathscr{S}(A, B, C)$.

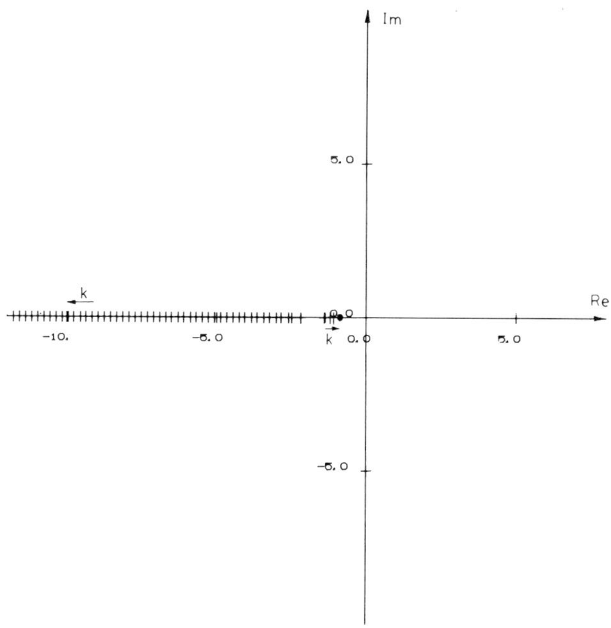

Figure 3 (b). The first Riemann sheet for $\mathscr{S}(A, B, C)$ showing the finite zero (denoted by ●) and the first-order asymptotic behaviour.

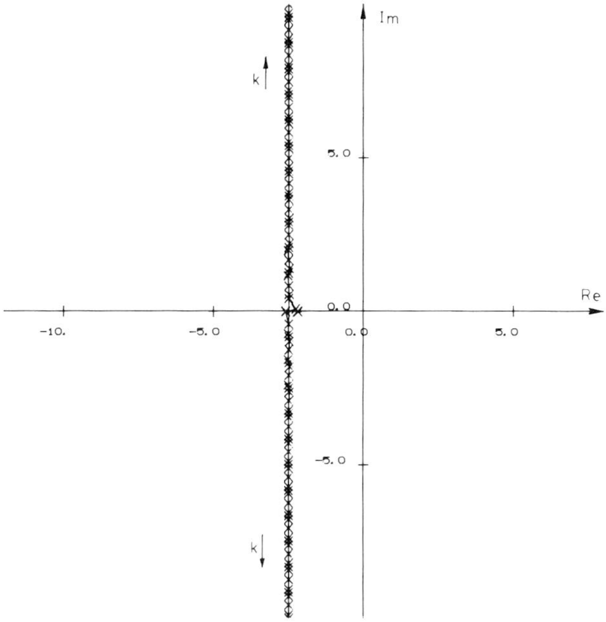

Figure 3 (*c*). The second Riemann sheet for $\mathscr{S}(A, B, C)$ showing the second-order asymptotic behaviour.

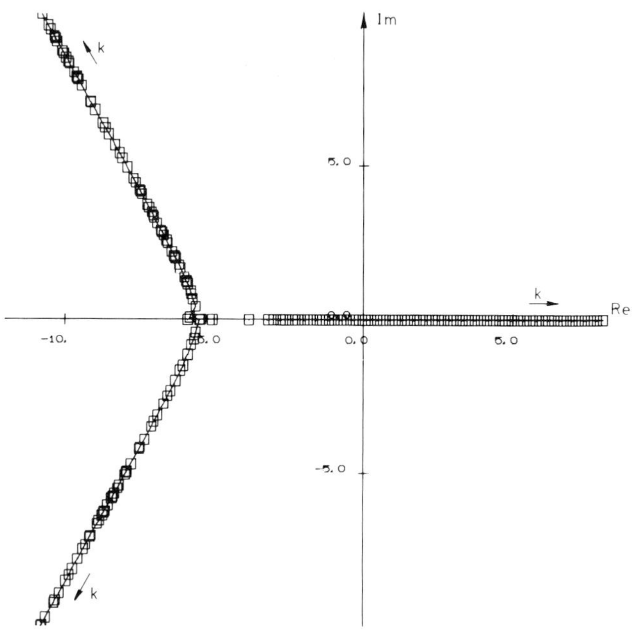

Figure 3 (*d*). The third Riemann sheet for $\mathscr{S}(A, B, C)$ showing the third-order asymptotic behaviour.

Projecting the next MP according to the maps N_1 and M_1 we obtain

$$\begin{aligned} G_2 &= N_1 C A B M_1 \\ &= \begin{bmatrix} 2 & 0 \\ 0 & 0 \end{bmatrix} \end{aligned} \tag{5.7}$$

The second PMP then is rank-deficient by $d_2 = 1$ degrees and as such leads to a further PMP defined by the annihilators of G_2, N_2 and M_2 :

$$N_2 = [0 \quad 1], \quad M_2 = \begin{bmatrix} 0 \\ 1 \end{bmatrix} \tag{5.8}$$

as

$$\begin{aligned} G_3 &= N_2 N_1 C A^2 B M_1 M_2 \\ &= -1 \end{aligned} \tag{5.9}$$

The above parameter however is full rank and therefore there will exist no further PMP.

We may now summarize the asymptotic behaviour of the RL of the system $\mathscr{S}(A, B, C)$ as follows :

(i) One of the seven closed-loop poles tends to the finite zero at $-\frac{3}{2}$.

(ii) Of the remaining closed-loop poles, one will tend to a first-order IZ along an asymptote which lies on the negative real axis, as dictated by the non-zero eigenvalue of G_1.

(iii) Of the remaining five closed-loop poles two will tend to a pair of second-order IZ along two asymptotes which, as indicated by the non-zero eigenvalue of G_2, make with the real axis angles of $\pm 90°$.

(iv) Finally the remaining three closed-loop poles vanish to infinity along asymptotes belonging to a Butterworth pattern defined by the three IZ of third order ; the orientation of the pattern of third-order asymptotes with respect to the real axis is dictated by G_3.

The asymptotic behaviour described above is demonstrated in Fig. 3 where the complete root-locus plots for $\mathscr{S}(A, B, C)$, as obtained by a digital computer, are displayed.

Suppose now we apply the same kind of analysis to the augmented system

$$\left.\begin{aligned} \hat{A} = \begin{bmatrix} A & BB^t \\ 0 & -A^t \end{bmatrix}, \quad \hat{B} = \begin{bmatrix} 0 \\ -C^t \end{bmatrix} \\ \hat{C} = [C \quad 0] \end{aligned}\right\} \tag{5.10}$$

where the matrices A, B, C are as defined by eqns. (5.1 *a–c*) and utilizing the modifications to the RL procedure outlined above and discussed in § 4.2 for the case of vanishing PMP. We then obtain the following results concerning the augmented system $\mathscr{S}(\hat{A}, \hat{B}, \hat{C})$:

(i) It has two finite zeros located at

$$\hat{z}_1 = -\tfrac{3}{2}, \quad \hat{z}_2 = \tfrac{3}{2} \tag{5.11}$$

(ii) There are no first-order IZ.

(iii) The asymptotic behaviour of order 2 is described by the two roots of

$$s = {}^2\sqrt{4k} \tag{5.12}$$

(iv) There are no third-order IZ.

(v) The asymptotic behaviour of order 4 is described by the four roots of

$$s = {}^4\sqrt{-4k} \tag{5.13}$$

(vi) There are no fifth-order IZ.

(vii) The remaining asymptotic behaviour is given by the six roots of

$$s = {}^6\sqrt{k} \tag{5.14}$$

In order to refer the above results to the ORL we have, of course, to consider the solutions of eqns. (5.11–14) that lie in the left-half of the complex plane only. Thus we have that there is one optimal finite zero at

$$z = -\tfrac{3}{2} \tag{5.15}$$

one second-order OIZ associated with an asymptote lying on the negative real axis, two fourth-order OIZ associated with two asymptotes which make an angle of $\pm 45°$ with the negative real axis and finally three sixth-order OIZ associated with three asymptotes, one lying on the negative real axis and the other two making angles of $\pm 60°$ with the negative real axis. This type of asymptotic behaviour is verified by the diagram obtained by a digital computer and displayed in Fig. 4.

The symmetrical example discussed above illustrates the results contained in the relevant subsections. Thus we see that there are as many OFZ as FZ for the system $\mathscr{S}(A, B, C)$, namely one, and, further, since $\mathscr{S}(A, B, C)$ is minimum phase the OFZ is exactly equal to the FZ of $\mathscr{S}(A, B, C)$. There are as many second-order OIZ as first-order IZ for $\mathscr{S}(A, B, C)$, namely, one, and this is given by the non-zero eigenvalue of the second PMP, $\hat{G}_2$:

$$\hat{G}_2 = \begin{bmatrix} 4 & 0 & 0 \\ 0 & 0 & 0 \\ 0 & 0 & 0 \end{bmatrix} = -G_1 G_1^{\mathrm{t}} \tag{5.16}$$

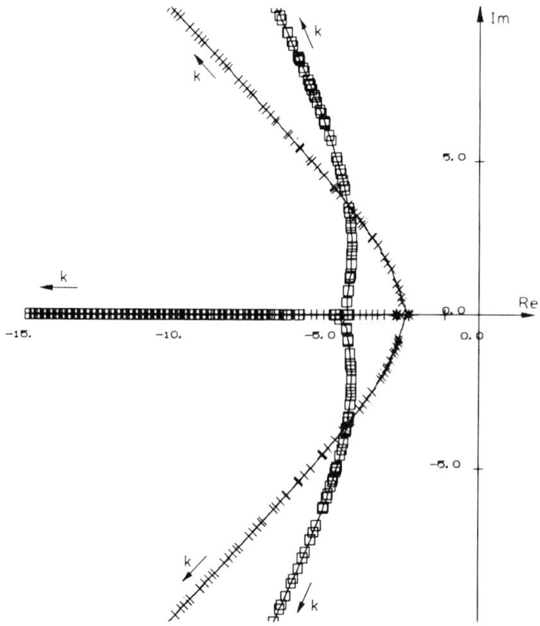

Figure 4 (*a*). The complete optimal root locus for $\mathscr{S}(A, B, C)$.

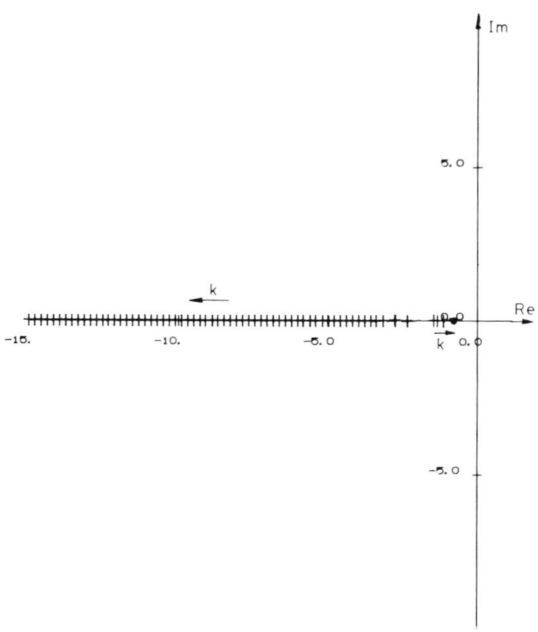

Figure 4 (*b*). The first Riemann sheet of the ORL showing the optimal finite zero (denoted by ●) and the second-order asymptotic behaviour.

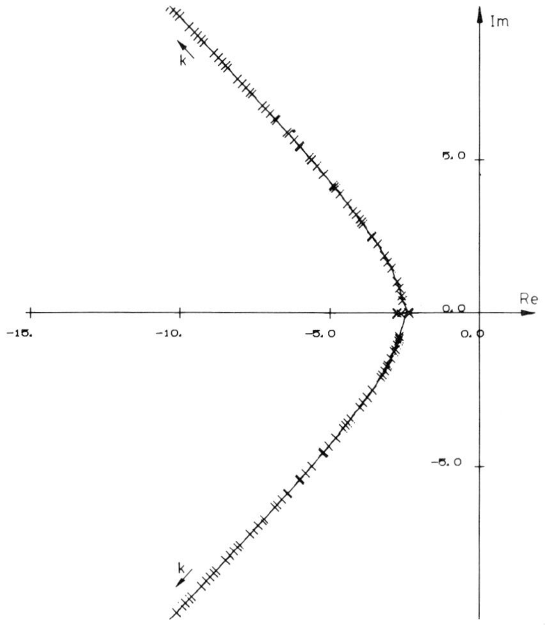

Figure 4 (c). The second Riemann sheet of the ORL showing the fourth-order optimal asymptotic behaviour.

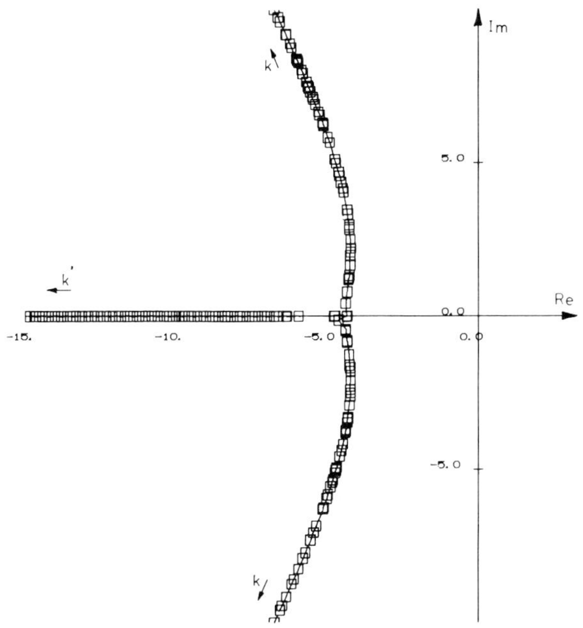

Figure 4 (d). The third Riemann sheet of the ORL showing the sixth-order optimal asymptotic behaviour.

The number of fourth-order OIZ is two, equal to the number of second-order IZ of $\mathscr{S}(A, B, C)$ and the location of these is dictated by the non-zero eigenvalue of the fourth PMP, $\hat{G}_4$:

$$\hat{G}_4 = \begin{bmatrix} 4 & 0 \\ 0 & 0 \end{bmatrix}$$

$$= [R_{N_1} G_2 R_{M_1}][R_{N_1} G_2 R_{M_1}]^{\mathrm{t}} \tag{5.17}$$

The sixth-order OIZ and the third-order IZ of $\mathscr{S}(A, B, C)$ are the same in number, that is three, and the location of these OIZ is dictated by the non-zero eigenvalue of the sixth PMP, $\hat{G}_6$:

$$\hat{G}_6 = -1$$

$$= -[R_{N_2} G_3 R_{M_2}][R_{N_2} G_3 R_{M_2}]^{\mathrm{t}}$$

Finally, as expected, no odd orders of OIZ exist.

References

Evans, W. R., 1948, *A, Inst. elect. electron. Engrs*, **67, 547.**

Gantmacher, F. R., 1959, *Theory of Matrices*, Vol. 1 (New York : Chelsea).

Kouvaritakis, B., and MacFarlane, A. G. J., 1976 a, *Int. J. Control*, **23,** 149 ; 1976 b, *Ibid.*, **23,** 167.

Kouvaritakis, B., and Shaked, U., 1976, *Int. J. Control*, **23,** 297.

Kwakernaak, H., 1976, *I.E.E.E. Trans. autom. Control*, **21,** 378.

Kwakernaak, H., and Sivan, R., 1972, *Linear Optimal Control of Systems* (New York : Wiley–Interscience).

MacFarlane, A. G. J., 1970 a, *Proc. Instn elect. Engrs*, **117,** 2037 ; 1970 b, *Ibid.*, **117,** 464 ; Personal correspondence.

Shaked, U., and Kouvaritakis, B., 1977, *I.E.E.E. Trans. autom. Control*, **22,** 597.

Walter, O. H. D., 1970, *I.E.E.E. Trans. autom. Control*, **15,** 486.

Wonham, W. M., 1974, *Linear Multivariable Control : A Geometric Approach* (New York : Springer-Verlag).

IV. DESIGN TECHNIQUES

Two different trends are clearly discernible in the currently developing literature on multivariable feedback control. One trend is towards a better mathematical formulation of the problems involved, to the development of fundamental principles, to the establishment of appropriate necessary and sufficient conditions for the solution of specific control problems, and to the development of synthesis techniques. Good examples of this approach are Wonham's work on a unified geometric approach to multivariable control (Wonham 1974) and the work of Wolovich (1974), Bengtsson (1973, 1977), Francis (1975, 1977), Davison (1976) and Pernebo (1978). The other trend is towards a more pragmatic and practical engineering approach in which a design problem is formulated in a way which may make good sense to an experienced engineer but which may appear distinctly vague and ill-defined from a more mathematical viewpoint. Such approaches to design problems tend to use a few key physically based concepts and to be critically dependent on the amount of physical insight which they can generate. Frequency-response techniques have always featured prominently in such approaches because of their direct link with experimental measurements. This particular trend has received a considerable boost from rapid developments in interactive computing (Rosenbrock 1974). When such engineering design approaches are used with an interactive computer graphic display, techniques which lend themselves to graphical interpretation have a built-in advantage and so it is natural that generalized Nyquist diagrams and generalized root-locus diagrams should emerge as useful design tools.

It is useful in this context to distinguish between the concepts of design and synthesis. Given an appropriately agreed meaning these terms become useful labels when classifying and discussing various trends and developments. In developing a synthesis technique one strives to formulate a sharply defined problem with a well founded mathematical solution, preferably expressible in the form of a workable and robust computer algorithm. In developing a design technique one strives to give a practising and experienced engineer a set of manipulative and interpretive tools which will enable him to build up, modify and assess a design put together on the basis of physical reasoning within the guidelines laid down by his engineering experience. The optimal control problem and its solution is a good example of a synthesis procedure. It, nevertheless, cannot avoid placing some burden on the engineering designer in the form of having to choose the performance index to be optimized.

The reprints reproduced in this section are concerned with the development of design techniques for multivariable feedback systems which are based on the use of generalized Nyquist diagrams and multivariable root loci. As mentioned in the Introduction they are meant to be used with a suitable interactive computer-aided design package (Edmunds 1978, 1979).

References

BENGTSSON, G., 1973, A theory for control of linear multivariable systems. Lund

Institute Technical Report 7341, Lund, Sweden ; 1977, Output regulation and internal models—a frequency domain approach. *Automatica*, **13,** 333–345.

Davison, E. J., 1976, The robust decentralized control of a general servomechanism problem. *I.E.E.E. Trans. autom. Control*, **21,** 529–534.

Edmunds, J. M., 1978, Cambridge Linear Analysis and Design Programs, Report No. CUED/F-CAMS/TR 170, Cambridge University Engineering Department ; 1979, Cambridge Linear Analysis and Design Programs. *Proceedings of the IFAC Symposium on Computer-Aided Control System Design*, Zurich.

Francis, B. A., 1975, The linear multivariable regulator problem. Memorandum ERL-M560, Electronics Research Laboratory, University of California, Berkeley ; 1977, The multivariable servomechanism from the input–output viewpoint. *I.E.E.E. Trans. autom. Control*, **22,** 322–328.

Pernebo, L., 1978, Algebraic control theory for linear multivariable systems. Lund Institute Technical Report LUTFD2/TFRT-1016/1-307, Lund, Sweden.

Rosenbrock, H. H., 1974, *Computer-aided Control System Design* (New York : Academic Press).

Wolovich, A. A., 1974, *Linear Multivariable Systems* (New York : Springer-Verlag).

Wonham, W. M., 1974, *Linear Multivariable Control : A Geometric Approach* (Berlin : Springer-Verlag).

A design technique for linear multivariable feedback systems

A. G. J. MacFARLANE and B. KOUVARITAKIS

A systematic approach is developed for the design of linear multivariable feedback control systems based on a manipulation of the set of frequency-conscious eigenvalues and eigenvectors of an open-loop transfer-function matrix. The key idea behind the approach used is that of an approximately-commutative controller. An algorithm for approximation of a frame of complex vectors by a frame of real vectors is developed and plays a basic role in the systematic design approach. An example, based on industrial plant data, is given showing how the design method is used.

1. Introduction

Recent work on the generalized Nyquist stability criterion for multivariable feedback systems and the related theory of matrix-valued functions of a complex variable (MacFarlane and Postlethwaite 1977) has provided a firm theoretical foundation for the design approach known as the Characteristic Locus Method (MacFarlane and Belletrutti 1973, Kouvaritakis 1974). The essence of the Characteristic Locus Method is the manipulation of the set of frequency-conscious eigenvalues and eigenvectors of an open-loop transfer-function matrix in such a way as to achieve a desired closed-loop response for a multivariable feedback system. The difficulty of the technique lies in the choice of controller transfer function matrix to achieve a desired set of eigenvector and eigenvalue behaviours for the cascade of plant and controller. It is the purpose of this paper to provide a systematic design approach to these manipulations. This is based on the concept of an approximately-commutative controller and gives a method of easily achieving desired eigenvalue and eigenvector modifications at specific values of angular frequency.

The basic concepts of matrix-valued functions of a complex variable are briefly reviewed in § 2, and § 3 deals with their use in the analysis of a standard multivariable feedback configuration. The key technique which forms the basis of the design approach, namely the construction of matrices which approximately commute with a given plant matrix at a specific frequency, is considered in § 4. The remaining two sections develop the proposed design technique and illustrate it by an example based on industrial data. The material of these last three sections is treated in more detail in Kouvaritakis (1974); the interested reader is referred there for a fuller account of the method together with additional examples of its application.

2. Matrix-valued functions of a complex variable

It has been shown (MacFarlane and Postlethwaite 1977) that the basic concepts of pole, zero, Nyquist diagram and Root Locus diagram used in classical frequency-response approaches to feedback system design can be rigorously extended to the appropriate multivariable case. The key to this extension is the association of a set of algebraic functions with an $m \times m$ square matrix-valued function of a complex variable, $Q(s)$ say, whose elements are rational functions in s. For any specific value of complex frequency $\bar{s}$, the matrix $Q(\bar{s})$ is a matrix of complex numbers and so has a set of complex eigenvalues. If we form the relevant characteristic equation

$$\det [q(s)I_m - Q(s)] \triangleq \Delta(q, s) = 0 \tag{2.1}$$

it is clear that a matrix $Q(s)$ will not normally have eigenvalues which are rational functions in s, since $\Delta(q, s)$ will not normally have factors linear in q when regarded as a polynomial in q with coefficients in the field of rational functions in s. In general $\Delta(q, s)$ will be reducible to a form

$$\Delta(q, s) = \Delta_1(q, s)\Delta_2(q, s) \ldots \Delta_p(q, s) \tag{2.2}$$

where the factors $\{\Delta_i(q, s) : i = 1, 2, \ldots, p\}$ are polynomials which are irreducible over the field of rational functions in s. Each irreducible factor will have the form

$$\Delta_i(q, s) = q_i^{t_i}(s) + a_{i1}(s)q_i^{t_i-1}(s) + \ldots + a_{it_i}(s), \quad i = 1, 2, \ldots, p \tag{2.3}$$

where the coefficients $\{a_{ij}(s)\}$ are rational functions in s. In almost all situations however where a transfer-function matrix arises from a practical plant, the associated expression $\Delta(q, s)$ will in fact be irreducible. If $b_{i0}(s)$ is the least common denominator of the coefficients $\{a_{ij}(s) : j = 1, 2, \ldots, t_i\}$ then eqns. (2.3) can be put in the form

$$b_{i0}(s)q_i^{t_i} + b_{i1}q_i^{t_i-1} + \ldots + b_{it_i}(s) = 0, \quad i = 1, 2, \ldots, p \tag{2.4}$$

where the coefficients $\{b_{ij}(s)\}$ are now polynomials in s. A function $q_i(s)$ defined by an equation of the form (2.4) is called an algebraic function (Bliss 1966). Algebraic functions are generalizations of the concept of a simple function of a complex variable. They have as domains of definition Riemann surfaces constructed by appropriately piecing together a number of copies of the complex plane equal to the order of the defining eqn. (2.4) : the algebraic function defines a single-valued analytic function from the Riemann surface on to the set of complex numbers. Finite poles and zeros of the algebraic function are defined by the roots of the equations

$$b_{i0}(s) = 0 \tag{2.5}$$

and

$$b_{it_i}(s) = 0 \tag{2.6}$$

respectively, and may be located on the relevant Riemann surface (MacFarlane and Postlethwaite 1977). The algebraic function has branch points wherever the defining eqn. (2.4) has repeated roots. In every simply-connected region of the complex plane which does not contain any branch points, the theory of algebraic functions shows that the values of the characteristic function form a set of locally distinct analytic functions. Each of these locally distinct analytic functions is called a *branch* of the algebraic function.

2.1. *Dyadic expansions*

It follows from the treatment of matrix-valued functions of a complex variable given by MacFarlane and Postlethwaite (1977) that at almost all points of the complex plane an $m \times m$ matrix $Q(s)$ whose elements are rational functions in s can be expressed in the dyadic form

$$Q(s) = \sum_{i=1}^{m} q_i(s) w_i(s) v_i^{\,t}(s) \tag{2.7}$$

where :

(i) The functions $\{q_i(s) : i = 1, 2, \ldots, m\}$ are a set of analytic functions whose values at any specific complex frequency $\bar{s}$ are the set of eigenvalues of the complex matrix $Q(\bar{s})$. These are a complete set of branches for all the algebraic functions associated with the matrix $Q(s)$ by the characteristic eqn. (2.1).

(ii) The vectors $\{w_i(s) : i = 1, 2, \ldots, m\}$ are an independent set of vector-valued functions of s whose values at $s = \bar{s}$ are eigenvectors of the complex matrix $Q(\bar{s})$. The individual component functions of these vectors are branches of analytic functions defined on the same Riemann surfaces as the appropriate eigenvalue functions $q_i(s)$.

(iii) The vectors $\{v_i^{\,t}(s) : i = 1, 2, \ldots, m\}$ are an appropriate set of dual-eigenvector functions for the eigenvector functions $w_i(s)$.

(iv) The expression (2.7) is well-defined at all points of the complex plane except the complete set of branch points for all the characteristic algebraic functions of $Q(s)$.

2.2. *Characteristic loci*

The characteristic loci of a matrix $Q(s)$ are the set of loci in the complex plane traced out by the eigenvalues of $Q(s)$ as s traverses the standard Nyquist contour in the standard (clockwise) direction ; they are discussed in detail in MacFarlane and Postlethwaite (1977).

2.3. *Generalized Nyquist stability criterion*

Let the multivariable feedback system shown in Fig. 1 have no open-loop uncontrollable and/or unobservable modes whose corresponding characteristic frequencies lie in the right-half complex plane. Then this feedback configuration will be closed-loop stable if, and only if, the net sum of anti-clockwise encirclements of the critical point $(-1 + j0)$ by the set of characteristic loci for the return-ratio matrix $L(s)$, corresponding to signals injected at point α and returned at point α', is equal to the total number of right-half plane poles of $G_1(s)$ and $G_2(s)$ (MacFarlane and Postlethwaite 1977).

2.4. *Multivariable root loci*

If the return-ratio matrix referred to in § 2.3 has a set of characteristic functions $\{l_i(s) : i = 1, 2, \ldots, p\}$ then the multivariable root loci for the feedback system of Fig. 1 are defined as the set of values of complex frequency which satisfy the equations

$$l_i(s) = -\frac{1}{k}, \quad i = 1, 2, \ldots, p \tag{2.8}$$

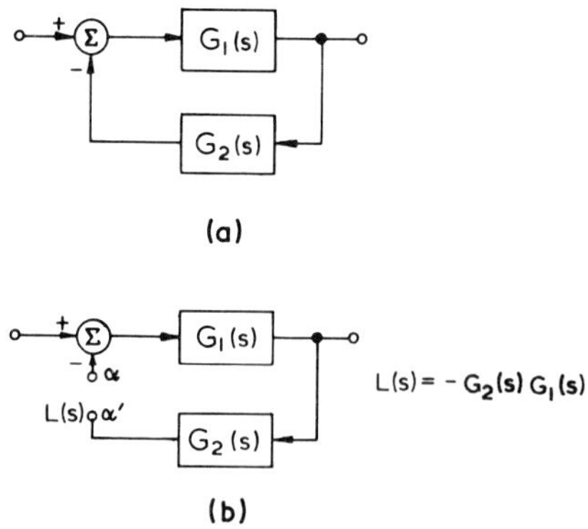

Figure 1. Feedback configuration for generalized Nyquist stability criterion. (*a*) Closed loop ; (*b*) open loop.

where k is a scalar gain parameter. They correspond to the paths traced out in the complex plane by the closed-loop characteristic frequencies of the feedback system shown in Fig. 2 as the scalar gain parameter k is varied from zero to infinity. In Riemann surface terms, the multivariable root loci are the 180° phase contours of the set of characteristic functions $\{l_i(s) : i=1, 2, \ldots, p\}$ on their appropriate Riemann surfaces. Note that eqn. (2.8) does not deal with single point root loci that may arise due to cancellations of poles by zeros or due to decoupling. For a treatment of this special type of root loci the interested reader is referred to MacFarlane and Postlethwaite (1977).

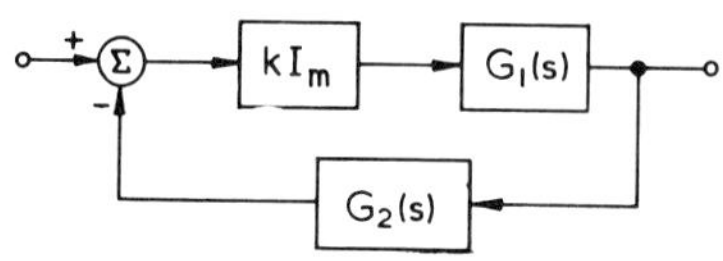

Figure 2. System for definition of multivariable root loci.

3. Analysis of standard multivariable feedback configuration

Consider the multivariable feedback configuration shown in Fig. 3, where $G(s)$ is an $m \times m$ plant transfer function matrix and $K(s)$ is an $m \times m$ controller transfer function matrix. Let

$$Q(s)=G(s)K(s) \tag{3.1}$$

so that the closed-loop transfer function matrix for this arrangement is

$$R(s)=[I_m+Q(s)]^{-1}Q(s) \tag{3.2}$$

It follows from standard relationships in matrix algebra (Marcus and Minc 1964) that every eigenvector $w_i(s)$ of $Q(s)$ is also an eigenvector of $R(s)$, and that for every eigenvalue $q_i(s)$ of $Q(s)$ there is a corresponding eigenvalue $[q_i(s)/1+q_i(s)]$ of $R(s)$. Since the set of complex numbers $\{q_i/1+q_i :$

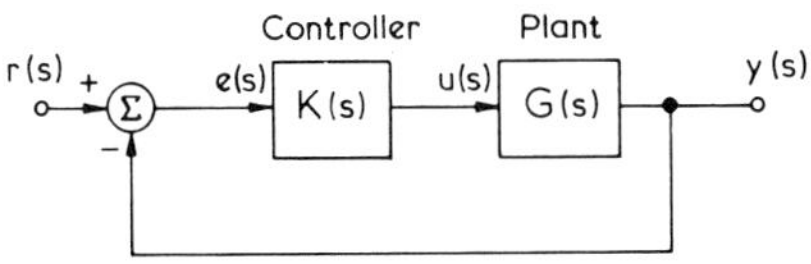

Figure 3. Multivariable feedback system.

$i=1, 2, \ldots, m\}$ will have two or more members with the same value if and only if the set of complex numbers $\{q_i : i=1, 2, \ldots, m\}$ has two or more members with the same value, it follows that the sets of branch points associated with the matrices $R(s)$ and $Q(s)$ are identical. The characteristic algebraic functions associated with the open-loop matrix $Q(s)$ and the closed-loop matrix $R(s)$ thus share the same Riemann surface, and these matrices also share the same set of eigenvectors. Therefore if the open-loop matrix $Q(s)$ has a dyadic expansion of the form

$$Q(s)=\sum_{i=1}^{m} q_i(s)w_i(s)v_i{}^t(s) \tag{3.3}$$

valid for almost all values of s (except at the set of branch points) then $R(s)$ will have a corresponding dyadic expansion of the form

$$R(s)=\sum_{i=1}^{m}\left[\frac{q_i(s)}{1+q_i(s)}\right] w_i(s)v_i{}^t(s) \tag{3.4}$$

valid in the same punctured plane as $Q(s)$ (complex plane with set of branch points deleted).

3.1. *Sinusoidal response of closed-loop system*

Consider the arrangement of Fig. 3, and let $r(s)$ be the input transform vector. Let a reference input vector sinusoid be applied to this closed-loop system and let $r(j\omega)$ denote the corresponding input phasor vector. Then the corresponding output phasor vector for the closed-loop system is given by :

$$\begin{aligned} y(j\omega) &= \sum_{i=1}^{m}\left[\frac{q_i(j\omega)}{1+q_i(j\omega)}\right] w_i(j\omega)v_i{}^t(j\omega)r(j\omega) \\ &= \sum_{i=1}^{m}[v_i{}^t(j\omega)r(j\omega)]\left[\frac{q_i(j\omega)}{1+q_i(j\omega)}\right] w_i(j\omega) \end{aligned} \tag{3.5}$$

This expression gives a clear geometrical picture of the way in which the closed-loop system responds to a vector sinusoidal input. In these terms the action of the closed-loop system may be described in the following way.

(i) The reference input phasor vector $r(j\omega)$ is split up into a sum of components

$$[v_i{}^t(j\omega)r(j\omega)]w_i(j\omega)$$

directed along the characteristic vector set $\{w_i(j\omega) : i=1, 2, \ldots, m\}$.

(ii) Corresponding to the eigenvector $w_i(j\omega)$, the ith component coordinate $v_i^t(j\omega)r(j\omega)$ is acted upon by a ' coordinate positioning ' closed-loop transference

$$\frac{q_i(j\omega)}{1+q_i(j\omega)}$$

to produce the corresponding ith component of the output phasor vector as

$$(i\text{th output coordinate}) = \left[\frac{q_i(j\omega)}{1+q_i(j\omega)}\right] \times (i\text{th input coordinate}) \tag{3.6}$$

(iii) The output phasor vector $y(j\omega)$ of the closed-loop system is then assembled as a sum of component phasor vectors lying along the characteristic phasor directions $\{w_i(j\omega)\}$, and of amounts determined by the outputs of the ' coordinate-positioning ' characteristic control gains :

$$\text{output phasor } y(j\omega) = \sum_{i=1}^{m} \left[\frac{q_i(j\omega)}{1+q_i(j\omega)}\right] [v_i^t(j\omega)r(j\omega)]w_i(j\omega) \tag{3.7}$$

The importance of this expression is that it shows how the closed-loop performance of the system of Fig. 3 may be analysed in terms of the *open-loop* characteristic loci and characteristic vector set of $Q(s)$. In particular if the moduli of all the characteristic values $\{q_i(j\omega) : i=1, 2, \ldots, m\}$ are sufficiently high, such that

$$\frac{q_i(j\omega)}{1+q_i(j\omega)} \simeq 1 \quad \text{for } i=1, 2, \ldots, m$$

then

$$y(j\omega) \simeq \left[\sum_{i=1}^{m} w_i(j\omega)v_i^t(j\omega)\right] r(j\omega)$$

so that

$$y(j\omega) \simeq r(j\omega)$$

3.2. *Interaction*

In general, if some specific reference input demand $r_i(s)$ is imposed on the ith closed-loop input then some form of response $\{y_j(s) : j=1, 2, \ldots, m\}$ will result from all the closed-loop system outputs. In most cases one will want one specific output $y_k(s)$ to respond to the input $r_i(s)$, and all other outputs $\{y_k(s) : k \neq i\}$ to remain small in some suitably defined sense. By the general term ' interaction ' is simply meant the body of relationships influencing the way in which a specific input $r_i(s)$ affects the set of outputs $\{y_k(s) : k \neq i\}$.

It follows from the discussion given in § 3.1 that interaction at low frequencies can in principle be suppressed to any required degree by simply ensuring that the moduli of all the characteristic loci $\{q_i(j\omega) : i=1, 2, \ldots, m\}$ are sufficiently large at the frequency concerned. At high frequencies, however, one cannot normally arrange for the characteristic loci to have arbitrarily large moduli, because of the power demands which this imposes on the system, and because the fact that the characteristic loci must satisfy the generalized Nyquist stability criterion usually forces them to have small moduli at high frequencies. Furthermore, for all practical forms of plant transfer function matrix, all the elements of the transfer function matrix and hence all the characteristic loci, will tend to zero as the modulus of s tends to infinity. Suppose therefore that at some specific high value of angular frequency, ω_h say, all the characteristic loci of $Q(s)$ have arbitrarily small modulus. Then

$$I_m + Q(j\omega_h) \to I_m$$

and so

$$R(j\omega_h) \to Q(j\omega_h)$$

One thus reaches the following general conclusion regarding high-frequency interaction : the high-frequency cross-couplings in $R(j\omega)$ are essentially the same as those in $Q(j\omega)$, and are not therefore influenced by any feedback action. Thus, in general terms, the only way to remove high-frequency interaction in the closed-loop system is by some suitable modification to the forward path transference $Q(j\omega)$.

A useful approach to the problem of high-frequency interaction is *via the characteristic direction set* of the operators involved. Consider the unity-feedback situation of Fig. 3. When one says that there must be little closed-loop interaction, this simply means that only the ith output of $R(j\omega)$ must respond to the ith input. This, in turn, simply means that one of the standard basic vectors $\{e_i\}$ where

$$e_1 = \begin{bmatrix} 1 \\ 0 \\ 0 \\ \vdots \\ 0 \end{bmatrix}, \quad e_2 = \begin{bmatrix} 0 \\ 1 \\ 0 \\ \vdots \\ 0 \end{bmatrix} \quad \cdots \quad e_m = \begin{bmatrix} 0 \\ 0 \\ \vdots \\ 0 \\ 1 \end{bmatrix}$$

must be a characteristic direction vector of $R(j\omega)$, and therefore of $Q(j\omega)$. It then follows that a sufficient condition for a small amount of high-frequency interaction is that the characteristic direction set of $Q(j\omega)$ must approach the standard basis direction set at high frequencies. Thus one may take as a convenient measure of interaction at high frequencies the angle between the vectors $w_k(j\omega)$ and the standard basis directions e_i for $i, k = 1, 2, \ldots, m$. This can be measured in terms of the angular misalignment between these vectors, given by

$$\cos \theta_l(j\omega) = \frac{|(w_k(j\omega), e_l)|}{\|w_k(j\omega)\|} \tag{3.8}$$

where the convention is adopted that k is chosen such that w_k is that eigenvector which makes the smallest angle θ_l with the standard basis vector e_l at the frequency ω.

In summary then, interaction is determined at low frequencies by θ_l and/or the moduli of the characteristic loci of the open-loop operator, and at high frequencies by the angular misalignment between the standard basis set and the characteristic direction set of the open-loop operator. Thus a convenient way to handle interaction over the whole required frequency range is via a set of graphical plots showing : (i) moduli of $q_i(j\omega)$ versus ω ; and (ii) the misalignment angles $\theta_l(j\omega)$, as given by eqn. (3.8), versus ω.

3.3. *Performance of the closed-loop system*

We conclude from the above analysis that the performance of the closed-loop system of Fig. 3 can be discussed in terms of the characteristic loci of the open-loop matrix $Q(s)$ and the corresponding set of frequency-conscious characteristic directions. For a satisfactory closed-loop performance the following conditions must be satisfied.

(i) The set of characteristic loci $\{q_i(j\omega) : i=1, 2, \ldots, m\}$ must satisfy the generalized Nyquist stability criterion.
(ii) The characteristic loci must all have a sufficiently high gain over a desired operating bandwidth to provide the required accuracy for the tracking of $r(j\omega)$ by $y(j\omega)$ at the frequency ω.
(iii) For the characteristic gains to have the required effect, the characteristic direction set should form a good spanning set for the m-dimensional complex signal space. If the characteristic direction set is excessively skew it will be necessary, as explained below, to use an alternative means (the principal gain set) of checking the system gains.
(iv) The high-frequency alignment to the standard basis direction set of the characteristic directions of $Q(j\omega)$ should be sufficiently good to keep high-frequency interaction within acceptable limits.

If these conditions are satisfied then we may regard the ith eigenvalue $q_i(j\omega)$, associated with the eigenvector $w_i(j\omega)$ which aligns best with the jth standard basis vector e_j, as a reasonable approximation (for the purposes of discussing performance) to the transference between jth input and jth output at the particular angular frequency ω. An assessment of closed-loop response of the jth output to the jth input can then be based on an inspection of the plots of $q_i(j\omega)$ at the frequencies of interest.

It follows from the expression (3.4) that if the eigenvectors of $Q(j\omega)$ are approximately orthogonal, or at least not excessively skew, then high gains for $\{q_i(j\omega) : i=1, 2, \ldots, m\}$ will ensure a satisfactory closed-loop response. However, if the eigenvectors are very skew or (as in certain cases of coincident eigenvalues) a complete set of eigenvectors does not exist, then we cannot use the eigenvectors of $Q(j\omega)$ in an assessment of closed-loop performance. In such a situation we can use the concepts of vector gain and principal gain direction. The *principal gains* of $Q(j\omega)$ are defined to be set the of square roots of the eigenvalues of Q^*Q (where Q^* is the complex conjugate transpose of Q). If the principal gains are denoted by $\{\nu_i : i=1, 2, \ldots, m\}$ then the following properties can be established.

(i) $\nu_i \geq 0$, $i=1, 2, \ldots, m$, where ν_i are real numbers.

(ii)
$$\nu_i = \frac{\|Qm_i\|}{\|m_i\|}, \quad i=1, 2, \ldots, m \tag{3.9}$$

where m_i is the eigenvector of Q^*Q corresponding to the eigenvalue ν_i^2.

(iii) If the ν_i are ordered so that

$$\nu_1 \leq \nu_2 \leq \ldots \leq \nu_m \tag{3.10}$$

then
$$\nu_1 \leq \frac{\|y\|}{\|e\|} \leq \nu_m \tag{3.11}$$

where

$$y = Qe$$

(iv) The vector set $\{m_i : i = 1, 2, \ldots, m\}$ are an orthogonal set ; they are called the principal gain directions of Q in its input space.

It is the fact that the principal gain directions are always an orthogonal set which makes principal gains useful in performance evaluation. Since the reference, error and output vectors are related by

$$r = e + y$$

we will have that

$$\frac{\|y\|}{\|r\|} = \frac{\|y\|}{\|e+y\|} = \frac{\dfrac{\|y\|}{\|e\|}}{\dfrac{\|e+y\|}{\|e\|}} \geqslant \frac{\dfrac{\|y\|}{\|e\|}}{1 + \dfrac{\|y\|}{\|e\|}} \tag{3.12}$$

so that using (3.11) we get that

$$\frac{\|y\|}{\|r\|} \geqslant \frac{\nu_1}{1 + \nu_m} \tag{3.13}$$

We will also obviously have that

$$\frac{\|y\|}{\|r\|} \rightarrow 1 \quad \text{as} \quad \frac{\|y\|}{\|e\|} \rightarrow \infty$$

These simple arguments show that the principal gains of $Q(j\omega)$ can be used for an assessment of closed-loop performance in the case when the eigenvectors of Q are badly skewed, or when a complete set of eigenvectors of $Q(j\omega)$ does not exist. A sufficiently high set of values for all the principal gains of the open-loop operator $Q(j\omega)$ will ensure that $y(j\omega)$ closely tracks $r(j\omega)$ regardless of the nature of the eigenvectors of $Q(j\omega)$.

4. Approximate commutativity and the alignment algorithm

The analysis of the previous section has shown how the behaviour of a closed-loop system can be discussed in terms of open-loop characteristic loci and open-loop characteristic vectors. One can thus translate a closed-loop performance specification into an equivalent specification of the frequency-dependent behaviour of the open-loop eigenvalues and eigenvectors. From this point of view the essential difficulty in choosing an appropriate form of controller matrix $K(s)$ lies in the fact that very little is known of the way in which the eigenvalues and eigenvectors of the product of two matrices $G(s)K(s)$ are related to the individual eigenvalues and eigenvectors of the matrices $G(s)$

and $K(s)$. There is one exception to this situation, however, which is that in which the matrices concerned commute, that is when

$$G(s)K(s)=K(s)G(s) \tag{4.1}$$

This happens if and only if $G(s)$ and $K(s)$ have a common set of eigenvectors. In such a situation the eigenvalues of the product $G(s)K(s)$ are simply appropriate products of the individual eigenvalues of the matrices $G(s)$ and $K(s)$; the eigenvalues which are multiplied together are those associated with a common eigenvector.

Since the eqn. (2.1) will normally be irreducible the eigenvalues and eigenvectors of the matrix $G(s)$ will not be expressible in terms of rational fractions. It will therefore not normally be possible to construct a realizable controller $K(s)$ which will exactly commute with $G(s)$ for all values of s. Even in the rare cases when this might be possible, such a controller could be unnecessarily complicated. A more practical and rewarding approach is to investigate the possibility of using controllers which are approximately commutative with the plant at appropriately chosen values of frequency.

In the design technique proposed here the controller matrix $K(s)$ is taken to have the specific form

$$K(s)=A\Lambda_K(s)B \tag{4.2}$$

where A and B are matrices with *real* elements and $\Lambda_K(s)$ is a *diagonal* matrix of rational functions in s. The matrix $K(s)$ can therefore be expressed in the dyadic form

$$K(s)=\sum_{j=1}^{m} k_j(s)a_jb_j{}^t \tag{4.3}$$

where the vector sets $\{a_j:\ j=1, 2, \ldots, m\}$ and $\{b_j:\ j=1, 2, \ldots, m\}$ are both *real*. The corresponding open-loop transfer function matrix for the combined plant and controller can then be written in the form

$$\begin{aligned} Q(s) &= \left[\sum_{i=1}^{m} g_i(s)w_i(s)v_i{}^t(s)\right]\left[\sum_{j=1}^{m} k_j(s)a_jb_j{}^t\right] \\ &= \sum_{i,j=1}^{m} g_i(s)k_j(s)[v_i{}^t(s)a_j][w_i(s)b_j{}^t] \end{aligned} \tag{4.4}$$

Now suppose that at some specific value of complex frequency, $\bar{s}$ say, we could choose an a_i and a $b_i{}^t$ which were good approximations to the eigenvector $w_i(\bar{s})$ and the reciprocal eigenvector $v_i{}^t(\bar{s})$, respectively, so that

$$a_i \simeq w_i(\bar{s})$$

and

$$b_i{}^t \simeq v_i{}^t(\bar{s})$$

and consider the effect of the corresponding open-loop transfer function matrix $Q(\bar{s})$ on the vector $w_i(\bar{s})$. The problem of how to make such an approximation is considered in the next section. We have that

$$\begin{aligned} Q(\bar{s})w_i(\bar{s}) &= \sum_{k,j=1}^{m} g_k(\bar{s})k_j(\bar{s})[v_k{}^t(\bar{s})a_j][w_k(\bar{s})b_j{}^t]w_i(\bar{s}) \\ &\simeq g_i(\bar{s})k_i(\bar{s})w_i(\bar{s}) \end{aligned} \tag{4.5}$$

since *both*

$$v_j^t(\bar{s})a_i \simeq 0$$

and

$$b_j^t \, w_i(\bar{s}) \simeq 0$$

for all $i \neq j$, and since both $v_i^t(\bar{s})a_i$ and $b_i^t \, w_i(\bar{s})$ will approximate to unity.

We conclude from this that if we can make the real vectors a_i and b_i^t sufficiently good approximations to the complex vectors $w_i(s)$ and $v_i^t(s)$ at some specific complex frequency $\bar{s}$, then we will have that :

(i) they will also be good approximations to an appropriate eigenvector and dual eigenvector pair for the compensated system ; and
(ii) the product $g_i(\bar{s})k_i(\bar{s})$ will be a good approximation to an open-loop eigenvalue of the compensated-plant matrix $G(\bar{s})K(\bar{s})$.

These considerations lead to the idea of manipulating the characteristic loci of a compensated plant into an appropriate form by choosing what we may call an *approximately-commutative controller* (Kouvaritakis, 1974)

$$K(s) = A\Lambda_K(s)B$$

whose constituent matrices A and B are real matrices chosen in such a way that $K(s)$ is approximately commutative with $G(s)$ at *some specific value* of complex frequency $\bar{s}$. At this frequency we will then have that

$$q_i(\bar{s}) \simeq g_i(\bar{s})k_i(\bar{s}), \quad i = 1, 2, \ldots, m \tag{4.6}$$

where $q_i(\bar{s})$, $g_i(\bar{s})$ are eigenvalues of $Q(\bar{s})$ and $G(\bar{s})$, respectively, and $k_i(\bar{s})$ are appropriate diagonal elements of $\Lambda_K(s)$ evaluated at $s = \bar{s}$. This set of relations gives an obvious basis for manipulating the characteristic loci of $Q(s)$ in a systematic manner. It also turns out, as discussed in § 5, that the approximately-commutative technique can be used to manipulate the eigenvectors of $Q(s)$.

Note that the effectiveness of this technique will depend on the degree of alignment, and hence the range of frequencies over which the approximate adjustment of the eigenvalues will be valid will depend on the rate of change of the eigenvector directions. In general one can expect the approximately commutative controller to be effective over a band of frequencies above and below $\bar{s}$. Difficulties may arise in the case of eigenvalue adjustments near the vicinity of branch points when the eigenvectors become skewed. It is hoped to do further work on this aspect of the analysis.

4.1. *The alignment procedure for approximating complex vector frames by real vector frames* (Kouvaritakis, 1974)

Let $\{w_i(s) : i = 1, 2, \ldots, m\}$ and $\{v_i^t(s) : i = 1, 2, \ldots, m\}$ be a set of eigenvectors and dual eigenvectors of the plant matrix $G(s)$, and let $W(s)$ and $V(s)$ be a pair of matrices having $\{w_i(s)\}$ and $\{v_i^t(s)\}$ as columns and rows, respectively, *with the further proviso that* the column vector set of $W(s)$ is normalized to have unit length. If $x_i(s)$ is any given vector of unit length, then $x_i(s)$ is exactly equivalent to $w_i(s)$ if and only if

$$V(s)x_i(s) = e_i \tag{4.7}$$

where e_i denotes, as usual, the ith column of the unit matrix I_m. As $x_i(s)$ and $w_i(s)$ become misaligned, the magnitude of the ith entry of the vector Vx_i will decrease, and all the other entries will come into play. We can therefore use the relationship between the magnitude of the ith entry of Vx_i and the magnitudes of all the other entries as a measure of the alignment of some given vector x_i with some desired vector w_i. This leads us to take as a measure of the misalignment between a pair of vectors a_i and w_i the quotient

$$\phi_i = \frac{|(v_i, a_i)|^2}{\sum_{\substack{j=1 \\ j \neq i}}^{m} |(v_j, a_i)|^2} \tag{4.8}$$

where $\{v_i^t : i = 1, 2, \ldots, m\}$ is the dual vector set to the set $\{w_i : i = 1, 2, \ldots, m\}$. Since the whole set of dual vectors $\{v_i^t\}$ is involved in this quotient, it follows that in maximizing ϕ_i we are carrying out a *frame alignment* and the successive choices of the vectors a_i which achieve stationary maximum values of ϕ_i for the vector set $\{v_1, v_2, \ldots, v_m\}$ can thus be thought of as generating a frame of vectors $\{a_i\}$ which is closely aligned with a given frame $\{w_i\}$.

Let us now examine in detail the problem of choosing a real vector a_i such that the quotient

$$\phi_i = \frac{|(v_i, a_i)|^2}{\sum_{\substack{j=1 \\ j \neq i}}^{m} |(v_j, a_i)|^2}$$

is maximized. We have, using an overbar to denote the complex conjugate of the quantity over which it is placed, that

$$\begin{aligned} |(v_j, a_i)|^2 &= \overline{(v_j, a_i)}(v_j, a_i) \\ &= [v_j^t a_i][\bar{v}_j^t a_i] \\ &= [a_i^t v_j][\bar{v}_j^t a_i] \end{aligned}$$

so that

$$|(v_j, a_i)|^2 = a_i^t v_j \bar{v}_j^t a_i \tag{4.9}$$

Let v_j be expressed in terms of a pair of real vectors α_j and β_j as

$$v_j = \alpha_j + j\beta_j \tag{4.10}$$

so that

$$\bar{v}_j^t = \alpha_j^t - j\beta_j^t$$

Equation (4.9) can then be re-written in the form

$$\begin{aligned} |(v_j, a_i)|^2 &= a_i^t(\alpha_j + j\beta_j)(\alpha_j^t - j\beta_j^t)a_i \\ &= (a_i^t \alpha_j + j a_i^t \beta_j)(\alpha_j^t a_i - j\beta_j^t a_i) \\ &= (a_i^t \alpha_j)^2 + (a_i^t \beta_j)^2 \\ &= (a_i^t \alpha_j)(\alpha_j^t a_i) + (a_i^t \beta_j)(\beta_j^t a_i) \\ &= a_i^t(\alpha_j \alpha_j^t + \beta_j \beta_j^t)a_i \end{aligned} \tag{4.11}$$

If we now put

$$C_i = \alpha_i \alpha_i^t + \beta_i \beta_i^t \tag{4.12}$$

and

$$D_i = \sum_{\substack{j=1 \\ j \neq i}}^{m} (\alpha_i \alpha_i^t + \beta_i \beta_i^t) \tag{4.13}$$

then the quotient ϕ_i becomes

$$\phi_i = \frac{a_i^t C_i a_i}{a_i^t D_i a_i} \tag{4.14}$$

For a maximum value of ϕ_i we must have that

$$\frac{d}{da_i}(\phi_i) = 0$$

which gives that

$$2C_i a_i (a_i^t D_i a_i) = 2D_i a_i (a_i^t C_i a_i)$$

from which we obtain

$$C_i a_i = \left[\frac{a_i^t C_i a_i}{a_i^t D_i a_i}\right] D_i a_i$$

$$= \phi_i D_i a_i \tag{4.15}$$

Thus if λ_i is the relevant maximum value of ϕ_i, it must be such that

$$C_i a_i = \lambda_i D_i a_i \tag{4.16}$$

and a_i will then be the corresponding vector for which such a maximum is obtained.

It follows from the definitions of the matrices C_i and D_i that they will both be real, symmetric and non-negative definite. In turn it follows from this that their individual eigenvalues will be real non-negative numbers (see Appendix 1).

We may now assume that D_i is positive-definite and therefore non-singular (see Appendix 1); it will then have a spectral decomposition of the form

$$D_i = R_d \Lambda_d^2 R_d^t \tag{4.17}$$

where R_d is an appropriate orthogonal matrix whose columns are eigenvectors of D_i, and Λ_d^2 denotes a diagonal matrix of positive real eigenvalues. Further assume that the columns of R_d are so normalized that

$$R_d R_d^t = R_d^t R_d = I \tag{4.18}$$

Equation (4.16) implies that, for a non-trivial solution, we must have that

$$\det [C_i - \lambda_i D_i] = 0 \tag{4.19}$$

so that, substituting for D_i from eqn. (4.17), we get

$$\begin{aligned}\det [C_i - \lambda_i D_i] &= \det [C_i - \lambda_i R_d \Lambda_d^2 R_d^t] \\ &= \det [R_d \Lambda_d (\Lambda_d^{-1} R_d^t C_i R_d \Lambda_d^{-1} - \lambda_i I) \Lambda_d R_d^t] \\ &= \det [R_d R_d^t] \det [\Lambda_d^2] \\ &\quad \times \det [\Lambda_d^{-1} R_d^t C_i R_d \Lambda_d^{-1} - \lambda_i I] \\ &= \det [\Lambda_d^2] \det [S_i - \lambda_i I] \\ &= 0 \end{aligned} \tag{4.20}$$

where

$$S_i = \Lambda_d^{-1} R_d^t C_i R_d \Lambda_d^{-1} \tag{4.21}$$

Since we may take D_i to be non-singular, it follows that $\det [\Lambda_d^2]$ will not be zero, and so the required value of λ_i will be given by

$$\det [S_i - \lambda_i I] = 0 \tag{4.22}$$

It thus follows that the maximum value of ϕ_i will be given by the maximum eigenvalue of S_i. The matrix S_i is easily seen to be real and symmetric; consequently, all its eigenvalues will be real. Let z_i then be the real eigenvector of S_i corresponding to the real eigenvalue λ_i, so that

$$S_i z_i = \lambda_i z_i \tag{4.23}$$

On substituting for S_i from eqn. (4.21) we then get

$$\Lambda_d^{-1} R_d^t C_i R_d \Lambda_d^{-1} z_i = \lambda_i z_i \tag{4.24}$$

so that

$$\begin{aligned} C_i (R_d \Lambda_d^{-1} z_i) &= \lambda_i R_d \Lambda_d z_i \\ &= \lambda_i R_d \Lambda_d \Lambda_d R_d^t R_d \Lambda_d^{-1} z_i \\ &= \lambda_i R_d \Lambda_d^2 R_d^t R_d \Lambda_d^{-1} z_i \\ &= \lambda_i D_i R_d \Lambda_d^{-1} z_i \end{aligned} \tag{4.25}$$

Therefore, on putting

$$p_i = R_d \Lambda_d^{-1} z_i \tag{4.26}$$

we get

$$C_i p_i = \lambda_i D_i p_i \tag{4.27}$$

and hence deduce that p_i is the required real vector which results in a maximum value of the quotient ϕ_i.

5. Characteristic locus design technique

The analysis presented in § 3 shows how the stability and performance of a multivariable feedback system may be discussed in terms of the characteristic loci and characteristic directions of the open-loop transfer function matrix $G(s)K(s)$. It is therefore natural to consider a design technique based upon methods of choosing the controller matrix $K(s)$ in such a way as to give the characteristic loci and characteristic directions a required set of properties. Such a design technique is presented here ; it is convenient before becoming involved in the details of the method to briefly summarize the desired properties of the characteristic loci and characteristic directions of $G(s)K(s)$ as follows.

(i) The characteristic loci must satisfy the generalized Nyquist stability criterion.
(ii) For high tracking performance, and therefore low interaction, at any frequency, the gains of all the characteristic loci must be suitably large. This is particularly relevant to the low frequency situation in which it is feasible to have high characteristic gains.
(iii) At high frequencies, however, it is not feasible to have large gains in the characteristic loci since this would tend to lead to a violation of the generalized Nyquist stability criterion.
(iv) To reduce interaction at high frequencies, therefore, one cannot in general deploy high gains for this purpose, and so one must instead make the characteristic direction set of the open-loop transference align with the standard basis set at high frequencies.

The principal components of this characteristic locus design method are therefore :

(*a*) a method of manipulating characteristic loci ; and
(*b*) a method of manipulating characteristic directions.

These are now examined in turn.

5.1. *Procedure for manipulation of characteristic loci*

A computer algorithm has been programmed to carry out the best alignment of a given complex frame by a real frame, according to the misalignment measure and using the results of the previous section. It is convenient to refer to this procedure by the name ALIGN. To manipulate the values of the characteristic loci at any given angular frequency, the following procedure may therefore be adopted.

(i) Compute $W(j\omega)$ and $V(j\omega)$ at the angular frequency involved.
(ii) Compute a real frame matrix A which approximates the frame matrix $W(j\omega)$, using ALIGN.
(iii) Compute a real frame matrix B which approximates the dual frame matrix $V(j\omega)$, using ALIGN.
(iv) Manipulate the loci via the compensating matrix

$$K_a = A\Lambda_k(s)B \tag{5.1}$$

where the diagonal elements $k_i(s)$ of $\Lambda_K(s)$ are chosen on the assumption that

$$q_i \simeq g_i k_i \tag{5.2}$$

5.2. *Procedure for manipulation of characteristic vectors*

If it were possible at any frequency to design a controller such that

$$K(j\omega) = G^{-1}(j\omega) \tag{5.3}$$

then this would :

(i) annihilate all the phase from the characteristic loci of $GK(j\omega)$, at that frequency ; and
(ii) align all the characteristic directions of $GK(j\omega)$ with the standard basis set.

Such a controller is, of course, not physically realizable since it would require, apart from its excessive phase-advancing characteristics, arbitrarily high characteristic gains at high frequencies. However, it is entirely feasible to approximate $G^{-1}(j\omega)$, regarded as a frame matrix, by a real frame matrix at some suitably chosen high frequency. Furthermore, since the frame vectors so obtained would be continuous functions of the angular frequency, it would be reasonable to assume that the approximating properties of the real frame chosen would persist over a range of frequencies.

Such an approximating real frame could be generated using the ALIGN algorithm, just as in the procedure for the manipulation of characteristic loci via an approximately-commutative controller. One further important factor is however involved with this use of the ALIGN procedure which was not present in the previous case. This is that the *signs* of the real vectors involved in the approximating real frame must be adjusted, after exit from the ALIGN procedure, in order to conform with the requirements on the characteristic loci for satisfaction of the generalized Nyquist stability criterion and to obtain satisfactory patterns of root-locus asymptotes (see § 6.1). Such an adjustment of signs is not required when the *two* frames A and B are used in an approximately-commutative controller arrangement, since A and B tend naturally to be approximately dual since they are individual approximants to the dual frames $W(j\omega)$ and $V(j\omega)$. The adjustment of the solution frames obtained from the ALIGN algorithm is discussed in Appendix 2.

5.3. *Basic design procedure*

We will first consider the case where the closed-loop performance specification does not require a very wide bandwidth for tight closed-loop control. The design can then be broken into two distinct parts corresponding, respectively, to high-frequency and low-frequency conditions.

(i) *High-frequency controller behaviour.* At high frequencies we wish to reduce the misalignment angles between the compensated system characteristic direction set and the standard basis set. This will be done by designing a real controller K_h which approximates the frame matrix of $G^{-1}(j\omega)$ at some appropriately high frequency ω_h.

(ii) *Low-frequency controller behaviour.* At low frequencies we wish to manipulate the characteristic loci in such a way that :

(*a*) the encirclement criteria are satisfied for stability, and
(*b*) the gain criteria are satisfied for performance.

This will be done by designing an appropriate approximately-commutative controller K_l by means of which the required characteristic loci manipulations may be made.

Since the controller design has been split up into two distinct frequency regions in this way, it is therefore necessary to devise a way of combining the effects of K_l and K_h so that each comes into proper operation in the appropriate frequency range. In particular, it is essential to ensure that K_h does not upset the stability characteristics imposed via K_l, and that K_l does not upset the reduction in high-frequency interaction achieved by K_h. The frequency-decoupling of the two controllers can be achieved in a simple way by making the low-frequency controller a matrix-integral controller. The complete controller then becomes of matrix-proportional-plus-integral form :

$$K(s) = \frac{\alpha}{s} K_l + K_h \tag{5.4}$$

where α is a constant chosen to achieve a suitable transition from low-frequency to high-frequency behaviour of $K(s)$. The presence of the integral term $1/s$ ensures that :

$$K(j\omega) \simeq \frac{\alpha}{j\omega} K_l \quad \text{at low frequencies,}$$

and

$$K(j\omega) \simeq K_h \qquad \text{at high frequencies.}$$

It also ensures that there will be zero steady-state error in the closed-loop steady-state response, since all the characteristic loci will have gains tending to infinity as the angular frequency ω tends to zero.

The case where wider closed-loop bandwidth is required will necessitate a more careful treatment of the intermediate frequency range ; this is considered in the following sub-section.

An alternative approach to the design arises from rewriting the combined controller expression as

$$\begin{aligned} K(s) &= K_h \left[\frac{\alpha}{s} K_h^{-1} K_l + I \right] \\ &= K_h \left[\frac{\alpha}{s} K_l' + I \right] \end{aligned} \tag{5.5}$$

where

$$K_l' = K_h^{-1} K_l \tag{5.6}$$

From this point of view, the high-frequency portion K_h would be designed in the same way as previously described, and then the product $G(s)K_h$ would be treated as a new plant (the high-frequency compensated plant) for which an appropriate approximately-commutative controller K_l' would then be designed.

5.4. *Extended procedure for the intermediate frequency range*

If dynamic compensation is required at intermediate frequencies, then a medium-frequency range controller factor $K_m(s)$ can be added to give an overall controller structure of the form

$$K(s) = \frac{\alpha}{s} K_l + K_h K_m(s) \tag{5.7}$$

The job of the controller factor $K_m(s)$ is to inject phase-advance compensation into the compensated-system characteristic loci over the intermediate frequency range. This will be again done by the approximately-commutative technique of using the ALIGN procedure to generate a real frequency-independent frame which approximates the characteristic direction frame of $G(s)$ at a suitable point in the intermediate frequency range. The frequency-conscious characteristic transfer functions of $K_m(s)$ will be those lead/lag transfer functions which will make the required gain and phase adjustments to the characteristic loci of $G(s)K_h$ at the intermediate frequency ω_m.

In choosing these compensating transfer functions, however, we must ensure that

$$K_m \simeq I \quad \text{for } \omega \geqslant \omega_h \tag{5.8}$$

so that, over the whole frequency range the composite controller will have the following pattern of behaviour :

$$K(s) = \frac{\alpha}{s} K_l \qquad \text{for low frequencies ;}$$

$$K(s) \simeq K_h K_m(s) \quad \text{for intermediate frequencies ;}$$

$$K(s) \simeq K_h \qquad \text{for high frequencies.}$$

The complete design procedure then becomes as follows. The overall controller can be written in the form :

$$K(s) = K_h K_m(s) \left[\frac{\alpha}{s} K_l'' + I \right] \tag{5.9}$$

where

$$K_l'' = K_m^{-1}(s) K_h^{-1} K_l \tag{5.10}$$

In interpreting this last term, it is useful to remember that $K_m(s)$ will tend to a constant matrix at those low frequencies at which K_l'' will be computed, so that no significant dynamic variation of K_l'' is implied and this matrix may therefore be treated as if it were a constant matrix. With this arrangement of the overall controller, we then proceed in the following stages.

(i) K_h is designed first along the lines previously set out.

(ii) The high-frequency-compensated plant $G(s)K_h$ is formed and an approximately-commutative controller $K_m(s)$ designed to provide the required amounts of phase-advance in the characteristic loci over the intermediate frequency range.

(iii) K_l'' is then designed using an approximately-commutative controller approach to achieve the following pair of objectives :

(*a*) manipulate the characteristic loci at low frequencies to obtain the required number of encirclements of the critical point necessary for stability ; and

(*b*) to balance up the characteristic gains in the low frequency range.

(iv) The final form of controller is then put together using eqn. (5.9).

The design technique which has been introduced is a natural generalization to the multivariable case of the classical Bode–Nyquist scalar frequency response approach. It is complete and systematic and aims at :

(i) the stabilization of the closed-loop system ;
(ii) the reduction of interaction ; and
(iii) the improvement of closed-loop dynamic behaviour.

Due to the approximately-commutative approach adopted, the resulting controller will have simple elements and be easily realizable.

A design example is next given, based on realistic industrial plant data.

6. Example

The example used here (Kouvaritakis, 1974) to illustrate the use of the design technique is based on a model of an open-loop unstable chemical reactor. Its state-space description is given by the set of matrices

$$\left.\begin{aligned} A &= \begin{bmatrix} 1\cdot 38 & -0\cdot 2077 & 6\cdot 715 & -5\cdot 676 \\ -0\cdot 5814 & -4\cdot 29 & 0 & 0\cdot 675 \\ 1\cdot 067 & 4\cdot 273 & -6\cdot 654 & 5\cdot 893 \\ 0\cdot 048 & 4\cdot 273 & 1\cdot 343 & -2\cdot 104 \end{bmatrix} \\ B &= \begin{bmatrix} 0 & 0 \\ 5\cdot 679 & 0 \\ 1\cdot 136 & -3\cdot 146 \\ 1\cdot 136 & 0 \end{bmatrix}, \quad C = \begin{bmatrix} 1 & 0 & 1 & -1 \\ 0 & 1 & 0 & 0 \end{bmatrix} \end{aligned}\right\} \tag{6.1}$$

corresponding to a transfer-function matrix

$$G(s) = \frac{1}{\Delta_0(s)} \begin{bmatrix} \Gamma_{11}(s) & \Gamma_{12}(s) \\ \Gamma_{21}(s) & \Gamma_{22}(s) \end{bmatrix} \tag{6.2}$$

where

$$\Delta_0(s) = s^4 + 11\cdot 67s^3 + 15\cdot 75s^2 - 88\cdot 31s + 5\cdot 514$$

$$\Gamma_{11}(s) = 29\cdot 2s + 263\cdot 3$$

$$\Gamma_{12}(s) = -3\cdot 146s^3 - 32\cdot 62s^2 - 89\cdot 83s - 31\cdot 81$$

$$\Gamma_{21}(s) = 5\cdot 679s^3 + 42\cdot 67s^2 - 68\cdot 84s - 106\cdot 8$$

$$\Gamma_{22}(s) = 9\cdot 43s + 15\cdot 15$$

The set of poles is $\{0\cdot 06318,\ 1\cdot 991,\ -5\cdot 057,\ -8\cdot 666\}$. Since there are two poles in the right-half plane, the generalized Nyquist stability theorem will require for stability that the characteristic loci must have a net sum of two anti-clockwise encirclements (if outputs are to be fed back through unit negative gains to correspondingly-numbered inputs). The state-space model parameters are such that

$$CB = \begin{bmatrix} 0 & -3\cdot 146 \\ 5\cdot 679 & 0 \end{bmatrix} \tag{6.3}$$

This is of full rank and so $G(s)$ will have a number of finite zeros equal to the number of states minus the number of outputs, which is two. These zeros are found as the eigenvalues of the matrix

$$NAM = \begin{bmatrix} -4{\cdot}296 & 1{\cdot}039 \\ 2{\cdot}221 & -1{\cdot}935 \end{bmatrix} \tag{6.4}$$

where N and M are defined by Kouvaritakis and MacFarlane (1976)

$$NB=0, \quad CM=0, \quad NM=I_{n-m} \tag{6.5}$$

The set of finite zeros is thus $\{-1{\cdot}192,\ -5{\cdot}039\}$. There are no finite zeros in the right-half plane and therefore no serious difficulty in exercising feedback control is anticipated.

The asymptotic behaviour of the multivariable root loci is determined by the eigenvalues of CB, which are $\pm j4{\cdot}227$. Thus these asymptotes will be parallel to the imaginary axis of the complex plane. Using the techniques for asymptotic location developed by Kouvaritakis and Shaked (1976), these asymptotes can be shown to intersect the real axis at the point $(-2{\cdot}72, 0)$.

The equation defining the characteristic gain function $g(s)$

$$\det\,[gI_2 - G(s)] = 0 \tag{6.6}$$

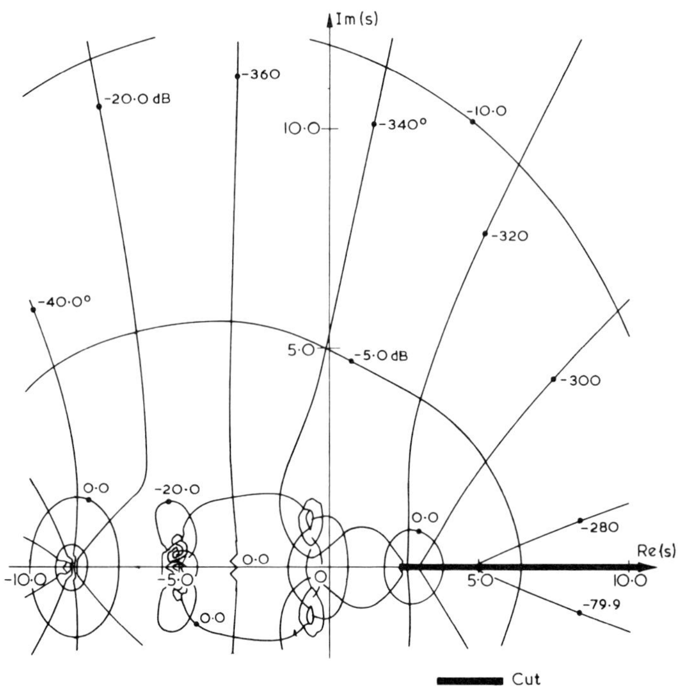

Figure 4. Sheet 1 of Riemann surface for system of eqn. (6.2). Only the cut along the positive real axis is shown.

gives

$$\Delta_0(s)g^2 - (38{\cdot}63s + 278{\cdot}45)g + 17{\cdot}866(s^2 + 6{\cdot}231s + 6{\cdot}006) = 0 \qquad (6.7)$$

Note that (MacFarlane and Postlethwaite 1977) the poles are the zeros of the polynomial coefficient of g^2 and the zeros are the zeros of the polynomial coefficient of g^0 in this expression. The polynomial on the left-hand side of eqn. (6.7) is irreducible over the field of rational functions in s and hence the characteristic gain function $g(s)$ is associated with a Riemann surface formed by joining together two copies of the s-plane. This is shown in Figs. 4 and 5 in the conventional manner of two copies of the s-plane joined together along appropriate cuts and overlaid with constant-modulus and constant-phase contours of $g(s)$. Since the cut shown runs from the branch point at $s_0 = 2{\cdot}374$ along the real axis to infinity it is obvious that (MacFarlane and Postlethwaite 1977) $G(s)$ will only have one characteristic locus. This locus, however, because the system is proper, will go through the origin of the image space at $s = \infty$. As a consequence, it has the appearance of two closed curves touching at a common point at $s = \infty$. This pair of curves will be called the characteristic loci of $G(s)$ and are partially shown in Figs. 6 and 7. These diagrams show only the

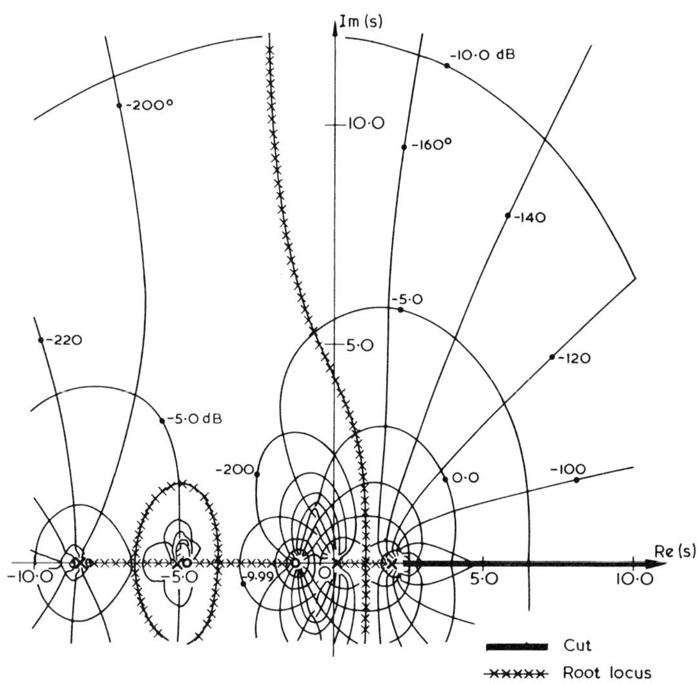

Figure 5. Sheet 2 of Riemann surface for system of eqn. (6.2). Only the cut along the positive real axis is shown. Zeros shown thus :○ and are terminal points of some branches of the loci. Open-loop poles shown thus : × and are the starting points of the loci.

image of the positive part of the imaginary axis ; the remainder of the loci are simply mirror images in the real axis of those parts shown. Figure 7 shows the behaviour of both loci in the vicinity of the critical point.

An inspection of the characteristic loci of the uncompensated plant shows that neither characteristic locus encircles the critical point. Thus this plant would be closed-loop unstable under direct negative feedback of outputs to correspondingly-numbered inputs. Further consideration of the characteristic loci shows that such a negative feedback with equal gains in each channel of

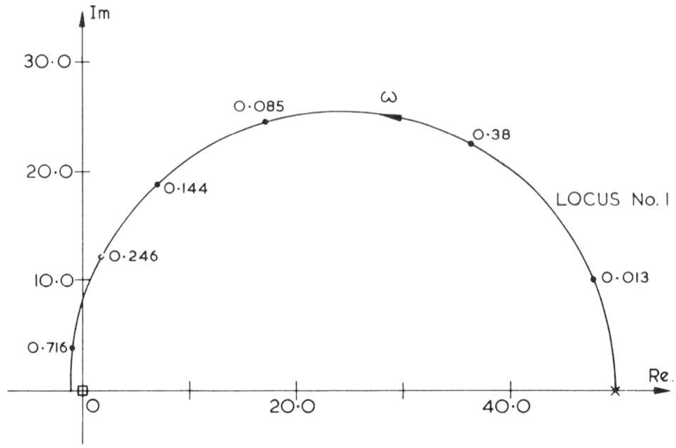

Figure 6. Characteristic locus No. 1 for system of eqn. (6.2).

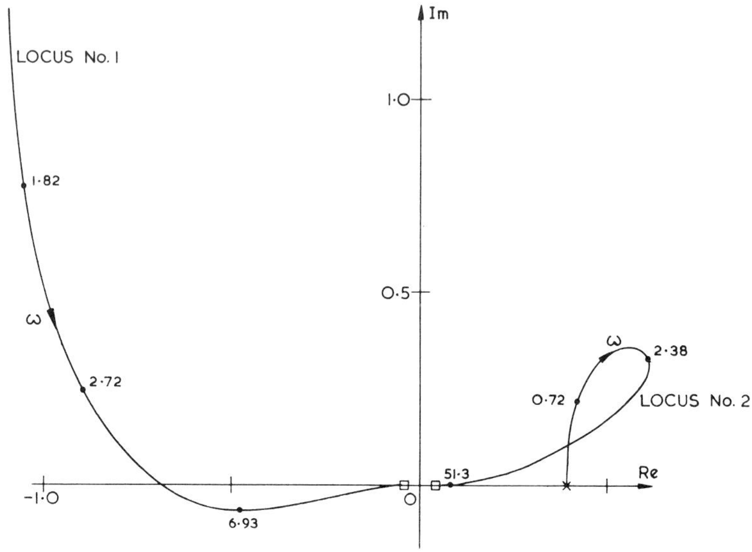

Figure 7. Characteristic loci for system of eqn. (6.2).

greater than about 1·5 would result in two anti-clockwise encirclements by one of the loci (see Fig. 7). Such a feedback scheme would, however, have fairly poor phase margins, and would at best give a highly oscillatory behaviour of the outputs in response to a step input or disturbance. This may be confirmed by an inspection of the multivariable root loci contained in Figs. 4 and 5. In these diagrams of the Riemann surface showing the behaviour of the characteristic gain function $g(s)$, the multivariable root loci are the 180° constant phase contours for $g(s)$ (shown by crosses). It can be seen that for high gains the feedback system is stable but that the root loci asymptotes, as predicted above by the analysis of their behaviour based on the matrix CB, run parallel to the imaginary axes of the frequency planes out of which the Riemann surface is built.

Therefore, although the system can be made stable under simple direct proportional feedback, it is obvious that some form of compensator is required. The need for compensation is also indicated by the high-frequency interaction caused by the form of $G(s)$. The plant characteristic direction misalignments are shown in Figs. 8 and 9. These show an unacceptably high degree of misalignment in the high-frequency range above 1 rs^{-1}. Hence a compensator should be chosen which makes a suitable adjustment of the high-frequency alignment of the eigenvectors of the compensated plant.

6.1. *High-frequency compensation*

The controller design process starts, as explained in § 5.4, by finding a real frame matrix K_h which approximates the complex matrix $G^{-1}(j\omega)$ at some suitably high values of frequency. This is done by using the ALIGN algorithm to approximate the complex columns of $G^{-1}(j\omega_h)$ by the real columns of K_h at the frequency ω_h. An investigation of the frequency range above 10 rs^{-1} showed that the best results were obtained at an $\omega_h = 30$ rs^{-1} where we found

$$K_h = \begin{bmatrix} 0 & 1 \\ 1 & 0 \end{bmatrix} \begin{bmatrix} \pm 1 & 0 \\ 0 & \pm 1 \end{bmatrix} \tag{6.8}$$

The sign ambiguity is a reminder that the column signs of K_h must be adjusted to give the best results from the point of view of closed-loop stability margins. A most useful way of uniquely defining the signs of the columns of K_h is via the asymptotic behaviour of the compensated system's root loci. To improve stability margins we must change the directions of the asymptotes from being parallel to the imaginary axis in the frequency plane to being parallel to the negative real axis in the frequency plane. For this to be the case the matrix product CBK_h must have positive real eigenvalues which is evidently so if we choose

$$K_h = \begin{bmatrix} 0 & 1 \\ -1 & 0 \end{bmatrix}$$

This choice will ensure that, under direct scalar feedback through equal gains, two of the system's closed-loop poles will migrate to the zeros at $-1{\cdot}192$ and $-5{\cdot}039$, and the remaining two characteristic frequencies will migrate to

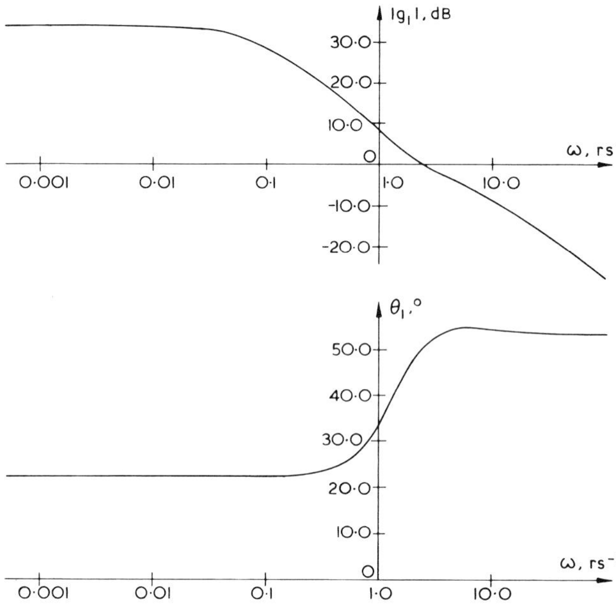

Figure 8. Misalignment angle plot for system of eqn. (6.2).

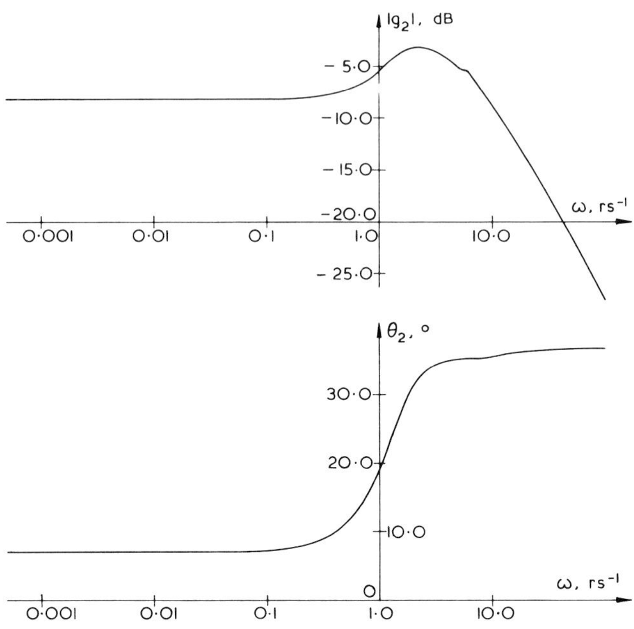

Figure 9. Misalignment angle plot for system of eqn. (6.2).

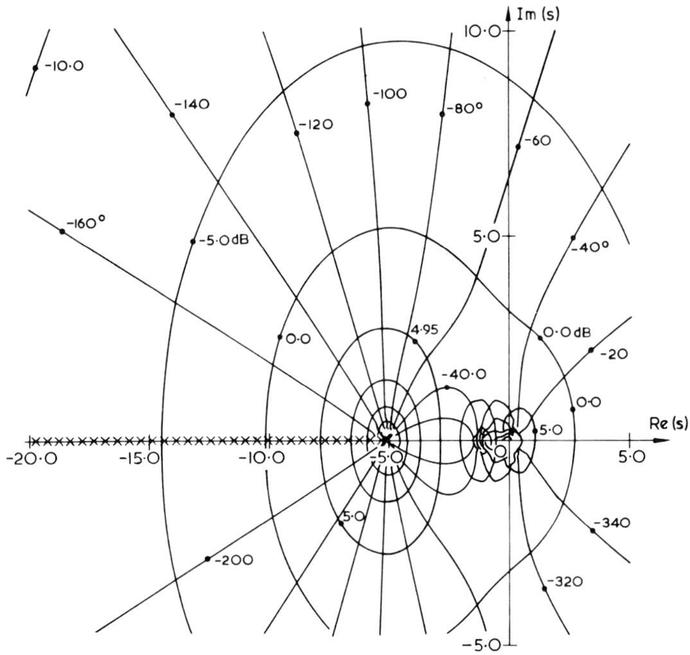

Figure 10. Sheet 1 of Riemann surface for system modified by pre-compensator $K_h = \begin{bmatrix} 0 & 1 \\ -1 & 0 \end{bmatrix}$. No cuts shown. Root locus shown by × × × ×.

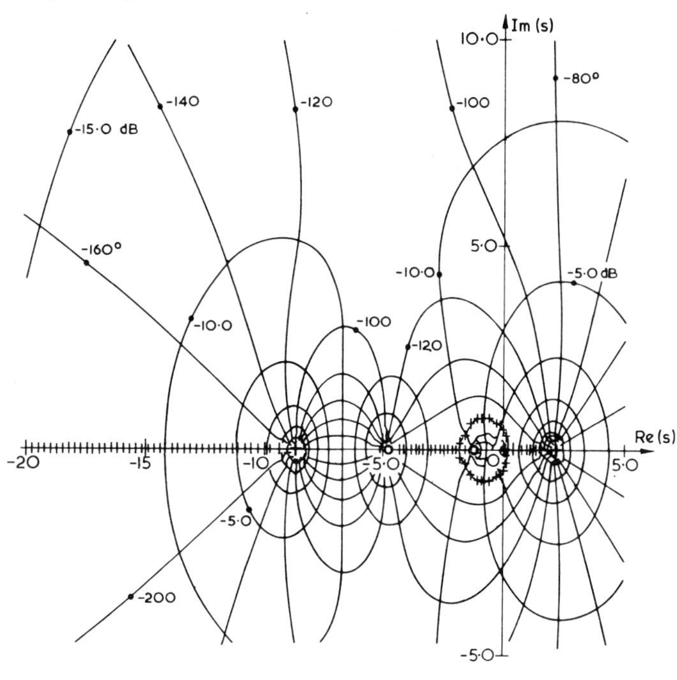

Figure 11. Sheet 2 of Riemann surface for system modified by pre-compensator $K_h = \begin{bmatrix} 0 & 1 \\ -1 & 0 \end{bmatrix}$. No cuts shown. Root locus shown by + + + +.

infinity along negative real axes of the frequency planes out of which the Riemann surface is built. That this is indeed the case can be seen by an inspection of the frequency-surface for the compensated plant shown in Figs. 10 and 11. In particular on one sheet we see a closed-loop characteristic frequency starting at the pole located at $-5{\cdot}057$ and moving off to infinity along a negative real axis with increasing value of the scalar gain parameter. On the other sheet we have the closed-loop characteristic frequency which starts from the pole located at $-8{\cdot}666$ moving off to infinity along a negative real axis, while the remaining two closed-loop characteristic frequencies starting from the poles at $0{\cdot}06318$ and $1{\cdot}991$ move over to the left-half frequency plane in a typical root-locus pattern to end up on the zeros at $-1{\cdot}192$ and $-5{\cdot}039$.

The improvement in stability margins achieved by the high-frequency compensator K_h must naturally be reflected in the forms of the characteristic loci of the compensated plant, and this is confirmed by an examination of the characteristic loci for $G(s)K_h$, partially shown in Figs. 12 and 13. An inspection of these loci shows that the generalized Nyquist criterion is satisfied for direct negative feedback through unit values of gain. Furthermore the characteristic loci have infinite gain margins, so that a controller matrix of the form kI_2 with arbitrarily large values of the scalar gain parameter k could be imposed on the compensated plant represented by $G(s)K_h$ while still maintaining the two anti-clockwise encirclements of the critical point required for closed-loop stability. Also the phase characteristics of the characteristic loci for the compensated plant $G(s)K_h$ are seen to be satisfactory over the intermediate frequency range. For this plant therefore, no dynamic compensation is required over the intermediate frequency range.

Before moving on to the last stages of the design we must investigate whether the alignment of the eigenvectors of $G(j\omega)K_h$ with the standard basis set at high values of ω is satisfactory or not. The relevant misalignment angles

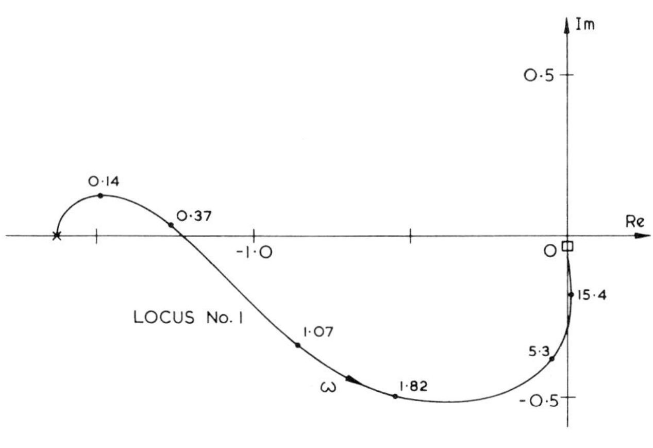

Figure 12. Characteristic locus No. 1 for system modified by pre-compensator

$$K_h = \begin{bmatrix} 0 & 1 \\ -1 & 0 \end{bmatrix}.$$

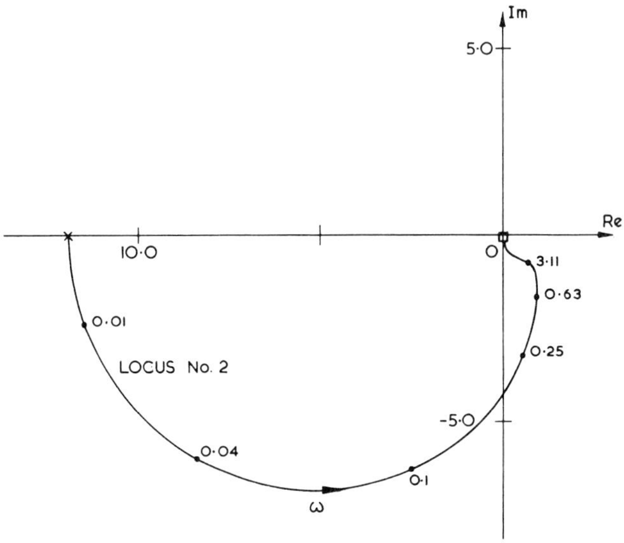

Figure 13. Characteristic locus No. 2 for system modified by pre-compensator

$$K_h = \begin{bmatrix} 0 & 1 \\ -1 & 0 \end{bmatrix}.$$

are displayed in Figs. 14 and 15. These show that the misalignment angles, θ_1° and θ_2°, tend to zero with increasing frequency and fall to acceptably low values for frequencies greater than 5 rs^{-1}. Hence we would expect the suppression of high-frequency interaction provided by the high-frequency compensator K_h to be adequate.

6.2. *Low-frequency compensation*

We now proceed to the final stage of the design. The discrepancy in size of the characteristic loci shows that it would be useful to balance up the gains of the characteristic loci at low frequencies. This will make the most efficient use of any final overall gain injection used to improve response speed, and to maintain an adequate level of interaction rejection at frequencies where the misalignment angles of the characteristic directions are becoming significant. The particular action taken here will be to design an approximately-commutative controller which roughly doubles the modulus of the smallest characteristic locus at a frequency of about $\omega = 1\ \mathrm{rs}^{-1}$. An application of the frame alignment algorithm to the frame matrix $G(s)K_h$ at a frequency of $1{\cdot}12\ \mathrm{rs}^{-1}$ gives the approximating frames

$$A(j1{\cdot}12) = \begin{bmatrix} 0{\cdot}584 & 0{\cdot}997 \\ -0{\cdot}812 & -0{\cdot}0754 \end{bmatrix} \tag{6.9}$$

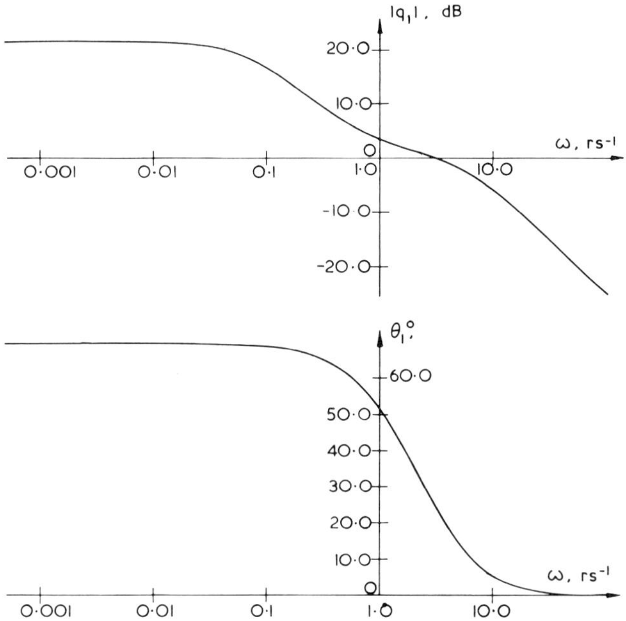

Figure 14. Misalignment angle plot for system modified by pre-compensator

$$K_h = \begin{bmatrix} 0 & 1 \\ -1 & 0 \end{bmatrix}.$$

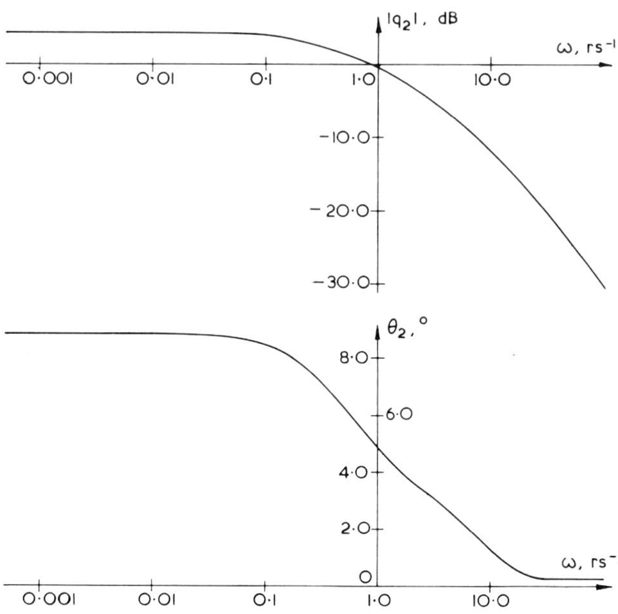

Figure 15. Misalignment angle plot for system modified by pre-compensator

$$K_h = \begin{bmatrix} 0 & 1 \\ -1 & 0 \end{bmatrix}.$$

and

$$B(j1{\cdot}12)=\begin{bmatrix} -0{\cdot}0754 & -0{\cdot}997 \\ 0{\cdot}812 & 0{\cdot}584 \end{bmatrix}$$

$$=0{\cdot}776A^{-1}(j1{\cdot}12) \qquad (6.10)$$

Re-scaling $B(j1{\cdot}12)$, so that it becomes the inverse of $A(j1{\cdot}12)$, then gives the required low-frequency controller term as

$$K_l = A(j1{\cdot}12)\begin{bmatrix} 1 & 0 \\ 0 & 2 \end{bmatrix} B(j1{\cdot}12)$$

$$=\begin{bmatrix} 2{\cdot}06 & 0{\cdot}81 \\ -0{\cdot}08 & 0{\cdot}94 \end{bmatrix} \simeq \begin{bmatrix} 2{\cdot}1 & 0{\cdot}8 \\ 0 & 0{\cdot}9 \end{bmatrix} \qquad (6.11)$$

The overall controller design is now concluded by an appropriate combination of the low and high frequency properties of K_l and K_h, respectively, in the form

$$K(s)=K_h\left(\frac{\alpha}{s}K_l+I\right)$$

where the best value of α is selected by a process of trial and error. The selection of a value of $\alpha=1$ gives

$$K(s)=\begin{bmatrix} 0 & 1 \\ -1 & 0 \end{bmatrix}\begin{bmatrix} \dfrac{s+2{\cdot}1}{s} & \dfrac{0{\cdot}8}{s} \\ 0 & \dfrac{s+0{\cdot}9}{s} \end{bmatrix}$$

$$=\begin{bmatrix} 0 & \dfrac{s+0{\cdot}9}{s} \\ -\dfrac{(s+2{\cdot}1)}{s} & -\dfrac{0{\cdot}8}{s} \end{bmatrix} \qquad (6.12)$$

and injecting a further gain of 10 in each loop results in a final controller design of the form

$$K(s)=\begin{bmatrix} 0 & \dfrac{10s+9}{s} \\ -\dfrac{(10s+21)}{s} & -\dfrac{8}{s} \end{bmatrix} \qquad (6.13)$$

6.3. *Survey of final design*

The final characteristic loci and characteristic vector misalignment diagrams are shown in Figs. 16, 17 and 18. These show that the compensated system now exhibits good gain and phase margins, has a good bandwidth over which characteristic gains are high, has well-balanced low-frequency gains and low

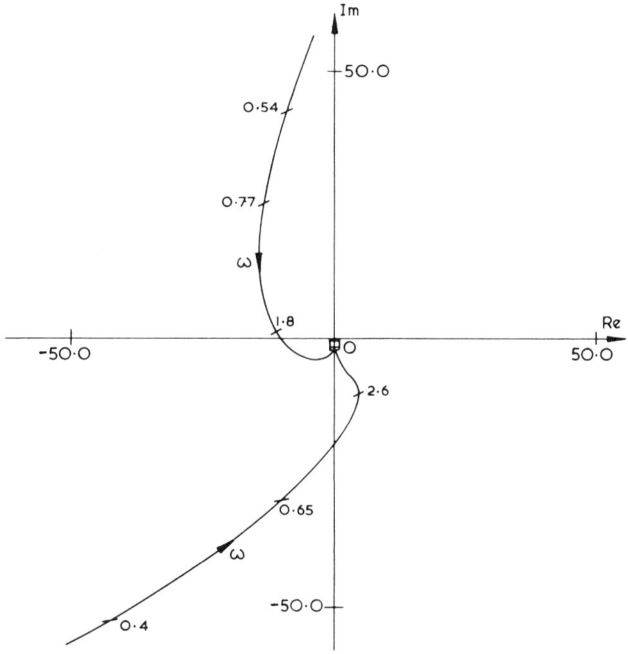

Figure 16. Final characteristic loci.

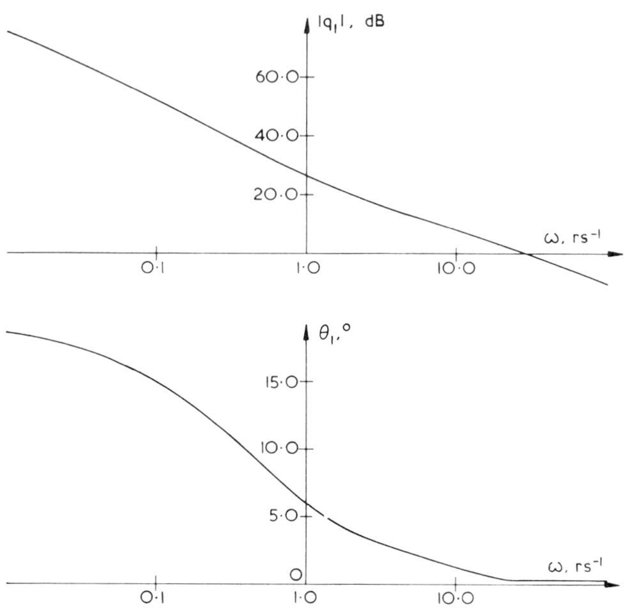

Figure 17. Final misalignment angle plot.

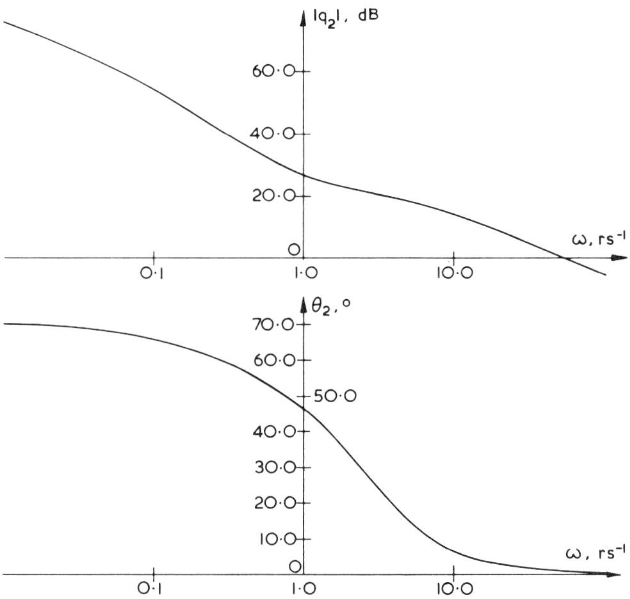

Figure 18. Final misalignment angle plot.

high-frequency misalignment angles. Hence the closed-loop system behaviour should be such that a fast and non-oscillatory step response is obtained for both reference inputs, with fairly low interaction during the transient regime, and accurate steady-state behaviour. This expectation is borne out by the digitally-simulated closed-loop step responses shown in Figs. 19 and 20.

Figure 19 shows the response of both system outputs to a unit step reference demand for the first controlled variable alone. Figure 20 shows the responses for a unit step demand on the second controlled variable only. There remains a certain amount of interaction in this second set of responses, and this is one aspect of the design which could be improved by further compensation if the improvement was felt to justify a more complicated form of compensator. The form of transient interaction shown corresponds to a predominant frequency term around the region where a transition from high to low values occurs in the misalignment angle θ_2 shown in Fig. 18. The associated characteristic gain falls below a level of 20 dB in this frequency region, and this is the root of the lack of interaction suppression. A simple remedy for this situation is to double the gain at each input, raising the final adjustment of controller gain referred to in § 6.2 above, and resulting in the $K(s)$ quoted in eqn. (6.13), from 10 to 20. The correspondingly altered form of controller thus becomes

$$K(s)=\begin{bmatrix} 0 & \dfrac{20s+18}{s} \\ -\dfrac{(20s+42)}{s} & -\dfrac{16}{s} \end{bmatrix} \tag{6.14}$$

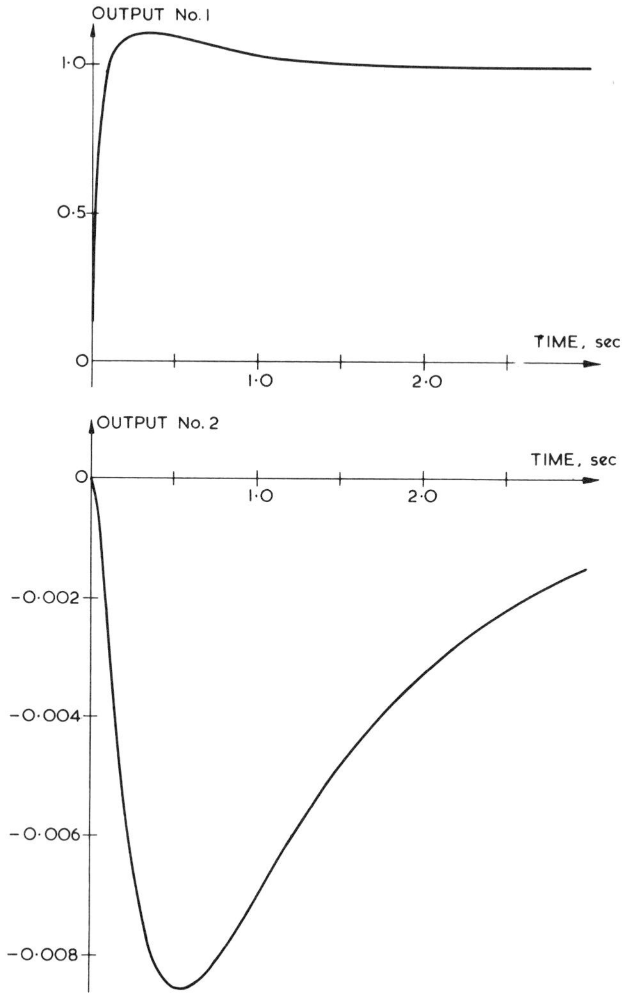

Figure 19. Response to step input on channel 1 with $K(s)$ given by eqn. (6.13).

This will raise both characteristic gains by 6 dB, and thus would compensate for the lack of alignment which is causing the interaction in the intermediate frequency range. Naturally this will also result in a speeding-up of the closed-loop transient response and a higher degree of accuracy as the steady state is approached. This is borne out by the corresponding simulated step-response behaviour which is shown in Figs. 21 and 22.

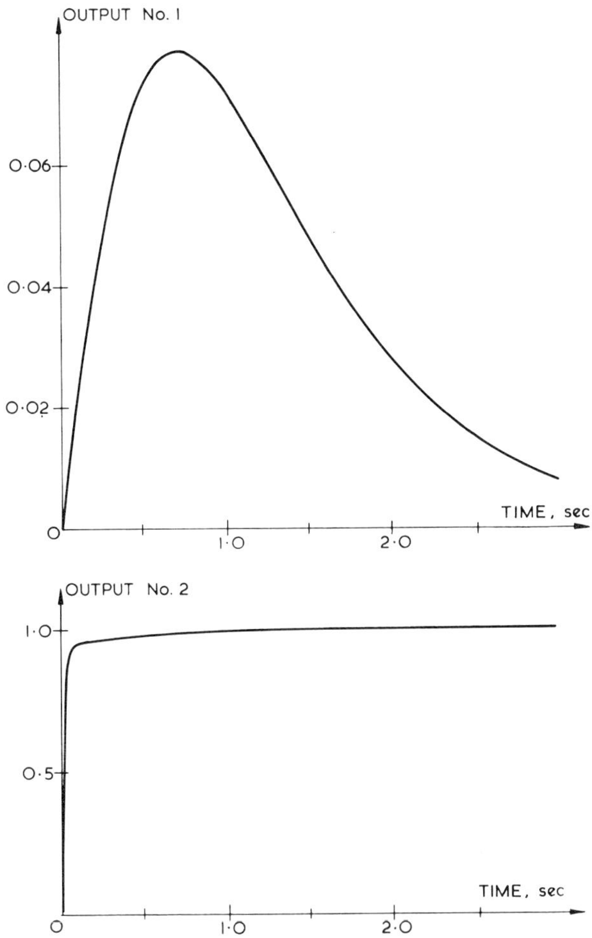

Figure 20. Response to step input on channel 2 with $K(s)$ given by eqn. (6.13).

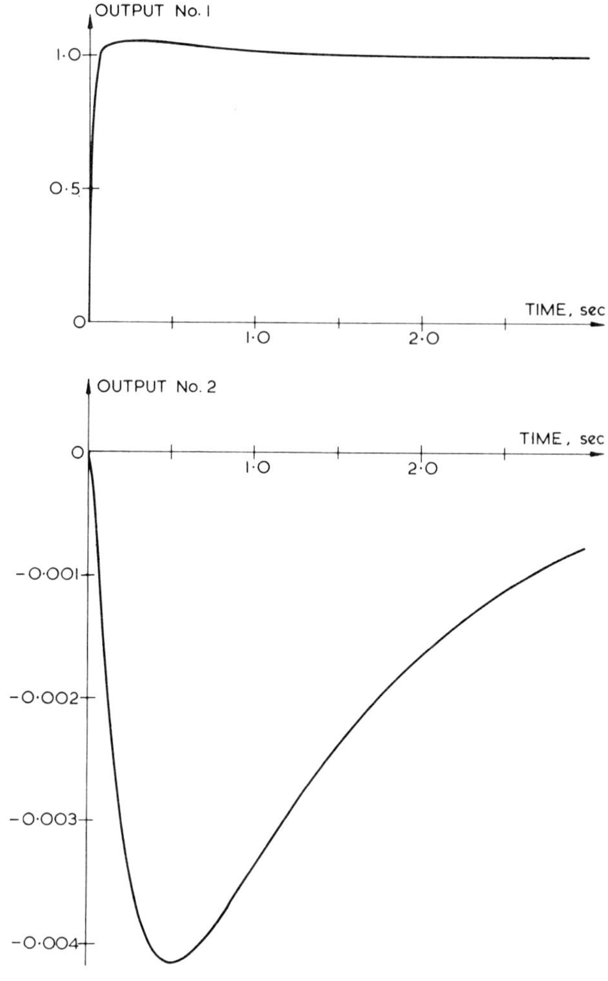

Figure 21. Response to step input on channel 1 with $K(s)$ given by eqn. (6.14).

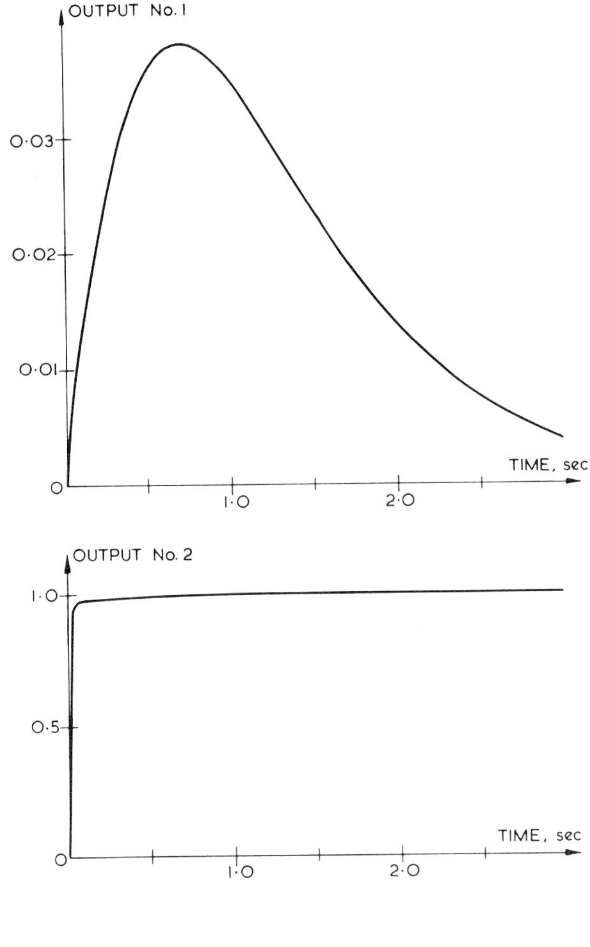

Figure 22. Response to step input on channel 2 with $K(s)$ given by eqn. (6.14).

Appendix 1

Consider the real non-zero vector μ and form the product $\mu^t C_i \mu$. Reversing the steps taken in deriving eqns. (4.11) and (4.12) from eqns. (4.9) and (4.10) in § 4.1 it readily follows that

$$\mu^t C_i \mu = |(v_i, \mu)|^2 \geqslant 0 \tag{A 1}$$

Now the matrix C_i is real symmetric and therefore it will have real eigenvalues, λ_k say, and real corresponding eigenvectors, u_k say. Letting $\mu = u_k$ in the above relation we obtain

$$\begin{aligned} u_k{}^t C_i u_k &= \lambda_k u_k{}^t u_k \\ &= |(v_i, u_k)|^2 \\ &\geqslant 0 \end{aligned} \tag{A 2}$$

The scalar product $u_k{}^t u_k$, however, is always positive and therefore the eigenvalues λ_k of C_i are non-negative.

Following similar arguments it can be easily shown that the eigenvalues of D_i are all non-negative. Thus for any real non-zeros vector μ we have that

$$\begin{aligned} \mu^t D_i \mu &= \sum_{\substack{j=1 \\ j \neq i}}^{m} |(v_j, \mu)|^2 \\ &\geqslant 0 \end{aligned} \tag{A 3}$$

Furthermore if μ is chosen to be one of the real eigenvectors of D_i, say $\tilde{u}_k$, corresponding to the real eigenvalue $\tilde{\lambda}_k$ then

$$\begin{aligned} \tilde{u}_k{}^t D_i \tilde{u}_k &= \tilde{\lambda}_k \tilde{u}_k{}^t \tilde{u}_k \\ &= \sum_{\substack{j=1 \\ j \neq i}}^{m} |(v_j, \tilde{u}_k)|^2 \\ &\geqslant 0 \end{aligned} \tag{A 4}$$

The eigenvalues of D_i therefore cannot be negative.

The equality sign in the above relation will only hold if the eigenvalue $\tilde{\lambda}_k$ associated with $\tilde{u}_k$ were zero. In such a case (A 4) would reduce to

$$\sum_{\substack{j=1 \\ j \neq i}}^{m} |(v_j, \tilde{u}_k)|^2 = 0 \tag{A 5}$$

which of course is equivalent to

$$(v_j, \tilde{u}_k) = 0 \quad \text{for } j = 1, 2, \ldots, m, \ j \neq i \tag{A 6}$$

The implication of this in turn is that $\tilde{u}_k$ is in perfect alignment with the corresponding eigenvector of $G(s)$, $w_i(s)$. The vector $\tilde{u}_k$, however, is assumed to be real and thus D_i will be rank deficient only if the eigenvector we are trying to approximate is already real. For this reason then D_i may be considered to be non-singular and hence the arguments used in § 4.1 in connection with the rank of D_i are fully justified.

Appendix 2

First we note that when ALIGN is being used for the design of approximately commutative controllers then, denoting by A_0 and B_0 a particular pair of solutions for the frames A and B as computed by the algorithm, we may generate a whole family of aligning manifolds by computing the columns and rows of matrices

$$\left.\begin{aligned} &A = A_0 \text{ diagonal } \{\lambda_i\} \\ \text{and} \\ &B = \text{diagonal } \{\mu_i\} \, B_0 \quad \text{for } i = 1, 2, \ldots, m \end{aligned}\right\} \tag{A 7}$$

respectively, where λ_i and μ_i are any non-zero constants (real or complex). It should be understood, however, that because of the commutativity of diagonal matrices, as far as the approximate commutative controller K is concerned it is not the individual choice of the λ_i's and μ_i's that matters but rather the choice of the one set of scaling factors with respect to the other. Thus we may write

$$\begin{aligned} K &= A \Lambda_k(s) B \\ &= A_0 \text{ diag } \{\lambda_i\} \, \Lambda(s) \text{ diag } \{\mu_i\} \, B_0 \\ &= A_0 \Lambda(s) \text{ diag } \{\nu_i\} \, B_0 \end{aligned} \tag{A 8}$$

where $\nu_i = \lambda_i \mu_i$ for $i = 1, 2, \ldots, m$. The implication of this of course is that without loss of generality we need only adjust the scaling of the rows of B_0, say, and accept the solution for A_0 with an arbitrary scaling (for example the scaling produced by the ALIGN algorithm). The choice of the scaling factors ν_i then clearly must be such that

$$\begin{aligned} A_0 \text{ diag } \{\nu_i\} \, B_0 &\simeq \text{diag } \{\nu_i\} \, B_0 A_0 \\ &\simeq I_m \end{aligned} \tag{A 9}$$

where I_m is a unit matrix of order m.

Now it turns out that for the 2×2 case ($m = 2$) a set of factors ν_i can always be found such that eqn. (A 9) is satisfied exactly. This is the case because for systems with two inputs and two outputs the first row vector of B_0 is always orthogonal to the second column vector of A_0 and the second row vector of B_0 is always orthogonal to the first column vector of A_0.

In the case of $m \geqslant 3$ it is no longer true to say that $B_0 A_0$ will always be diagonal as this depends on the original scaling of the eigenvectors of $G(s)$. The exhaustive treatment of this point is involved and is not given here. Instead we propose two alternative methods of overcoming this difficulty. First the designer may normalize the eigenvectors of $G(s)$, then use ALIGN to obtain the frame A_0 that forms the best real approximation to the eigenvectors and finally select the operators A and B of the commutative controller to be $A = A_0$ and $B = A_0^{-1}$. Alternatively we may wish to initialize the synthesis of the approximately commutative controller by selecting the eigenvector scaling factors in such a way as to obtain a product $B_0 A_0$ which is near diagonal. This could be achieved by either some trial and error method or by some computer-automated hill-climbing technique. The final choice for A and B would be the result of appropriate scaling of the rows of B_0 or the columns of A_0 or both, so that $B_0 A_0$ becomes nearly equal to the identity matrix.

References

Bliss, G. A., 1966, *Algebraic Functions*, reprint of 1933 original (New York : Dover).
Kouvaritakis, B., 1974, Ph.D. Thesis, University of Manchester.
Kouvaritakis, B., and MacFarlane, A. G. J., 1976, *Int. J. Control*, **23**, 149.
Kouvaritakis, B., and Shaked, U., 1976, *Int. J. Control*, **23**, 297.
MacFarlane, A. G. J., and Belletrutti, J. J., 1973, *Automatica*, **9**, 575.
MacFarlane, A. G. J., and Postlethwaite, I., 1977, *Int. J. Control*, **25**, 81.
Marcus, M., and Minc, H., 1964, *A Survey of Matrix Theory and Matrix Inequalities* (Boston : Allyn & Bacon).

Characteristic frequency–gain design study of an automatic flight control system

B. KOUVARITAKIS, W. MURRAY
and A. G. J. MacFARLANE

Recent work on the application of complex variable theory to the analysis of linear multivariable systems has led to a design approach that integrates the multivariable frequency-response and root-locus techniques. A brief discussion of a two-stage design method that derives from this approach is first undertaken and its application to an automatic flight control system is then considered. The design starts with the introduction of an inner-loop compensator for which root-locus techniques are employed for the adjustment of closed-loop characteristic frequencies. This is then followed by an outer-loop gain injector design where frequency-response techniques are employed for the adjustment of open-loop characteristic gains. The overall scheme results in a controlled system whose response is fast, non-oscillatory, accurate in the steady state and largely non-overshooting and non-interactive.

1. Introduction

This paper considers the application of recently developed complex-variable design methods to an automatic flight control system. The system model is given in terms of a linearized state–space equation set describing the motion of an aircraft in the vertical plane when small perturbations about a datum flight condition are considered. The model has three control inputs, three commanded outputs and five states, all of which are accessible for measurement. Generalized frequency-response and root-locus techniques are used in a two-stage design approach ; the two design stages are associated with an inner-loop compensator and an outer-loop gain injector. The methods underlying such an approach have been described in detail by MacFarlane, Kouvaritakis and Edmunds (1977), Kouvaritakis and Edmunds (1977), and MacFarlane and Kouvaritakis (1977) ; a brief review of their main features is given here. Difficulties in multivariable feedback control usually arise from a limitation on the amount of information available to the controller. It is essentially because of this lack of information that one is faced with frequency-response plots which exhibit excessive phase-lag characteristics or, equivalently, by root-locus diagrams with branches deflected towards the right half of the complex plane. It follows then that the availability of a number of measured variables in excess of the number of commanded outputs should lead to a design which makes a more efficient use of gain. The use of such extra measurements however involves a ‘ squaring-down ’ procedure, that is a way of combining all the available measured variables in order to form a new set of outputs (available for feedback purposes) equal in number to the number of inputs. Squaring-down operations are known to generate multivariable zeros (Kouvaritakis and MacFarlane 1976) and to affect vitally the asymptotic behaviour of the multivariable root loci as described by a set of infinite zeros (Kouvaritakis and Shaked 1976). Some frequency-response design techniques (Mayne 1973, Rosenbrock 1969) are not equipped with a systematic procedure for dealing with this problem of squaring-down. By contrast, the multivariable root locus techniques used here permit the systematic placement of zeros (both finite and infinite) and hence provide a means of shaping the asymptotic behaviour of root loci when forming ' squared-down ' feedback configurations.

2. The two-stage design approach

The configuration considered is shown in Fig. 1. A given plant has l inputs and m outputs, and it is assumed that $m \geqslant l$. The plant output vector y consists of two sub-vectors c and z.

$$y = \begin{bmatrix} c \\ \\ z \end{bmatrix} \tag{2.1}$$

where c is an l-dimensional vector of variables whose behaviour is to be controlled and z is an $(m-l)$-dimensional vector of extra available measurements. The design process is carried out in two main stages, corresponding to the arrangement shown in Fig. 2.

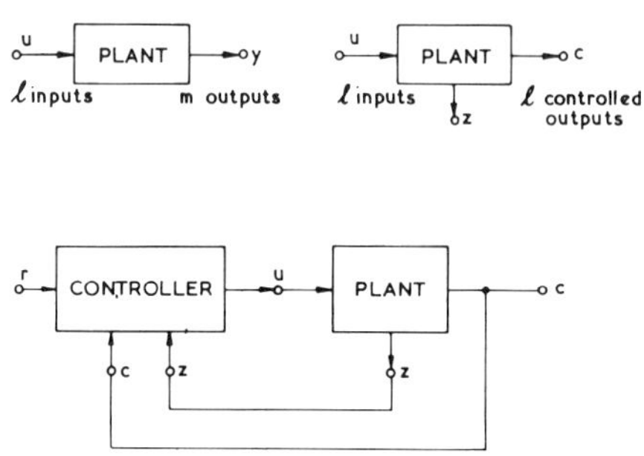

Figure 1.

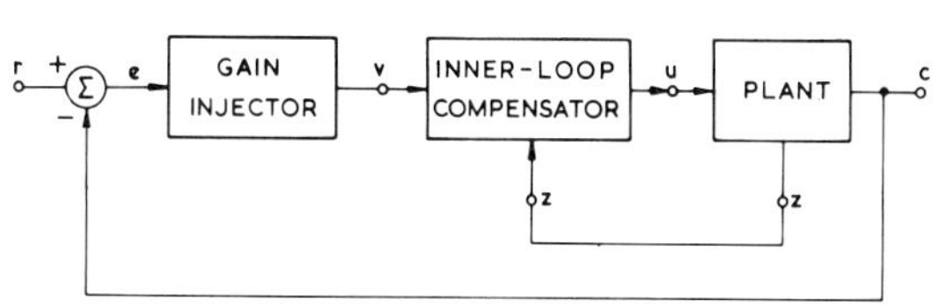

Figure 2.

Stage 1—Inner-loop compensator design. The purpose of the inner-loop compensator is to make use of the extra information contained in the vector of measurements z to create a new l-dimensional input point, v, which gives a suitable form of transmittance from v to the set of controlled outputs c. A suitable form of this transmittance is taken to be one which allows acceptably large amounts of feedback gain to be applied around the modified plant by the gain-injector block.

Stage 2—Outer-loop gain injector design. Having used all of the available measurements to the best advantage in arranging for a suitable transmission characteristic between v and c, the design is completed by choosing a gain-injector block which will inject the suitably high feedback gains required to meet the closed-loop performance specification over the operating bandwidth. The gain-injector block will normally contain both integral action terms, to remove low-frequency errors, and phase-compensating terms to improve overall closed-loop damping.

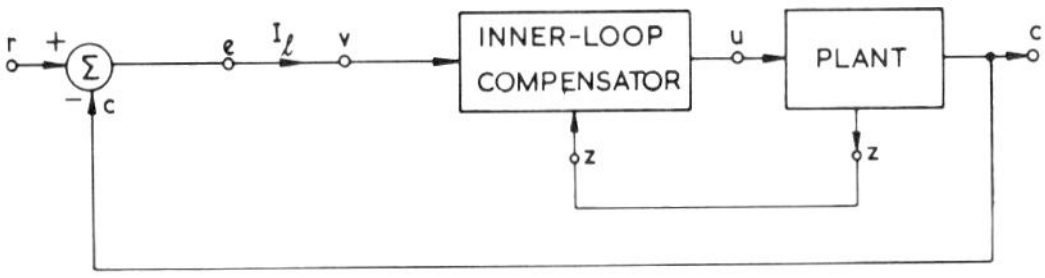

Figure 3.

The basic idea behind this two-stage design approach is best seen from a consideration of Figs. 2 and 3 ; the arrangement of Fig. 3 is obtained from that of Fig. 2 by replacing the gain-injector block by a unit-transmittance operator. The first stage of the design process is to design the compensator using the arrangement shown in Fig. 3 ; suppose for the moment that this stage of the design has been carried out. The set of closed-loop characteristic frequencies for the arrangement of Fig. 3 will depend on the nature of the characteristic gain loci for the transmittance $G_c(s)$ between v and c shown in Fig. 4. If the closed-loop characteristic frequencies for the arrangement in

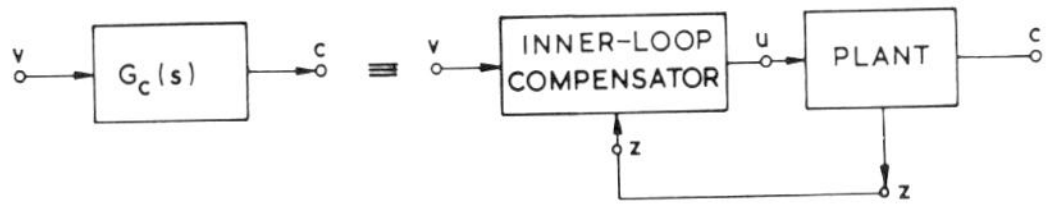

Figure 4.

Fig. 3 have good damping properties then the characteristic gain loci of the transmittance matrix $G_c(s)$ will have good gain and phase margins, and *vice versa* : good gain and phase margins for all the characteristic gain loci of $G_c(s)$ will correspond to a well damped set of closed-loop characteristic frequencies for the arrangement of Fig. 3. Hence the manipulation of the closed-loop frequencies for the arrangement of Fig. 3 is equivalent to the manipulation of the characteristic gain loci for the compensated plant $G_c(s)$. Therefore the inner-loop compensator is designed for the arrangement of Fig. 3 in such a way that a suitable set of closed-loop characteristic frequencies is obtained. Suitability of these frequencies is judged on a basis of damping and speed of response ; essentially one is implicitly attempting in the design of the inner-loop compensator to get the best possible set of characteristic gain loci for $G_c(s)$, since a good set of these characteristic gain loci will provide a suitable point of departure for the design of the gain injector at the next stage.

2.1. *Inner-loop compensator design*

The ideas behind the design of the inner-loop compensator are illustrated by Figs. 5, 6 and 7. The form of the characteristic gain loci for the compensated plant $G_c(s)$, that is for the transmittance between v and c in Fig. 7

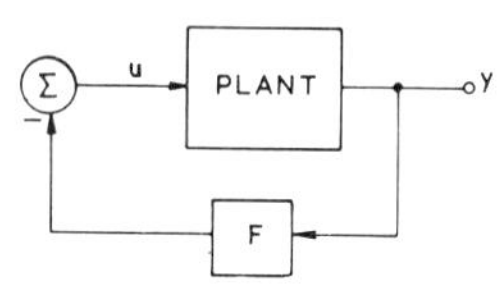

Figure 5.

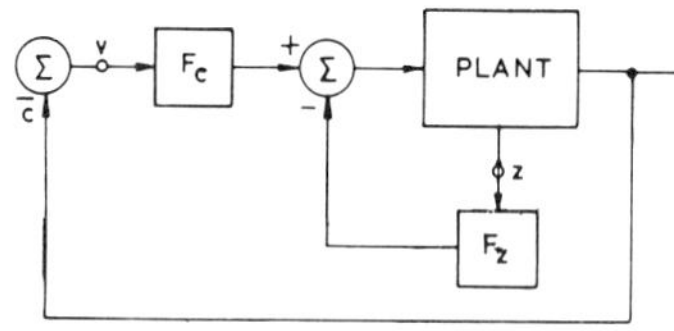

Figure 6.

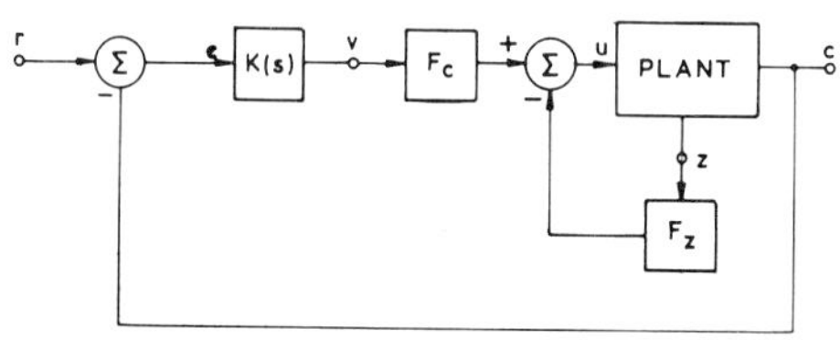

Figure 7.

is directly related to the nature of the closed-loop characteristic frequencies for the arrangement of Fig. 6 which in turn is equivalent to Fig. 5. Thus the problem of choosing the component blocks F_c and F_z of the inner-loop compensator is essentially the problem of choosing the feedback matrix F in the configuration of Fig. 5 in such a way that a suitable set of closed-loop characteristic frequencies is obtained. Since F has m inputs and l outputs where $l<m$ the choice of F is often referred to as a 'squaring-down' procedure. The choice of F is essentially based on generalized root-locus ideas and involves the placement of finite zeros and the manipulation of root-locus asymptotes. The final objective of the 'inner-loop compensator' design is to get suitable locations for the poles of the system seen by the gain injector $K(s)$ in Fig. 7. A satisfactory achievement of this objective requires some dynamical insight and engineering judgement since, in the usual case when all the plant states are not available, it will not normally be possible to place finite zeros in arbitrarily specified locations in the complex plane. Nevertheless the basic ideas underlying the technique of inner-loop compensator design are quite straightforward: one places finite zeros and manipulates root-locus asymptote patterns in such a way that, by finally setting a gain parameter at a suitably chosen value, the poles of the system seen by the gain injector $K(s)$ are pulled into suitable locations in the complex frequency plane. In some circumstances one may wish to pull such poles well over into the left-half of the complex plane in order to speed up closed-loop response and to allow $K(s)$ to inject a desired amount of gain before the overall (inner + outer loops) system has its closed-loop characteristic frequencies driven into unacceptable regions of the complex frequency plane. In other circumstances one may wish to 'freeze' the location of major poles associated with certain open-loop plant characteristic frequencies by placing finite zeros near them. Often a judicious mixture of both approaches will be required.

2.1.1. *Multivariable root loci*

Let a linear, continuous, time-invariant system have the set of state–space equations.

$$\left.\begin{aligned} \dot{x} &= Ax + Bu \\ y &= Cx \end{aligned}\right\} \tag{2.2}$$

where x, u and y are the n-, l- and m-dimensional state, input and output vectors respectively and A, B and C are matrices of constants of appropriate dimensions. Assume for the moment that the system $\mathscr{S}(A, B, C)$ is square, that is assume $l=m$. Then, the multivariable root loci of $\mathscr{S}(A, B, C)$ are defined as the paths traced by the system closed-loop characteristic frequencies with unity negative feedback and a positive scalar gain k introduced at each of the system inputs, while k is varied from zero to infinity. The main results of the multivariable root locus method deal with the asymptotic behaviour of the multivariable root loci as k tends to infinity. It is shown that, as in the case of scalar systems, a number of closed-loop poles tend to finite limits while the remainder vanish to infinity. The finite limits, called finite zeros, are identified as the system invariant zeros given by the finite solutions of the equation (Kouvaritakis and MacFarlane 1976 b)

$$\det [sNM - NAM] = 0 \tag{2.3}$$

where N and M are the full-rank left and right annihilators of maximum dimension of the matrices B and C respectively. The poles that diverge to infinity are attracted by a set of zeros located at the point at infinity, called infinite zeros (IZ). The manner in which the root loci approach the infinite zeros as k tends to infinity is described by equations of the type

$$s^i = \lambda_j^{(i)} k, \quad j = 1, 2, \ldots, d_{i-1} - d_i$$
$$i = 1, 2, \ldots, \nu \tag{2.4}$$

where i denotes the order of divergence which is associated with the ith parameter G_i from the chain of the projected Markov parameters (PMP) defined below; $\lambda_j^{(i)}$ denotes the jth non-zero eigenvalue of G_i; d_i denotes the rank defect of G_i with $d_0 = m$; and ν denotes the smallest integer for which G_ν is full rank. The projected Markov parameters are defined in an iterative manner. The first PMP, G_1, is defined to be equal to the first Markov parameter and has the spectral decomposition

$$G_1 = CB = [U_1 \ \ M_1] \begin{bmatrix} \Lambda_1 & 0 \\ 0 & 0_{d1} \end{bmatrix} \begin{bmatrix} V_1 \\ N_1 \end{bmatrix} \tag{2.5}$$

The second PMP is defined as the projection of the second Markov parameter onto the null space of G_1 so that

$$G_2 = N_1 CABM_1 \tag{2.6}$$

The remaining PMP's, G_i, may be defined by a continued projection of the Markov parameters onto the null spaces of the relevant PMP's. Appropriate correction to the Markov parameter chain, however, may be necessary for systems with three or more different orders of divergence (Kouvaritakis and Edmunds 1979). It is the non-zero eigenvalues of the parameters G_i that determine the asymptotic directions along which the infinite zeros are approached. Thus the infinite zeros corresponding to eqn. (2.4) have asymptotes that make with the positive real axis angles equal to the phase of the ith roots of minus the corresponding non-zero eigenvalue of G_i; these infinite zeros are called ith order due to the rate of divergence of the closed-loop poles. With every such eigenvalue therefore, there will exist i asymptotes all symmetrically arranged in the complex plane, forming a Butterworth pattern.

The above may be summarized as follows. A system $\mathscr{S}(A, B, C)$ with PMP defined by a suitable extension of eqns. (2.5) and (2.6) has exactly

$$n_z = n - \sum_{i=0}^{\nu-1} d_i$$

finite zeros given by the finite roots of eqn. (2.3) and

$$\sum_{i=0}^{\nu-1} d_i$$

infinite zeros described by the non-zero eigenvalues of the PMP. For k tending to infinity all poles tend to zeros (finite and infinite). Each finite zero attracts one pole, while the infinite zeros can be separated into groups with asymptotes grouped in Butterworth patterns. Thus each non-zero eigenvalue of the ith PMP is associated with i infinite zeros of order i acting as terminating pole locations for a set of i symmetrically arranged asymptotes to the root loci, one infinite zero corresponding to each asymptote. The description of the asymptotic root locus behaviour in terms of the spatial arrangement of asymptotes and infinite zeros into Butterworth patterns, and the orientation of these patterns with respect to the real axis, is incomplete since no indication as to the exact location of the patterns is given. To remedy this one can define a set of points, the ' pivots ' which are points of radiation of all the asymptotes belonging to one and the same pattern. Unlike the scalar case, a multivariable system may have more than one pivot, just as it may have more than one order of asymptotic behaviour. Moreover the pivots may be located on the real axis or they may appear in complex conjugate pairs. Explicit expressions for these points in terms of the system parameters A, B, C may be obtained and for further detail the reader is referred to (Kouvaritakis 1978). Further aspects of the root locus behaviour such as ' angles of approach and departure ', and the intersection of the root loci with the imaginary axis have also been investigated and the relevant results can be found in the existing literature (Shaked 1976, 1977).

2.2. *Outer-loop design*

The key to the extension of the classical Nyquist–Bode approach to the multivariable case (MacFarlane and Kouvaritakis 1977) is to be found in the association of a set of algebraic functions with the eigenvalues of the $m \times m$ Laplace transfer matrix, $G(s)$, which for the system of eqn. (2.2) is given by

$$G(s) = C(sI - A)^{-1}B \tag{2.7}$$

From the relevant characteristic gain equation

$$\det [gI - G(s)] = \Delta(g, s) = 0 \tag{2.8}$$

it is seen that $\Delta(g, s)$ will usually be irreducible in g, so that the eigenvalues of $G(s)$, $g_i(s)$, will not usually be rational in s. To overcome the problems of multivaluedness, appropriate Riemann surfaces are constructed and are used as domains of definition over which eqn. (2.8) defines an algebraic function, $g(s)$, which is single valued. It can be shown that almost everywhere (except at branch points, that is points where $\Delta(g, s)$ has repeated roots) the values of $g(s)$ form a set of m locally distinct analytic functions $g_i(s)$ called the eigenfunctions of $G(s)$ or the characteristic gain functions of $G(s)$. To each eigenfunction, $g_i(s)$, there corresponds a vector-valued function of s, $w_i(s)$, satisfying the standard eigenvalue–eigenvector equation for $G(s)$; the $w_i(s)$, known as the characteristic directions of $G(s)$, will often simply be referred to as the eigenvectors of $G(s)$.

The importance of the characteristic transfer functions and directions of $G(s)$ in feedback systems analysis stems from the fact that unity feedback leaves the eigenframe unaltered while it results in a set of eigenfunctions that are related to the open-loop eigenfunctions in the same simple manner that open- and closed-loop transfer functions of scalar systems are related. Thus expressing both $G(s)$ and the closed-loop transfer-function matrix $R(s)$ in dyadic form we have that

$$G(s) = \sum_{i=1}^{m} g_i(s) w_i(s) v_i^t(s) \tag{2.9}$$

and

$$R(s)=\sum_{i=1}^{m}\frac{g_i(s)}{1+g_i(s)}\,w_i(s)v_i{}^t(s) \tag{2.10}$$

where $\{v_i{}^t(s),\ i=1,\ 2,\ \dots,\ m\}$ is a dual vector set to the eigenvector set $\{w_i(s)\}$. The above two equations yield a lot of insight into the way that open- and closed-loop systems operate on input signals in order to generate output signals. The eigenfunctions can now be seen as functions which 'co-ordinate' the input signals as seen through the eigenvector frame, while a referral of the co-ordinated signals back to the original frame produces the output vector. In practice one is often particularly interested in the response of systems to sinusoidal signals. For this reason, in what follows, one usually concentrates on values of s which lie on the classical Nyquist D-contour. Thus, the argument s may be replaced by $j\omega$ which in turn, for convenience, will frequently be omitted altogether. Computing and plotting the eigenvalues of $G(j\omega)$ along the imaginary axis gives a set of closed curves called characteristic gain loci. (Note that in general different characteristic gain functions are associated with one Riemann surface and therefore in certain cases it may be necessary to join their respective frequency-response plots together in order to obtain closed curves.) From the discussion above one is led to the conclusion that the gain loci contain in them information about closed-loop stability, and that the gain loci together with a measure of the eigenvector spatial arrangements characterize most aspects of closed-loop behaviour. A brief summary of the main results is given below.

(i) *Stability*. Closed-loop stability occurs if and only if the net sum of counter-clockwise encirclements of the critical point $(-1+j0)$ by the set of characteristic gain loci is equal to the number of open-loop unstable poles.

(ii) *Interaction*. The term interaction here signifies the property of all system outputs y_j to respond to a signal applied at the ith input for $i\neq j$. The system exhibits low interaction if the ith input stimulates the ith output while the response of the other outputs remains below an acceptable level. Low interaction at all frequencies requires that at any frequency, the moduli of the characteristic gains be large and/or the characteristic directions be well aligned with the standard basis vectors. For the purpose of assessing interaction therefore it is convenient to display a measure of alignment, usually taken to be the angles defined by the complex cosine between eigenvectors and standard basis vectors, called misalignment angles, as a function of frequency together with a plot of the modulus of the corresponding characteristic gain locus against frequency.

(iii) *Performance*. The characteristic gain loci of a system which satisfies the requirements for low interaction may be used for the qualitative assessment of closed-loop performance in exactly the same manner as the gain versus frequency response of a scalar system. The classical notions of gain/phase margins, bandwidth etc. may be be applied to the characteristic gain loci with equal ease in order to assess the performance in the multivariable case.

(iv) *Accuracy*. From eqn. (2.10) it is clear that a system with arbitrarily large characteristic gain loci will give perfect tracking. Generally speaking it is easy to show that such large loci will yield good closed-loop accuracy. An exception to this can be found in the case of badly skewed eigenvectors ; for an assessment of accuracy in a case like this the method reverts to a different set of frequency response plots suitably defined and called the principal gain loci (MacFarlane and Kouvaritakis 1977).

2.3. *Overall design procedure*

Squaring-down is known to create finite zeros and influence the order of the infinite zeros, and hence F should be designed so that it yields an acceptable pattern of finite zeros and reduces the order of the infinite zeros. Further gain margin improvement may be achieved by the introduction of a square constant controller K, as shown in Fig. 8 ; the function of K is to manipulate the projected Markov parameters in order to adjust the asymptotic directions of the root loci and place the corresponding pivots if necessary. Judicious placement of the finite zeros and infinite zeros will, for an increasing loop gain k (see Fig. 8), conduct the characteristic frequencies of the system towards preferred areas of the complex plane thus stabilizing the system and speeding up its response. The inner-loop design is concluded by selecting a suitably high value for k.

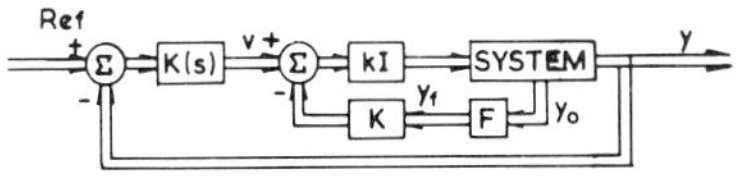

Figure 8.

The outer-loop design is initiated upon closure of the inner-loop. A transfer-function matrix representation, $G_c(s)$, for the system from commanding inputs, v, (see Fig. 8) to the commanded outputs, y with the inner-loop feedback connections made, may now be derived. The purpose of the second stage of the design is to construct a compensator $K(s)$ such that the characteristic quantities of the compensated transfer-function matrix, $Q(s)$, given by $Q(s) = G_c(s)K(s)$, possess a set of prescribed properties. For the adjustment of the characteristic loci, use is made of the approximately commutative controller technique according to which $K(s)$ is written as

$$K(s) = \hat{U}\Lambda_k(s)\hat{V} \tag{2.11}$$

where $\hat{U}$ and $\hat{V}$ are two $l \times l$ matrices of real constants and where scalar rational functions $k_i(s)$ form the elements of the diagonal matrix $\Lambda_k(s)$. The frames $\hat{U}$ and $\hat{V}$ are real frame matrix approximants to the complex eigenvector and dual eigenvector frame matrices of $G_c(s)$ at a specific frequency ω_0, respectively, suitably computed by a minimization algorithm called ALIGN (MacFarlane and Kouvaritakis 1977). For a successful approximation it can be shown that in the vicinity of ω_0 the eigenfunctions of $Q(s)$, $q_i(s)$, and those of $G_c(s)$, $g_i(s)$, are related in a simple manner

$$q_i(s) \simeq g_i(s)k_i(s) \tag{2.12}$$

The ALIGN algorithm may also be used for the adjustment of the characteristic directions. Thus real frame approximants to $G_c^{-1}(s)$ evaluated at some specific frequency ω_h, can be used as precompensators, K_h, for the purpose of improving the alignment of the eigenvectors of $Q(s)=G_c(s)K_h$, with the standard basis vectors in the vicinity of ω_h. The overall design procedure then may be summarized as follows.

(i) Calculate a squaring-down compensator F, to place the finite and infinite zeros of the inner-loop.

(ii) Choose a suitable inner-loop gain, and close the loops with this gain, in order to suitably place the closed-loop frequencies of the inner-loop system. Now proceed to the outer-loop design stages.

(iii) Compute K_h at some suitably high frequency ω_h in order to reduce the misalignment angles in the vicinity of ω_h. The role of K_h is to suppress high-frequency interaction.

(iv) Design an approximately commutative controller $K_m(s)$, to gain–phase compensate the characteristic loci of $G_c(s)K_h$ in the vicinity of a suitable intermediate frequency, ω_m, where the proximity of the characteristic loci to the critical point is most pronounced. One should ensure that $K_m(s)$ tends to the identity operator as s tends to $j\omega_h$ so that the high-frequency misalignment angles remain unaltered. The compensating action of $K_m(s)$ will improve closed-loop stability, suppress oscillatory behaviour and speed up the overall closed-loop response.

(v) Introduce an integral-plus-proportional controller of the form

$$\left[\frac{a}{s}K_l+I\right]$$

in order to suppress low-frequency interaction and improve the steady-state accuracy of the compensated system $Q(s)=G_c(s)K_hK_m(s)$. The low-frequency controller K_l could be an approximately commutative controller designed to balance the low frequency gains while the tuning coefficient a may be used to control the frequency range over which the integral action is to extend.

(vi) Introduce an additional gain k in each loop and test the resulting closed-loop time responses to assess the performance of the final overall controller configuration.

3. Flight control system example

An example is now given for a flight control system described by a state–space model whose A, B and C matrices are given by

$$\left.\begin{aligned} A&=\begin{bmatrix} 0 & 0 & 1\cdot132 & 0 & -1\cdot0\\ 0 & -0\cdot0538 & -0\cdot1712 & 0 & 0\cdot0705\\ 0 & 0 & 0 & 1\cdot0 & 0\\ 0 & 0\cdot0485 & 0 & -0\cdot8556 & -1\cdot013\\ 0 & -0\cdot2909 & 0 & 1\cdot0532 & -0\cdot6859 \end{bmatrix}\\ B&=\begin{bmatrix} 0 & 0 & 0\\ -0\cdot0012 & 1\cdot0 & 0\\ 0 & 0 & 0\\ 0\cdot4419 & 0 & -1\cdot6646\\ 0\cdot1575 & 0 & -0\cdot0732 \end{bmatrix} \quad C=\begin{bmatrix} 1 & 0 & 0 & 0 & 0\\ 0 & 1 & 0 & 0 & 0\\ 0 & 0 & 1 & 0 & 0 \end{bmatrix} \end{aligned}\right\} \qquad (3.1)$$

This model (Murray 1977) constitutes a linearized form of the equations of an aircraft motion in the vertical plane when small perturbations about a datum flight condition are considered. The elements of the state vector $x=(x_1, x_2, x_3, x_4, x_5)^t$ are as follows :

x_1 : height error relative to ground or guidance aid expressed in metres ;
x_2 : forward speed expressed in metres/second ;
x_3 : pitch angle expressed in degrees ;
x_4 : rate of change of pitch angle expressed in degrees/second ;
x_5 : vertical speed expressed in metres/second.

While the elements of the input vector $u=(u_1, u_2, u_3)^t$ are :

u_1 : spoiler angle expressed in degrees ;
u_2 : forward acceleration due to engine thrust expressed in metre/(second)^2 ;
u_3 : elevator angle expressed in degrees.

For this example all state variables are measurable. From the form of the output map C it is readily seen that the elements of the output vector over which we wish to exercise control are simply the first, second and third state variables, respectively. The objective of the design study is to ensure that the response of these three output variables to step changes in their reference set points are fast, non-oscillatory, non-overshooting, accurate in the steady-state and generally non-interactive.

A preliminary analysis of the open-loop system shows that the uncompensated characteristic frequencies are located at $\{0+j0,\ -0{\cdot}018\pm j0{\cdot}182$ and $-0{\cdot}78\pm j1{\cdot}03\}$. Open-loop responses therefore are expected to be slow and oscillatory. This type of behaviour is shown by the open-loop system responses to unit impulses being applied to the first, second and third input in turn, as shown in Figs. 9 (*a*), 9 (*b*) and 9 (*c*), respectively. The need for an adjustment of the system characteristic frequencies is obvious.

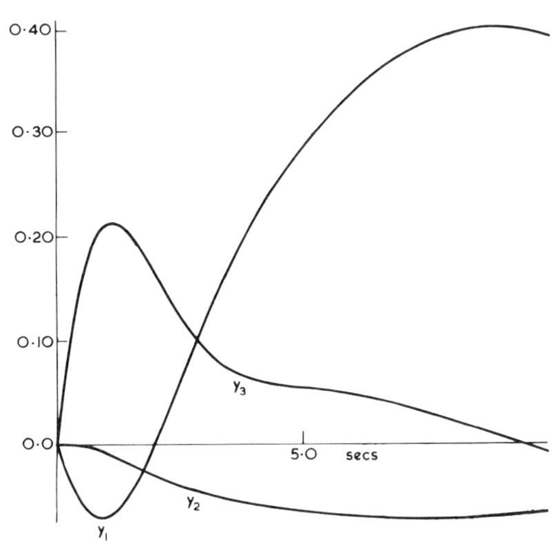

(*a*) *figure continued over*

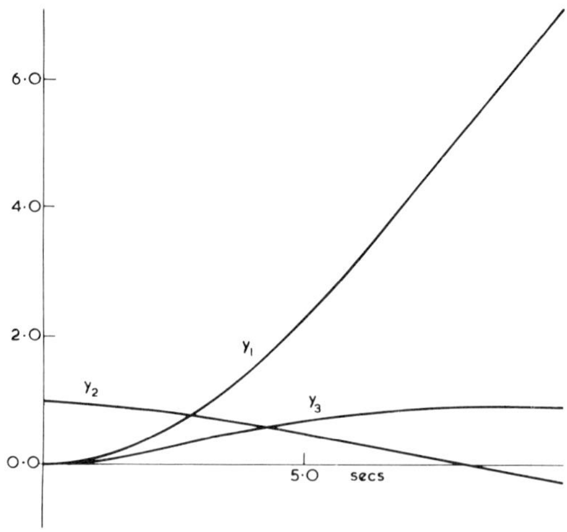

(*b*)

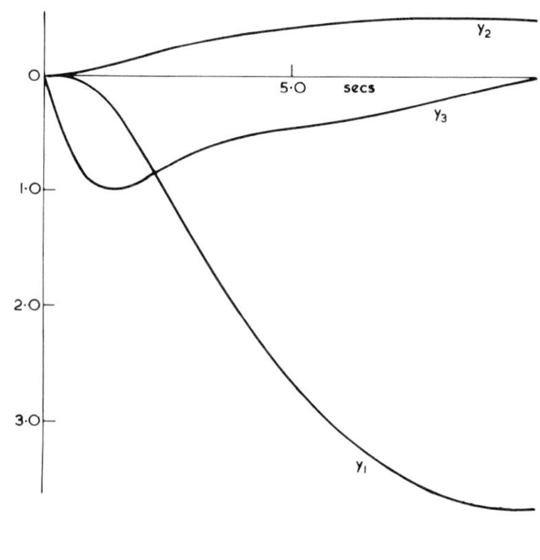

(*c*)

Figure 9.

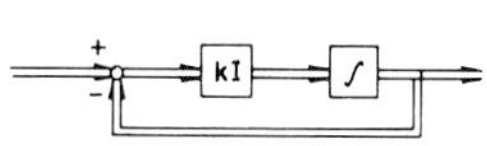

Figure 10.

Let us first consider what can be achieved by a simple application of proportional gain to the difference between the three controlled outputs and their reference inputs, as shown in Fig. 10. (That is we first see how things stand if no inner-loop compensation is used.) The asymptotic behaviour of the root loci that vanish to the point at infinity is dictated by the eigenproperties of the PMP's as defined by eqns. (2.5) and (2.6). The PMP for the system of eqn. (3.1) are

$$G_1 = CB = [U_1 \ \ M_1] \begin{bmatrix} \Lambda_1 & 0 \\ 0 & 0_{d_1} \end{bmatrix} \begin{bmatrix} V_1 \\ N_1 \end{bmatrix} = \begin{bmatrix} 0 & 0 & 0 \\ -0{\cdot}0012 & 1 & 0 \\ 0 & 0 & 0 \end{bmatrix}$$

$$= \begin{bmatrix} 0 & 1 & 0 \\ 1 & 0{\cdot}0012 & 0 \\ 0 & 0 & 1 \end{bmatrix} \begin{bmatrix} 1 & 0 & 0 \\ 0 & 0 & 0 \\ 0 & 0 & 0 \end{bmatrix} \begin{bmatrix} -0{\cdot}0012 & 1 & 0 \\ 1 & 0 & 0 \\ 0 & 0 & 1 \end{bmatrix} \tag{3.2}$$

and

$$G_2 = N_1 CABM_1 \tag{3.3}$$

$$= \begin{bmatrix} 1 & 0 & 0 \\ 0 & 0 & 1 \end{bmatrix} \begin{bmatrix} -0{\cdot}1575 & 0 & 0{\cdot}0732 \\ 0{\cdot}0112 & -0{\cdot}0538 & 0{\cdot}0052 \\ 0{\cdot}4419 & 0 & -1{\cdot}6646 \end{bmatrix} \begin{bmatrix} 1 & 0 \\ 0{\cdot}0012 & 0 \\ 0 & 1 \end{bmatrix}$$

$$= \begin{bmatrix} -0{\cdot}1575 & 0{\cdot}0732 \\ 0{\cdot}4419 & -1{\cdot}6646 \end{bmatrix} \tag{3.4}$$

The second PMP, G_2, is full rank so that the nested projection is terminated at this point. An inspection of the PMP chain now reveals that the system has one first-order infinite zero, two pairs of second-order infinite zeros and no finite zeros. Since the non-zero eigenvalue of G_1 is positive, the first-order asymptote lies on the negative half of the real axis. The eigenvalues of G_2 are both negative (given by $\lambda = -0{\cdot}1363$ and $\lambda = -1{\cdot}6858$) so that the two second-order Butterworth patterns will each have one asymptote on either half of the real axis. It follows that as the gain k increases in the configuration of Fig. 10 all five system poles tend to the point at infinity ; three do so along asymptotes on the negative half of the real axis and two do so on the positive half. The root locus diagram plotted in Fig. 11 (*a*) verifies this type of behaviour and furthermore it shows that the system goes unstable for k larger than about 0·1.

The poor gain margins of the feedback system in Fig. 10 are due to the orientation of the second-order asymptotes. Naturally the situation may be improved by a manipulation of the second PMP for correct placement of the second-order infinite zeros. Thus designing a precompensator which reverses the sign of eigenvalues of G_2, while at the same time retaining the positive sign of non-zero eigenvalue of G_1, will maintain the orientation of the first-order asymptote, and will rotate the second-order patterns of asymptotes by 90°. One such design study is carried out in a later section of this paper, and the resulting root locus diagram is in Fig. 18 (*a*). The gain margins of the system now are greatly improved, since the system remains stable for all values of scalar gain less than about 5. Further improvements could be

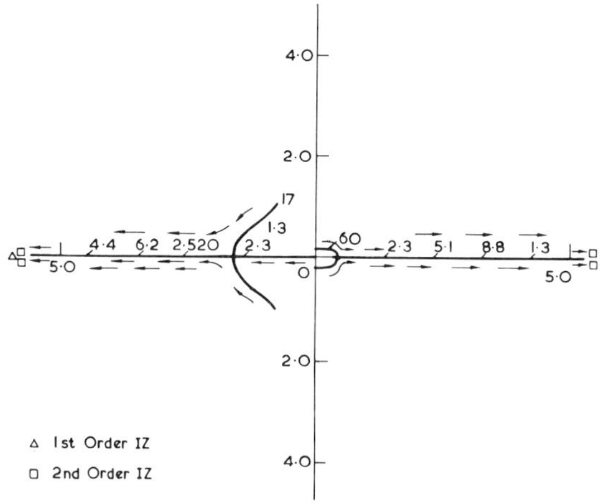

Figure 11 (a).

obtained by the placement of the pivots of both second-order Butterworth patterns of asymptotes on the negative half of the real axis.

Despite all possible improvements of this sort to the root locus asymptotic behaviour no amount of physically realizable compensation in this simple output-feedback controller structure, constant or dynamic, is capable of removing the second-order infinite zeros. It follows therefore that using the controlled output vector alone for the purpose of feedback will result in a configuration which does not permit the use of high gains since closed-loop poles will either migrate to the unstable half of the complex plane or become oscillatory with a large imaginary part. This shows the need for an inner-loop design where the feedback connections are made around not only the controlled-output signals but around all measurable signals.

3.1. *Inner-loop compensator design*

In the discussion of the model of the previous section it was noted that all the states of system (3.1) can be measured. The inner-loop configuration of Fig. 8 therefore makes feedback connections from the system states to the command input vector v by forming suitable combinations of the variables $\{x_i,\ i=1, 2, \ldots, 5\}$ to obtain the new set of 3 outputs contained in the vector y_f which may now be written as

$$y_f = Fx \tag{3.5}$$

The resulting system from the aircraft input vector u to the vector y_f has state–space equations

$$\left.\begin{aligned} \dot{x} &= Ax + Bu \\ y_f &= Fx \end{aligned}\right\} \tag{3.6}$$

and the function of F is to ensure that the system $\mathscr{S}(A, B, F)$ has desirable root locus properties. As was explained earlier F performs a squaring-down operation and hence F should be designed to influence primarily these root locus properties that remain invariant under the standard input, output and feedback operations after squaring-down has been completed. The two most important such properties are the order of the infinite zeros, and the location of finite zeros. We consider the order of the infinite zeros first.

The need for an inner-loop compensator design arose from the presence of second-order infinite zeros in the uncompensated system. It follows therefore that an inner-loop scheme would prove advantageous if the resulting $\mathscr{S}(A, B, F)$ system exhibited at the most first-order infinite zeros. This condition is secured if and only if F can be chosen such that Q_1, the first PMP of $\mathscr{S}(A, B, F)$ given by

$$Q_1 = FB \tag{3.7}$$

can be made full rank. If now f_i and $b_i{}^t$ for $i = 1, 2, \ldots, 5$ denote the ith column and row of F and B respectively, then from eqns. (3.1) and (3.7) we have that

$$Q_1 = F'B' \tag{3.8}$$

where $F' = [f_2, f_4, f_5]^t$ and $B' = [b_2, b_4, b_5]^t$. The matrix B' however can be seen to be full rank so that selecting any three independent vectors for the second, fourth and fifth column vectors of F will ensure that $\mathscr{S}(A, B, F)$ has infinite zeros of order 1 only. The number of these infinite zeros will of course be 3, and the orientation of the corresponding asymptotes will depend on the phase of the eigenvalues of Q_1. Placement of the asymptotes on the negative real axis requires the eigenvalues of Q_1 to be positive. If in addition we require the rate of divergence to the infinite zeros to be the same for all corresponding root loci, then F' must be of the form

$$F' = a[B']^{-1} \tag{3.9}$$

where a is a positive scalar. For the special choice of $a = 1$, F' becomes

$$F' = \begin{bmatrix} 0 & -0{\cdot}319 & 7{\cdot}244 \\ 1 & 0 & 0 \\ 0 & -0{\cdot}685 & 1{\cdot}923 \end{bmatrix} \tag{3.10}$$

The next task that the squaring-down operator must accomplish is to place the finite zeros at suitable locations in the left-half of the complex-frequency plane. Since the first PMP of $\mathscr{S}(A, B, F)$ is full rank the number of finite zeros according to the relevant result of § 2.1.1 is simply $n - m = 5 - 3 = 2$. Now columns 1 and 3 of the feedback compensator have not been specified yet, and therefore some degrees of design freedom can be drawn from these for the placement of zeros. Thus denoting by $f_{i,j}$ the element of F in the ith row and the jth column and selecting the second, fourth and fifth columns of F in accordance with eqn. (3.10) it is easy to show that an M, the right annihilator of F, satisfying the condition

$$FM = 0_{m, n-m} \tag{3.11}$$

may be written as

$$M=\begin{bmatrix} 1 & 0 \\ -f_{2,1} & -f_{2,3} \\ 0 & 1 \\ (-0\cdot442f_{1,1}+1\cdot665f_{3,1}) & (-0\cdot442f_{1,3}+1\cdot665f_{3,3}) \\ (-0\cdot158f_{1,1}+0\cdot073f_{3,1}) & (-0\cdot158f_{1,3}+0\cdot073f_{3,3}) \end{bmatrix} \quad (3.12)$$

Furthermore, if we define N, a left annihilator of B, according to the equation

$$NB=0_{n-m,\,n} \quad (3.13)$$

then it may be readily shown that a solution for N is given by

$$N=\begin{bmatrix} 1 & 0 & 0 & 0 & 0 \\ 0 & 0 & 1 & 0 & 0 \end{bmatrix} \quad (3.14)$$

so that by eqn. (2.3) we conclude that the finite zeros of the system $\mathscr{S}(A, B, F)$ are given the eigenvalues of the matrix product

$$NAM=\begin{bmatrix} (-0\cdot158f_{1,1}+0\cdot073f_{3,1}) & (-0\cdot158f_{1,3}+0\cdot073f_{3,3}+1\cdot132) \\ (0\cdot442f_{1,1}-1\cdot665f_{3,1}) & (0\cdot442f_{1,3}-1\cdot665f_{3,3}) \end{bmatrix} \quad (3.15)$$

The set of finite zeros therefore are defined by roots of the equation

$$s^2+(0\cdot442f_{1,3}+0\cdot073f_{3,1}-1\cdot665f_{3,3}-0\cdot158f_{1,1})s+ \\ +(0\cdot5f_{1,1}-1\cdot885f_{3,1}+0\cdot23f_{1,1}f_{3,3}-0\cdot231f_{3,1}f_{1,3})=0 \quad (3.16)$$

An inspection of the degrees of freedom available here shows that by a suitable choice of the four elements of F we may achieve an arbitrary placement of the system finite zeros. Ideally, these zeros should be placed well to the left on the negative real axis. However, in the face of practical limitations on the amount of feedback gain one may use, it was decided to place the zeros between the points at $(-4\cdot0+j0\cdot0)$ and $(-5+j0\cdot0)$.

Any one pair of values for the location of zeros leads to a whole family of solutions for the elements $f_{1,1}$, $f_{1,3}$, $f_{3,1}$, $f_{3,3}$. Consideration of the amount of gain used by each set of solutions measured, say as the sum of the squares of the four elements, provides guidelines for the selection of suitable solutions that make an efficient use of gain.

The particular choice for placement of zeros in the vicinity of $(-4\cdot5+j0\cdot0)$, after appropriate rounding up, was taken as

$$f_{1,1}=0, \quad f_{1,3}=6, \quad f_{3,1}=-6, \quad f_{3,3}=-4 \quad (3.17)$$

which resulted in a pair of zeros at the points

$$z_1=-4\cdot2 \quad \text{and} \quad z_2=-4\cdot7 \quad (3.18)$$

The remaining two elements of F, namely $f_{2,1}$ and $f_{2,3}$ were then chosen with the aim of manipulating the angles of departure (Shaked 1976, Kouvaritakis and Edmunds 1977) and the overall inner-loop compensator derived in this manner was finally obtained as

$$F=\begin{bmatrix} 0 & 0 & 6 & -0\cdot319 & 7\cdot244 \\ 4 & 1 & 7 & 0 & 0 \\ -6 & 0 & -4 & -0\cdot687 & 1\cdot923 \end{bmatrix} \tag{3.19}$$

This squaring-down operator F results in a system $\mathscr{S}(A, B, F)$ which has no second-order infinite zeros. Instead it has two finite zeros located near the $(-4\cdot5+j0\cdot0)$ point and three first-order infinite zeros with associated asymptotes which lie on the negative real axis. Furthermore, the closed-loop poles which vanish to the point at infinity with increasing gain do so in such a way that they approach the respective infinite zeros at equal rates. For this particular choice of F therefore the inner-loop system of Fig. 8 (with $K=I$) is expected to exhibit good gain margins, and the injection of a scalar gain into the feedback loop is expected to conduct the characteristic frequencies to preferred sectors of the complex plane, namely the part of the real axis to the left of the $(-4\cdot2+j0\cdot0)$ point. This type of behaviour is verified by the appropriate root locus diagram shown in Fig. 11 (b).

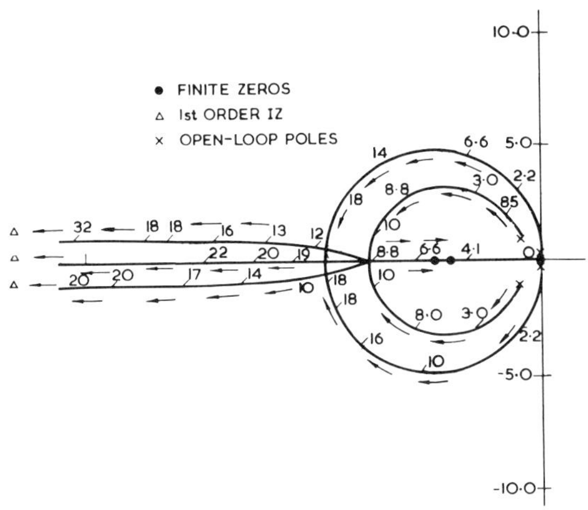

Figure 11 (b).

The inner-loop compensator design may be concluded by selecting an appropriate scalar gain k. From engineering considerations it is desirable to keep feedback gains as low as practicable. A suitable choice for k therefore appears to be $k=20$ since this is approximately the 'critical' value of gain above which all the closed loop characteristic frequencies become almost purely real. For this particular value of inner-loop gain, and for the squaring-down operator F of eqn. (3.19), the improved closed-loop characteristic frequency pattern is given by

$$\lambda_1=-4\cdot77, \quad \lambda_2=-8\cdot34, \quad \lambda_3=-11\cdot4, \quad \lambda_{4,5}=-18\cdot54\pm j1\cdot02 \tag{3.20}$$

These now form the set of poles for the system from the command input vector v to output vector y in Fig. 8. Thus the inner-loop compensator has improved the dynamics seen by the outer-loop system's gain injector.

3.2. *The outer-loop design*

The purpose of the inner-loop design is to improve the system dynamics by making use of all the five available measurements. The ultimate objective of the overall design study however is to exercise independent control over three of the five measured variables, namely x_1, x_2 and x_3. With this in mind we now rearrange the inner-loop configuration of Fig. 8 in a manner that will facilitate the outer-loop design. The new configuration is shown in Fig. 12. Note that in Fig. 12 the controlled output y_c is a sub-vector of the vector y which represents all the available outputs. For this reason the matrix F is shown as split into a pair of matrices F_1 and F_2 which are not the same as the matrices F_c and F_z shown in Figs. 6 and 7. The basic procedure being followed however is exactly as explained in § 2.1.

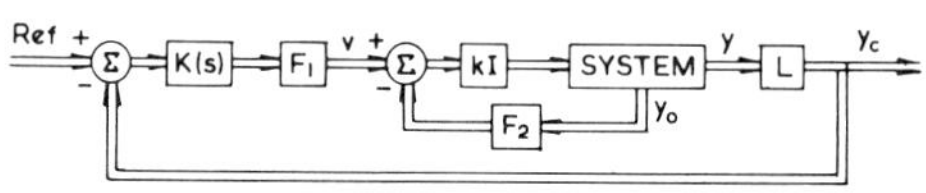

Figure 12.

The part of the controller of eqn. (3.19) that operates on x_1, x_2 and x_3 described by the matrix F_1, say, is moved around the loop to the forward path of the overall feedback scheme. Thus in effect we remove columns 1, 2 and 3 from F to form the operator F_1, and replace them in F by zero vectors to form the operator F_2 which remains in the feedback path. Note that with the outer-loop controller, $K(s)$, removed ($K(s)=I$), the resulting root loci for varying k are identical to those obtained before rearrangement. The new configuration (Fig. 12) however makes more efficient use of gain, since it closes the loops around x_1, x_2 and x_3 only once through the identity operator using the commanded output vector y_c in direct comparison with the reference input vector.

The second stage of the design procedure, the outer-loop design, consists of the application of the characteristic locus method to the system with the inner-loop compensator in place. For this purpose the Laplace transform transfer-function matrix representation, $G_c(s)$, for the system from the inputs of the compensator F_1 to the commanded outputs, y_c, of Fig. 12, with the inner-loop gain set at 20 is

$$G_c(s)=C(sI-A+20BF_2)^{-1}20BF_1 \tag{3.21}$$

The steps of the procedure outlined in § 2.2 are now carried out for the synthesis of the controller $K(s)$, which is intended to adjust the characteristic gains of $G_c(s)$ and thus yield a compensated system whose transfer-function matrix $Q(s)=G_c(s)\,K(s)$ has an eigenvalue–eigenvector set with the prescribed properties discussed earlier in the same section.

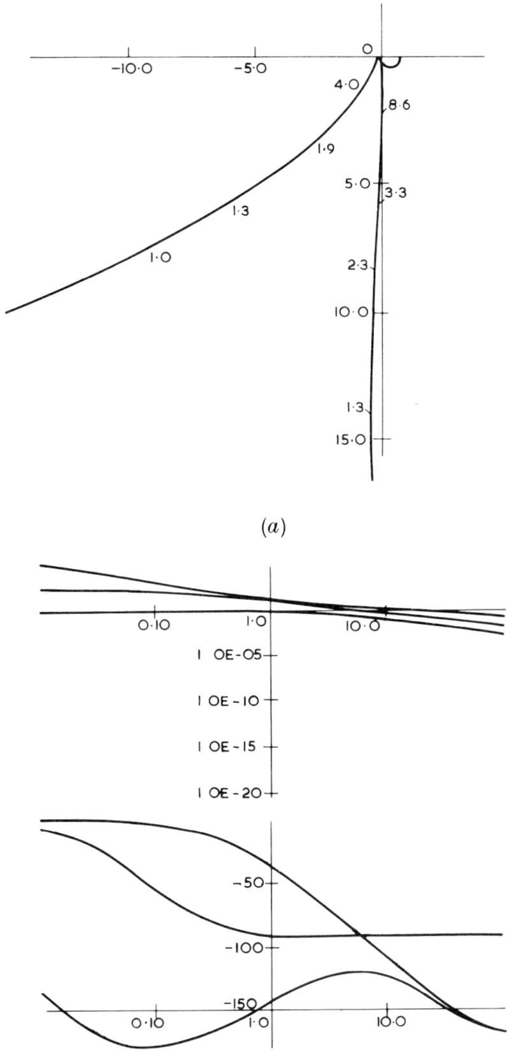

(*a*)

(*b*)

Figure 13.

Before entering into this characteristic locus design procedure however it is useful at this point to assess the effectiveness of the inner-loop design in terms of frequency response criteria. To do this we draw comparisons between the Nyquist and Bode plots of the characteristic loci of the transfer function matrix above shown in Fig. 13 (*a*) and 13 (*b*) respectively on the one hand, and the corresponding plots for the transfer function matrix

$$G(s) = C(sI - A)^{-1}B \tag{3.22}$$

which describes the input–output relationship of the system without the inner-loop compensator on the other. The latter set of frequency-response plots is given in Figs. 14 (*a*) and 14 (*b*). Note that the system with the inner-loop compensator exhibits infinite gain margins and good phase margins in contrast to the very poor gain/phase margins of the system without the inner-loop compensator. In the latter case, closed-loop stability requires a reduction of each loop gain by about ten times if scalar gain operators $K=kI$

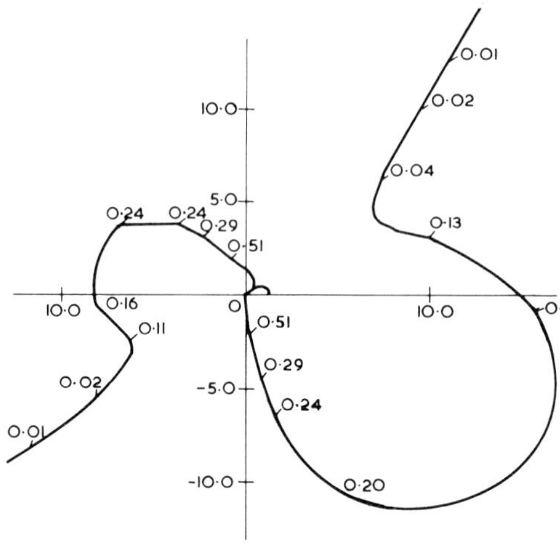

(*a*)

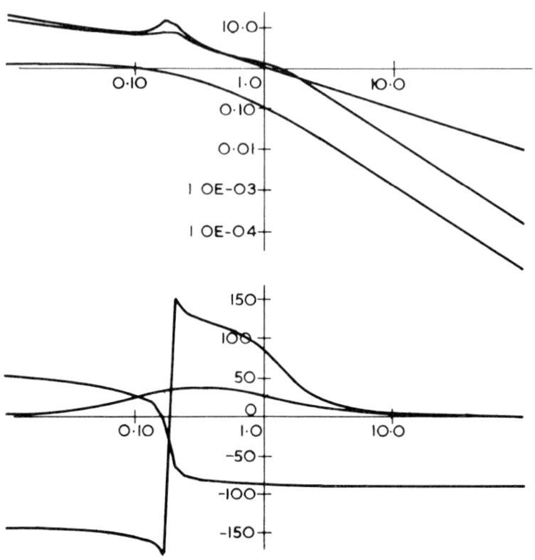

(*b*)

Figure 14.

alone are to be considered. Naturally this reduction of gain would have a detrimental effect on the speed of response of the system which already, from the frequency calibration of Figs. 14 (*a*) and 14 (*b*), is expected to be much slower for the feedback scheme with no inner-loop connections.

These arguments refer mostly to feedback configurations in which the only form of non-dynamic compensation employed is of the scalar type described earlier. For the sake of completeness, a further assessment of the effectiveness of the inner-loop compensator is undertaken in a later section where feedback compensation is derived from scalar as well as non-scalar and non-diagonal controllers of constant gains. For the moment however it should be noted that the total phase of the system as expressed by the sum of the phases of the characteristic loci exhibits much less phase-lag over the critical frequency range (from about 0·5 to 5 rad/s) when the inner-loop compensator is used. This is of course as expected due to the improved system poles obtained by the inner-loop given by eqn. (3.20). Controllers of constant gains may redistribute the phases of the individual characteristic loci but may not alter the total phase of the system. It follows then that the inner-loop design has reduced the amount of phase compensation required for the injection of feedback gain by the outer-loop.

3.2.1. *High frequency alignment*

A preliminary analysis of the eigenproperties of the transfer function matrix $G_c(s)$ of eqn. (3.22) establishes the need for an outer-loop design. Thus the characteristic loci shown in Figs. 13 (*a*) and 13 (*b*) display a disparity of characteristic gains and an imbalance of characteristic phases. Furthermore an inspection of the misalignment angles given by Fig. 15, together with the associated characteristic gains, shows that with increasing frequency two of

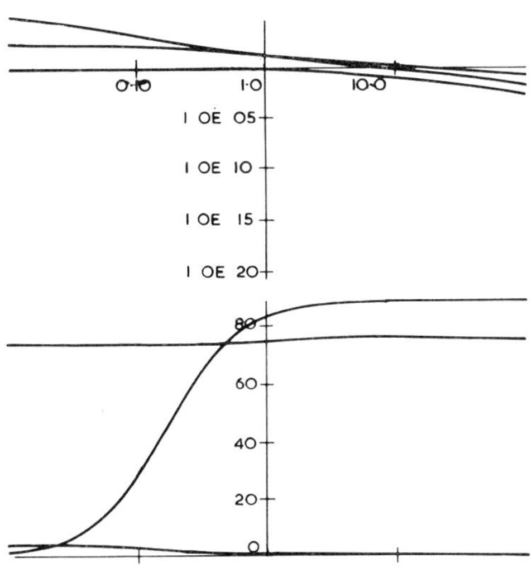

Figure 15.

the characteristic directions of $G(jw)$ remain badly aligned with the corresponding standard basis vector. Lack of alignment therefore, together with low characteristic gains (due to the system high-frequency attenuation), predicts high-frequency interaction. The outer-loop design is started by introducing a controller of constant gains for the supression of this interaction.

An inspection of the characteristic loci of $G_c(s)$ together with consideration of closed-loop responses defines the high frequency range of interest to lie between 1 rad/s and 10 rad/s. Application of the ALIGN algorithm at a few frequency points in this range shows that a satisfactory reduction of high-frequency misalignment angles is obtained by the controller K_h given by :

$$K_h = \begin{bmatrix} 2\cdot33 & 0\cdot21\times10^{-3} & 0\cdot32 \\ 8\cdot25 & 0\cdot105 & -0\cdot068 \\ -2\cdot51 & -0\cdot9\times10^{-3} & -0\cdot174 \end{bmatrix} \tag{3.23}$$

computed at $\omega_h = 2$ rad/s. Note that the frame matrix solutions produced by ALIGN are abitrary to within scaling of the columns of K_h ; appropriate scaling factors are therefore chosen with a view to improving stability and balancing the moduli of the characteristic gain loci.

The new transfer function matrix, with the introduction of the high-frequency controller in the forward path, is given by

$$Q(s) = G_c(s)K_h \tag{3.24}$$

The Nyquist and Bode plots for the characteristic gains of $Q(s)$ are displayed in Figs. 16 (*a*) and 16 (*b*) respectively and the misalignment angle plots are shown in Fig. 16 (*c*). From the latter it can be seen that with increasing frequency the new characteristic directions come to a satisfactory alignment with the standard basis vectors. Furthermore both the characteristic gains

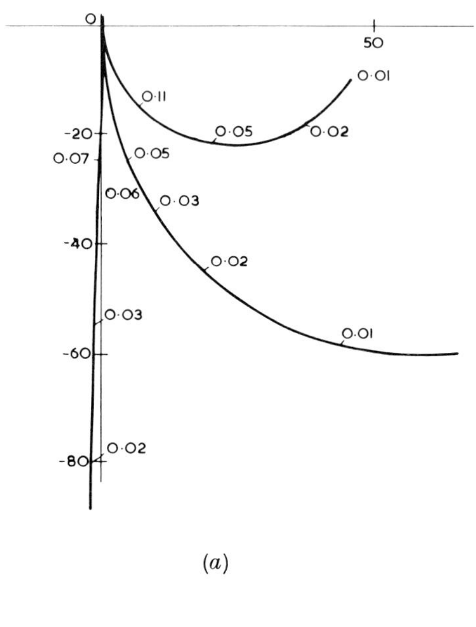

(*a*)

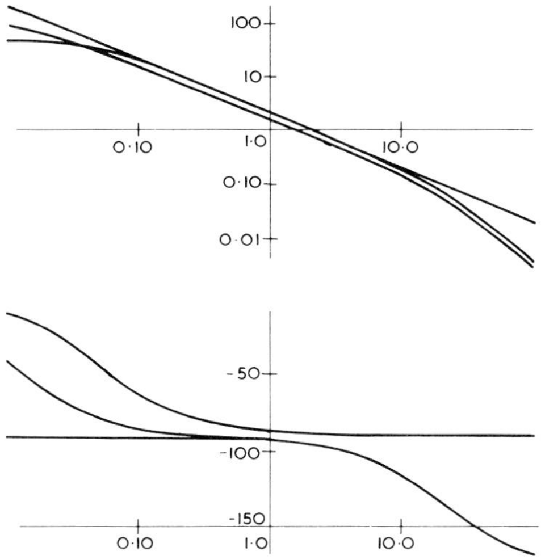

(*b*)

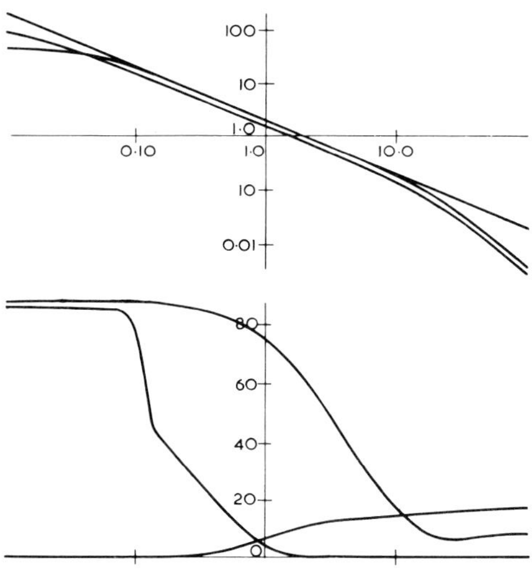

(*c*)

Figure 16.

now appear to be reasonably balanced in the frequency range from 0·2 rad/s to about 5 rad/s, thus affording the system good gain and phase margins. It follows then that the balancing effect that K_h is intended to have on the characteristic loci at 2 rad/s also extends to the intermediate frequencies where the proximity of the plots to the critical point is most pronounced, and hence there is no need for any further phase compensation. The outer-loop design may now be concluded by the introduction of some integral action,

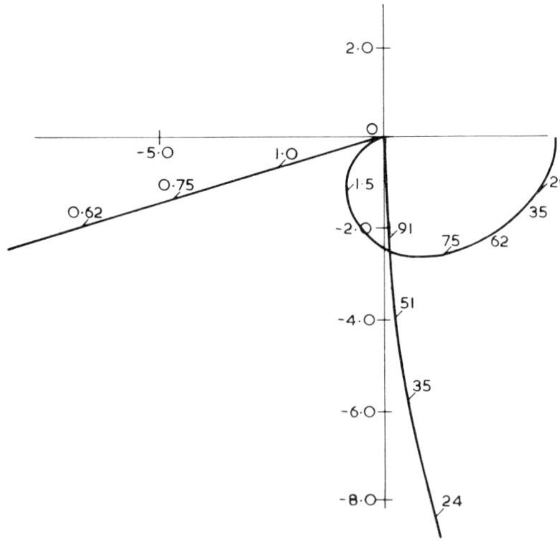

(*a*)

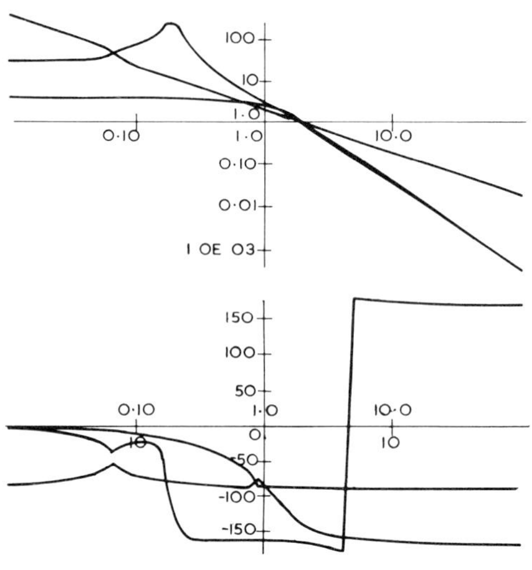

(*b*)

Figure 17.

with the aim of increasing and balancing the low-frequency gains ; high gains at low-frequencies are of course a prerequisite for good steady state accuracy.

The design of the high-frequency controller, K_h, completes the non-dynamic part of the outer-loop design. This therefore is a convenient point to reassess the usefulness of the inner-loop compensator. We do this by applying the ALIGN algorithm at high-frequencies to the transfer-function matrix of the system without the inner-loop compensator. By trial and error, $\omega = 2$ rad/s

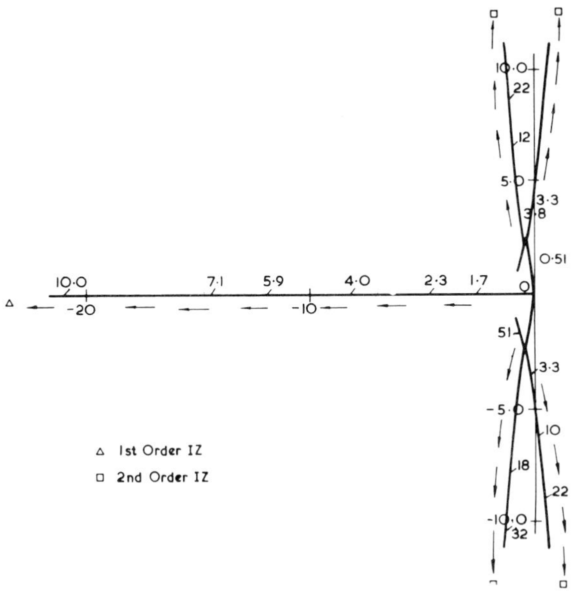
10·0
7·1
5·9
4·0
2·3
1·7
-20
-10
10·0
22
12
5·0
3·3
3·8
0·51
51
3·3
-5·0
10
18
22
-10·0
32
Δ 1st Order IZ
□ 2nd Order IZ

(*a*)

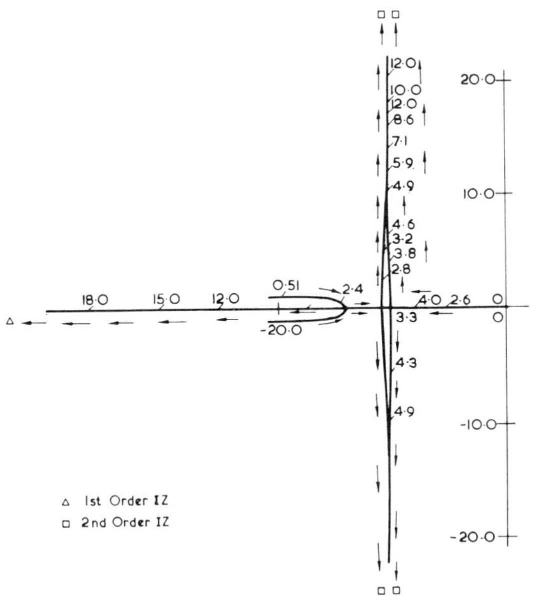
12·0
10·0
12·0
8·6
7·1
5·9
4·9
4·6
3·2
3·8
2·8
20·0
10·0
18·0
15·0
12·0
0·51
2·4
4·0
2·6
-20·0
3·3
4·3
4·9
-10·0
-20·0
Δ 1st Order IZ
□ 2nd Order IZ

(*b*)

Figure 18.

is selected as a suitable frequency point for the high-frequency alignment and this yields the controller

$$K'_h = \begin{bmatrix} -24\cdot55 & 0\cdot121 & -5\cdot712 \\ -0\cdot031 & 2\cdot002 & -0\cdot119 \\ -6\cdot118 & -0\cdot004 & -3\cdot66 \end{bmatrix} \quad (3.25)$$

The characteristic gain loci for the compensated transfer-function matrix $Q'(s) = G'(s)K'_h$ are shown in Figs. 17 (*a*) and 17 (*b*). A comparison of these plots with the loci of Figs. 16 (*a*) and 16 (*b*) shows that the removal of the inner-loop compensator causes a substantial deterioration of the system gain/phase margins as well as causing the speed of the system response to slow down. Further evidence of the beneficial effects of the inner-loop compensator is furnished by the root locus diagrams corresponding to the systems described by $Q'(s)$ and $Q(s)$ and plotted in Figs. 18 (*a*) and 18 (*b*) respectively. From these it can be seen that the inner-loop compensator, although unable to eliminate the second-order infinite zeros from the outer-loop system, has a very marked effect on the system gain margins. Thus whereas the use of an outer-loop scalar gain of 5 would render the system without the inner-loop compensator unstable, the same gain used in conjunction with the inner-loop compensator results in a set of poles with fast decay rates and damping ratios better than $\zeta = \sqrt{2}/2$.

3.2.2. *Low-frequency compensation*

The aim of the last stage in the characteristic locus design method is to boost and balance the system low-frequency characteristic gains. This may be conveniently done by the introduction of integral action through an approximate commutative controller K_l embedded in a composite integral plus-proportional controller, $K(s)$

$$K(s) = I + \frac{\alpha}{s} K_l \quad (3.26)$$

where α is a scalar parameter chosen to achieve a suitable transition from low-frequency to high-frequency behaviour of $K(s)$. Note that the identity operator I in eqn. (3.26) ensures that the proportional part of $K(s)$ does not modify the high-frequency effects of K_h. The approximately commutative controller K_l has the form of eqn. (2.11) so that

$$K_l = \hat{U} \Lambda_k \hat{V} \quad (3.27)$$

with

$$\Lambda_k = \text{diagonal}\,[k_1, k_2, k_3]$$

where k_1, k_2 and k_3 are the scaling factors required for balancing up the low-frequency gains.

The frequency point at $\omega = 0\cdot1$ rad/s was chosen as a suitable point for the application of the approximately commutative controller. The eigenframe approximants at this frequency were computed as

$$\hat{U} = \hat{V}^{-1} = \begin{bmatrix} 1\cdot97 & -0\cdot65 \times 10^{-1} & 0\cdot523 \\ -0\cdot16 \times 10^{-1} & 0\cdot683 & 0\cdot464 \\ 0\cdot34 \times 10^{-2} & -0\cdot4 \times 10^{-1} & -0\cdot67 \times 10^{-1} \end{bmatrix} \quad (3.28)$$

and the scaling factors k_1, k_2 and k_3 were selected to be

$$k_1 = 3, \quad k_2 = 5 \quad \text{and} \quad k_3 = 4{\cdot}5 \tag{3.29}$$

For the particular choice of $\alpha = 0{\cdot}1$, the composite controller of eqn. (3.27) becomes

$$K(s) = \hat{U}\Lambda(s)\hat{V}$$

where

$$\Lambda(s) = \text{diagonal}\left[\frac{s+0{\cdot}3}{s}, \frac{s+0{\cdot}5}{s}, \frac{s+0{\cdot}45}{s}\right] \tag{3.30}$$

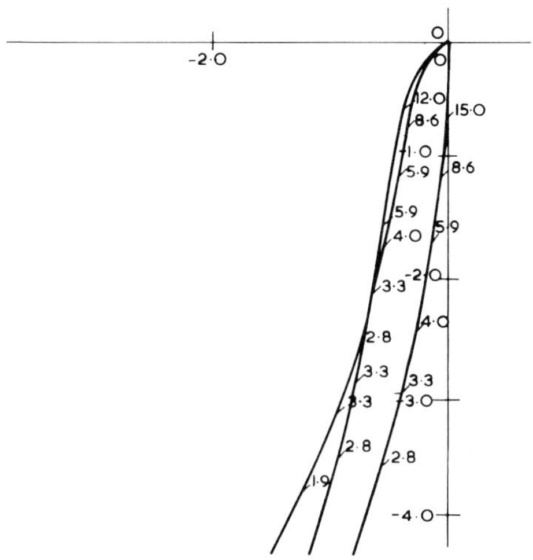

(*a*)

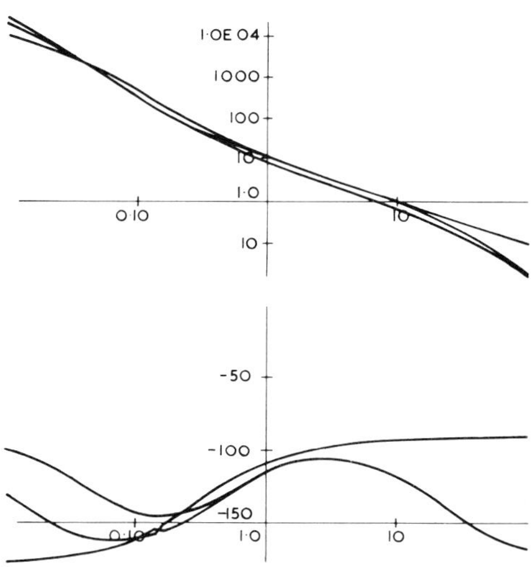

(*b*) *figure continued over*

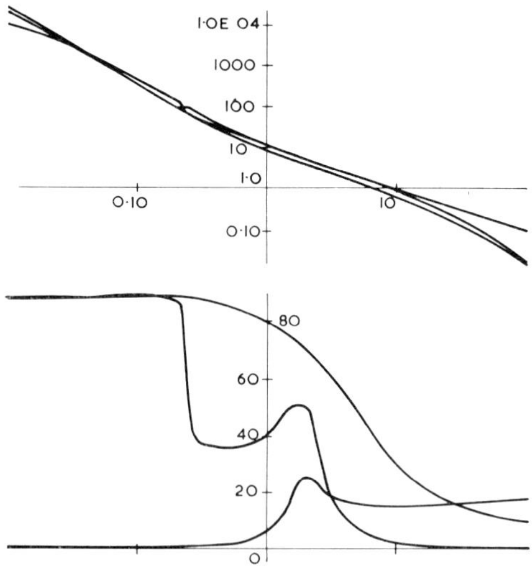

(*c*)

Figure 19.

The Nyquist and Bode plots of the characteristic gain loci, and the misalignment angle plots for the system with the high- and low-frequency compensation incorporated in the feedback scheme are shown in Figs. 19 (*a*), 19 (*b*) and 19 (*c*) respectively. From these it can be seen that the system has retained its good gain/phase margins, while the misalignment angles have remained substantially unaltered in the high-frequency range. At the same time, as expected, the low-frequency characteristic gains are larger and more balanced. Thus the overall feedback configuration is now considered to see if this results in a closed-loop system whose response is fast, non-oscillatory, non-overshooting and accurate in the steady state. An inspection of Fig. 19 (*c*) however shows that, although at low-frequencies the gains are high and at high-frequencies the misalignment angles are small, there exists an intermediate frequency range between 0·5 rad/s to about 5 rad/s where neither of these two conditions is satisfied. Thus at 1 rad/s the moduli of the characteristic gain loci do not exceed the value of 2 while the misalignment angle associated with the third standard basis vector is about 78°. It follows then that although loops 1 and 2 are expected to be non-interactive, the third loop will exhibit appreciable interaction at frequencies in the above range.

Most of this interaction may be reduced to an acceptable level by inserting into the feedback scheme an additional scalar gain of 5. Such an extra gain would restrict the frequencies for which neither the characteristic gains are sufficiently high nor the misalignment angles sufficiently small to the top end of the range from 0·5 to 5 rad/s, so that only a residual amount of high-frequency interaction will remain in the third loop. Note that, due to the good gain/phase margins, the increase in the loop gains is not going to harm the non-oscillatory nature of the expected system step response. The gain increase will of course speed up the system response further, resulting in

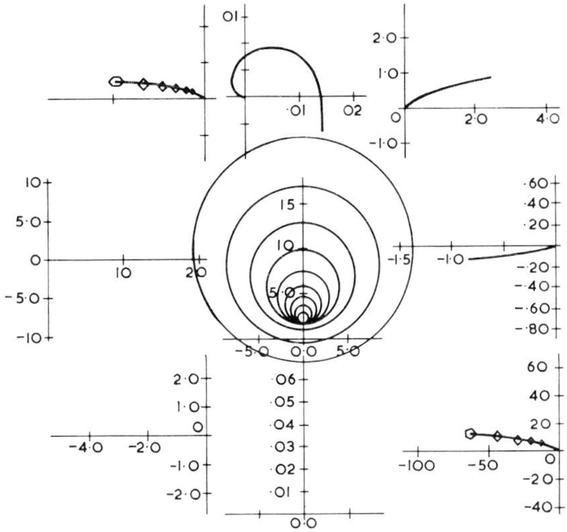

(*a*)

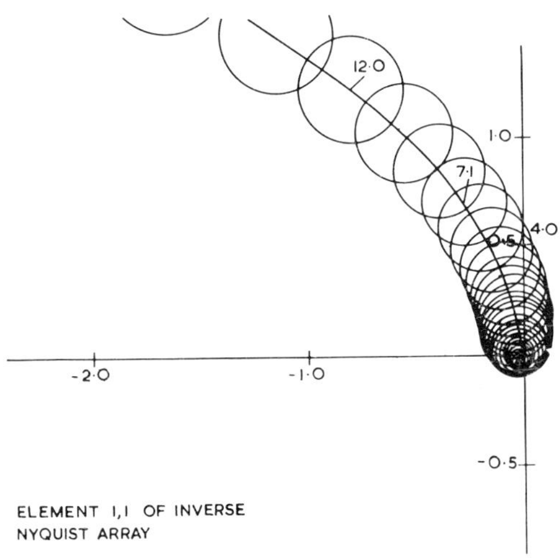

(*b*)

Figure 20.

estimated rise times which are better than 1 second, while the amount of expected overshoot (again estimated by calculations carried out on the individual characteristic gain loci) is better than 10% in all three loops. Before testing the system response it is perhaps of interest to observe that the final control scheme discussed above yields a system which is not diagonally dominant (Rosenbrook 1969) as is clearly demonstrated by the Nyquist array plots displayed in Figs. 20 (*a*) and 20 (*b*).

Digital computer simulations of the actual response of the overall feedback configuration to unit step inputs being applied in turn to the first, second and third reference inputs are shown in Figs. 21 (*a*), 21 (*b*) and 21 (*c*), respectively. Each one of these three figures displays the demand input signal together with the response of the three system outputs. The system response can indeed be seen to be fast, accurate in the steady state and non-oscillatory, with no significant overshoots. Furthermore the system response exhibits no appreciable interaction in all but the third loop where an acceptably low level of high-frequency interaction is present. Note that if the main reason for

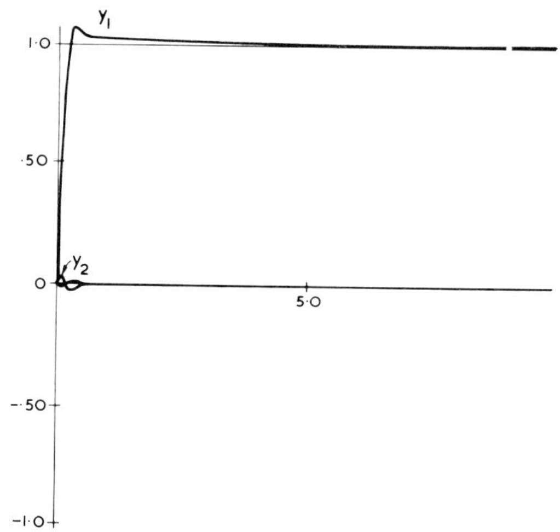

(*a*)

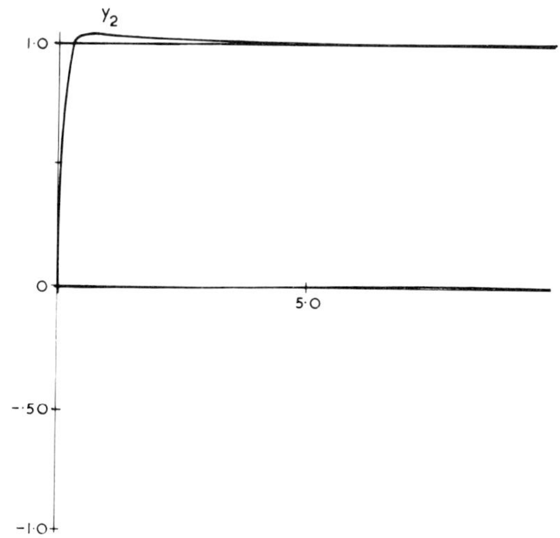

(*b*)

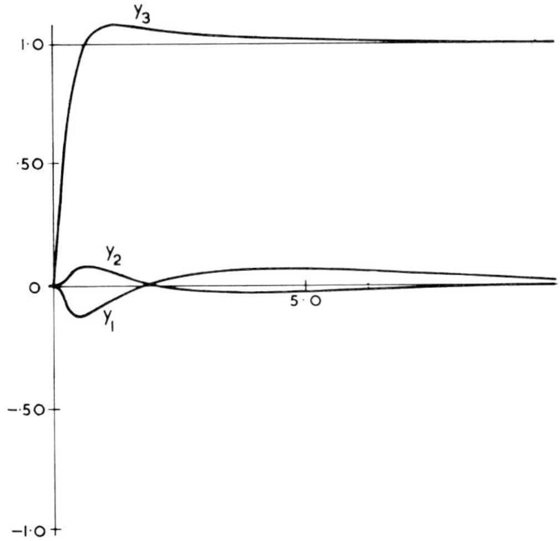

(c)

Figure 21.

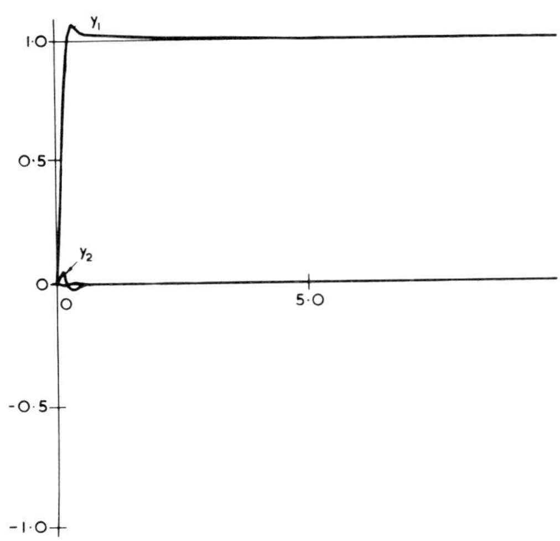

(a)

figure continued over

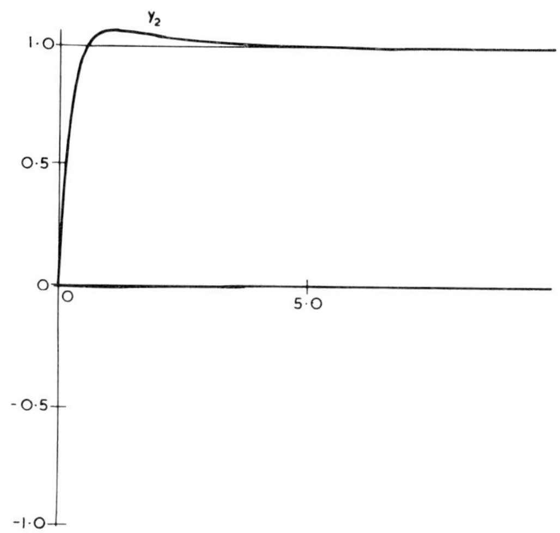

(b)

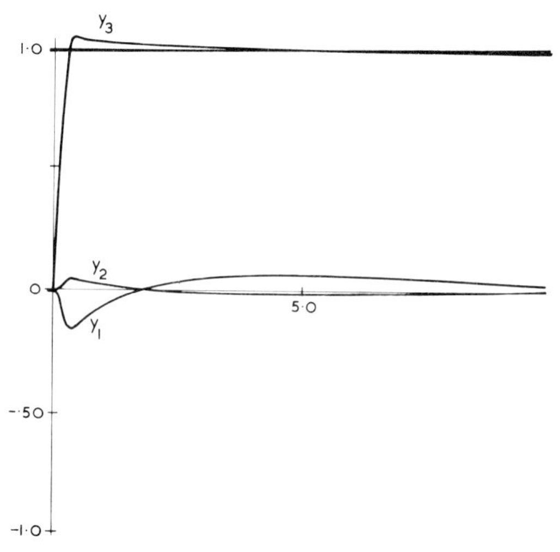

(c)

Figure 22.

introducing the extra gain of 5 is the suppression of interaction, and one did not require an equal speed of response in all loops, then a more efficient use of gain could be obtained by the use of non-scalar gain operators as for instance

$$K = \text{diagonal } [5,\ 2{\cdot}5,\ 2{\cdot}5] \tag{3.31}$$

where only half the amount of gain is used in loops 2 and 3. The time simulations for these values of gain are shown in Figs. 22 (a), 22 (b) and 22 (c).

The system step response has remained virtually unaltered, the main difference being that the rise time of loops 2 and 3 is about twice its previous value.

Acknowledgments

We are grateful to Mr. K. F. Goddard of the Royal Aircraft Establishment who provided the model of the flight control system. This work was supported by a grant from the Science Research Council.

References

Kouvaritakis, B., and MacFarlane, A. G. J., 1976 a, *Int. J. Control*, **23,** 149 ; 1967 b, *Ibid.*, **23,** 167.
Kouvaritakis, B., and Shaked, U., 1976, *Int. J. Control*, **23,** 297.
Kouvaritakis, B., 1978, *Int. J. Control*, **27,** 705.
Kouvaritakis, B., and Edmunds, J., 1977, *Proceedings of the National Engineering Consortium's International Forum on Alternatives for Linear Multivariable Control*, Chicago ; 1979, *Int. J. Control* (to be published).
MacFarlane, A. G. J., and Kouvaritakis, B., 1977, *Int. J. Control*, **25,** 837.
MacFarlane, A. G. J., Kouvaritakis, B., and Edmunds, J., 1977, *Proceedings of the National Engineering Consortium's International Forum on Alternatives for Linear Multivariable Control*, Chicago.
Mayne, D. Q., 1973, *Automatica*, **9,** 201.
Murray, W., 1977, Post-graduate diploma dissertation, University of Cambridge.
Rosenbrock, H. H., 1969, *Proc. Instn elect. Engrs*, **116,** 1929.
Shaked, U., 1976, *Int. J. Control*, **23,** 445 ; 1977, *Ibid.*, **25,** 603.
Postlethwaite, I., 1977, *Int. J. Control*, **25,** 677.

Extensions of the frame alignment technique and their use in the characteristic locus design method

J. EDMUNDS and B. KOUVARITAKIS

Frame alignment, performed by the ALIGN algorithm, plays an important role in the characteristic locus design method. A number of extensions to ALIGN intended to enhance the effectiveness of the design technique are proposed. In particular, the application of ALIGN at a number of frequency points is considered while an alternative least squares approach to frame alignment is developed. The problem of scaling both the frames to be aligned as well as the resulting frame solutions is examined and procedures for optimizing the use of the approximate commutative controller technique are suggested.

1. Introduction

The characteristic locus method studies system behaviour through the characteristic properties of the system transfer function matrix, $G(s)$ (MacFarlane and Kouvaritakis 1977, MacFarlane and Postlethwaite 1977). Thus the frequency response of the eigenfunctions and eigenvector functions of $G(s)$ are used for the assessment of closed loop stability, accuracy and interaction. For this reason the design technique that emanates from this method attempts to compensate and modify system performance by means of adjusting the eigenproperties of $G(s)$. An important role in achieving this is played by a maximization algorithm called ALIGN (Kouvaritakis 1974) which generates real frames that are aligned as closely as possible with given complex frames. ALIGN is then used to approximate the eigenframes of $G(s)$ at a specific frequency, s_0, so that simple controllers which exhibit a degree of commutativity with $G(s)$ at s_0 may be constructed for the adjustment of the eigenvalues of $G(s)$. Eigenvector adjustment may also be achieved by using ALIGN to approximate the column structure of the inverse of $G(s)$.

The application of ALIGN as described in § 2 has been shown to be effective in the design of multivariable systems. Here we consider some of the problems in its application and we suggest a number of extensions which help overcome these difficulties and which increase the flexibility of the characteristic locus design technique. Thus in § 3 a method for frame alignment at a number of frequency points is developed while an alternative approach to alignment, based on a least squares minimization algorithm is considered in § 4. The question of scaling the row or column vector members of a frame is investigated in § 5 while finally the problem of scaling the resulting frames themselves in order to improve the approximately commutative controller technique is considered in § 6.

2. The ALIGN algorithm

Consider the problem of characteristic gain adjustment (MacFarlane and Kouvaritakis 1977) for a system described by its $m \times m$ transfer function matrix $G(s)$. It is assumed that the spectral decomposition of $G(s)$ is given by (MacFarlane and Postlethwaite 1977)

$$G(s) = W(s)\Lambda_G(s)V(s) \tag{2.1}$$

where $W(s)$, $V(s)$ are the eigenvector and dual eigenvector matrices containing as their column and row vectors the characteristic directions (CD) $\mathbf{w}_i(s)$ and their dual, $\mathbf{v}_i^{\mathrm{T}}(s)$, respectively and where $\Lambda_G(s)$ is the diagonal eigenvalue matrix comprising the branches $g_i(s)$ of the characteristic gain (CG) of $G(s)$. Then the controller

$$K(s) = A\Lambda_K(s)B \tag{2.2}$$

where $A = [\mathbf{a}_1, \mathbf{a}_2, \ldots, \mathbf{a}_m]$, $B^{\mathrm{T}} = [\mathbf{b}_1, \mathbf{b}_2, \ldots, \mathbf{b}_m]$ and $\Lambda_K(s) = \mathrm{diag}\,\{k_i(s)\}$ will perform the desired CG adjustment yielding a new CG with branches

$$g_i(s) = k_i(s)g_i(s) \tag{2.3}$$

if $K(s)$ commutes with $G(s)$, namely if A and B are the eigenvector matrices $W(s)$ and $V(s)$ of $G(s)$.

On account of physical considerations (MacFarlane and Kouvaritakis 1977, Kouvaritakis 1974) one often wishes the frames A and B to be real and frequency independent. Since, however, both the $\mathbf{w}_i(s)$ and $\mathbf{v}_i^{\mathrm{T}}(s)$, in general, are complex and vary with s, one is confronted with the problem of selecting vectors $\mathbf{a}_i$ and $\mathbf{b}_i^{\mathrm{T}}$, $i = 1, 2, \ldots, m$ which approximate as well as possible the values of $\mathbf{w}_i(s)$ and $\mathbf{v}_i^{\mathrm{T}}(s)$ at a specific frequency of interest. It is noted that if $\mathbf{a}_i = \mathbf{w}_i(s)$ then $V(s)\mathbf{a}_i$ would be the ith standard basis vector $\mathbf{e}_i$; as $\mathbf{a}_i$, however, comes out of alignment with $\mathbf{w}_i(s)$ so the ith element of $V(s)\mathbf{a}_i$ decreases in relation to the remainder which now assume non-zero values. This simple argument indicates that an appropriate choice for $\mathbf{a}_i$ is the one that maximizes the ratio of the magnitude of the ith element to the sum of the magnitudes of the rest, namely the ratio

$$\phi_i = \frac{\overline{(\mathbf{v}_i^{\mathrm{T}}\,\mathbf{a}_i)}(\mathbf{v}_i^{\mathrm{T}}\,\mathbf{a}_i)}{\sum\limits_{\substack{j=1\\ j\neq i}}^{m} \overline{(\mathbf{v}_j^{\mathrm{T}}\,\mathbf{a}_i)}(\mathbf{v}_j^{\mathrm{T}}\,\mathbf{a}_i)} = \frac{|(\mathbf{v}_i, \mathbf{a}_i)|^2}{\sum\limits_{\substack{j=1\\ j\neq i}}^{m} |(\mathbf{v}_j, \mathbf{a}_i)|^2} \tag{2.4}$$

Note that for convenience the argument s in $\mathbf{v}_i(s)$ has been omitted. Simple algebraic manipulations will show that if $\boldsymbol{\alpha}_i$ and $\boldsymbol{\beta}_i$ denote the real and imaginary part of $\mathbf{v}_i$ then

$$|(\mathbf{v}_i, \mathbf{a}_i)|^2 = \mathbf{a}_i^{\mathrm{T}}\,C_i\mathbf{a}_i \quad \text{and} \quad \sum_{\substack{j=1\\ j\neq i}}^{m} |(\mathbf{v}_j, \mathbf{a}_i)|^2 = \mathbf{a}_i^{\mathrm{T}}\,D_i\mathbf{a}_i \tag{2.5}$$

where

$$C_i = \boldsymbol{\alpha}_i\boldsymbol{\alpha}_i^{\mathrm{T}} + \boldsymbol{\beta}_i\boldsymbol{\beta}_i^{\mathrm{T}} \quad \text{and} \quad D_i = \sum_{\substack{j=1\\ j\neq i}}^{m} C_j \tag{2.6}$$

All possible solutions for $\mathbf{a}_i$ may now be obtained by introducing the above into eqn. (2.4) and by subsequently setting the derivative of ϕ_i with respect to $\mathbf{a}_i$ equal to zero in order to obtain (MacFarlane and Kouvaritakis 1977)

$$C_i\mathbf{a}_i=\phi_i D_i\mathbf{a}_i \quad \text{or} \quad D_i^{-1}C_i\mathbf{a}_i=\phi_i\mathbf{a}_i \tag{2.7}$$

assuming of course D_i to be non-singular.

The best real approximation $\mathbf{a}_i$ to the complex eigenvector $\mathbf{w}_i(s)$ is given by the eigenvector of $D_i^{-1}C_i$ that corresponds to the maximum eigenvalue. It is easy to show that the algorithm for obtaining $\mathbf{a}_i$ described above, which is referred to as the ALIGN algorithm, always generates real solutions for $\mathbf{a}_i$ corresponding to positive values of ϕ_i (Kouvaritakis 1974). Furthermore, because of the special form of D_i and C_i, which are real symmetric matrices, the algorithm leads to a fast and efficient computer implementation. Note that frame alignment can be also achieved using the pseudo-diagonalization routine (Hawkins 1972) which is normally applied by the inverse Nyquist array designer. However, this routine, rather than maximizing the ratio of the numerator to the denominator of ϕ_i, simply minimizes the denominator. A moment's reflection will show that, on account of this difference, pseudo-diagonalization cannot produce better alignment than ALIGN and indeed in most applications it is likely to produce worse results.

The problem of approximating the dual set of eigenvectors $\mathbf{v}_i$ by real vectors $\mathbf{b}_i$ is dual to the one just examined since one is looking for solutions $\mathbf{b}_i$ such that $\mathbf{b}_i^{\mathrm{T}} W$ is as close as possible to the transpose of the ith standard basis vector $\mathbf{e}_i$. A simple transposition of this problem will readily bring it into the form of the ALIGN problem.

Apart from its use in the adjustment of the CG of $G(s)$ the ALIGN algorithm may also be employed for the manipulation of the CD. Thus, it is often desirable to improve the alignment of the CD with the standard basis vectors at high frequencies since this, in most practical applications, turns out to be the only acceptable means of suppressing high frequency interaction. A sensible way of achieving this is to produce controllers, K_h, that imitate the column structure of $G^{-1}(s)$. If in addition, it is now required that K_h be real, then the task of designing K_h leads to a frame alignment problem. Thus the ALIGN algorithm may be used once again, the only difference being that the vectors $\boldsymbol{\alpha}_i^{\mathrm{T}}$ and $\boldsymbol{\beta}_i^{\mathrm{T}}$ now must be chosen to be the real and imaginary part of the ith, $i=1, 2, \ldots, m$, row vector of $G(s)$ at the preselected frequency s.

3. The multi-frequency application of ALIGN

The effectiveness of ALIGN in design has been established in a number of case studies (MacFarlane and Kouvaritakis 1977, Kouvaritakis and Edmunds 1978). It will be remembered, however, that the algorithm in its present form may only be applied at one frequency, s_0, and thus the usefulness of the resulting controller will be limited to the range of frequencies about s_0 for which alignment is preserved. This restriction could cause concern in cases where the eigenframe $W(s)$ or the row structure of $G(s)$ undergo rapid changes as s moves in the vicinity of s_0. For example, consider the system with transfer function matrix

$$G(s)=\frac{1}{s^3+3\cdot5s^2+3\cdot5+1}\begin{bmatrix}-s^2+s-3 & s^2+s+7\\ 0\cdot5s^2-s-3\cdot5 & 0\cdot5s^2+s+6\cdot5\end{bmatrix} \tag{3.1}$$

with characteristic locus moduli, $|g_1(j\omega)|$ and $|g_2(j\omega)|$ and misalignment angles $\theta_1(j\omega)$ and $\theta_2(j\omega)$ as shown in Fig. 1. Note that misalignment angles give a measure of the angular displacement between eigenvectors and standard basis vectors and are as defined by MacFarlane (1972). An attempt to use ALIGN in order to reduce θ_1 and θ_2 around the frequency $\omega_0=2$ rad/s would produce good but very localized results. This is clearly demonstrated by Figs. 2 (*a–c*) which display the values of θ_1 and θ_2 after the application of ALIGN at 1 rad/s, 2 rad/s and 3 rad/s, respectively. This of course is as expected since $G(s)$ was especially chosen to possess a row structure which exhibits high sensitivity to frequency variations about $\omega_0=2$ rad/s.

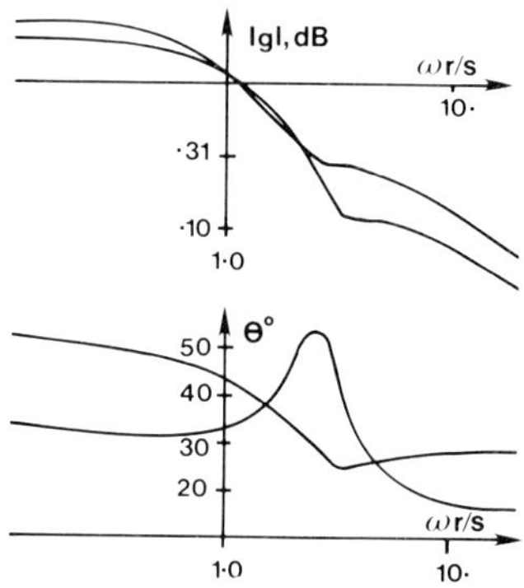

Figure 1

In circumstances like this the designer must avoid the critical frequency range thus being forced to select a frequency much lower or higher than his original choice, ω_0. Such a dilemma could be undesirable as the application of ALIGN at $\omega_1 \ll \omega_0$ or $\omega_2 \gg \omega_0$ would not cater for the misalignment angles at $\omega \geqslant \omega_0$ or $\omega \leqslant \omega_0$, respectively. This is certainly the case with the example of eqn. (3.1) as depicted by Figs. 3 (*a*, *b*) which correspond to a choice of $\omega_1=0\cdot5$ rad/s and $\omega_2=10$ rad/s.

To overcome this difficulty we extend the ALIGN algorithm so that it produces real frames which approximate complex frames considered over a number of frequency points, t. To do this we argue that if $\mathbf{a}_i$ were exactly equal to the ith eigenvector $\mathbf{w}_i(j\omega)$ at the frequencies $\omega_1, \omega_2, \ldots, \omega_t$ then each of the vectors $V(j\omega_k)\mathbf{a}_i$ for $k=1, 2, \ldots, t$ would be equal to the ith standard basis vector $\mathbf{e}_i$ and hence the vector $\boldsymbol{\xi}_i=\sum_{k=1}^{t} p_k V(j\omega_k)\mathbf{a}_i$ itself would be parallel to $\mathbf{e}_i$; the constants p_k are appropriate, positive and real weighting factors. However, because the elements of $\boldsymbol{\xi}_i$ are the geometric sum of the complex quantities $p_k\mathbf{v}_l^{\mathrm{T}}(j\omega_k)\mathbf{a}_i$, it is not possible to claim the converse, namely

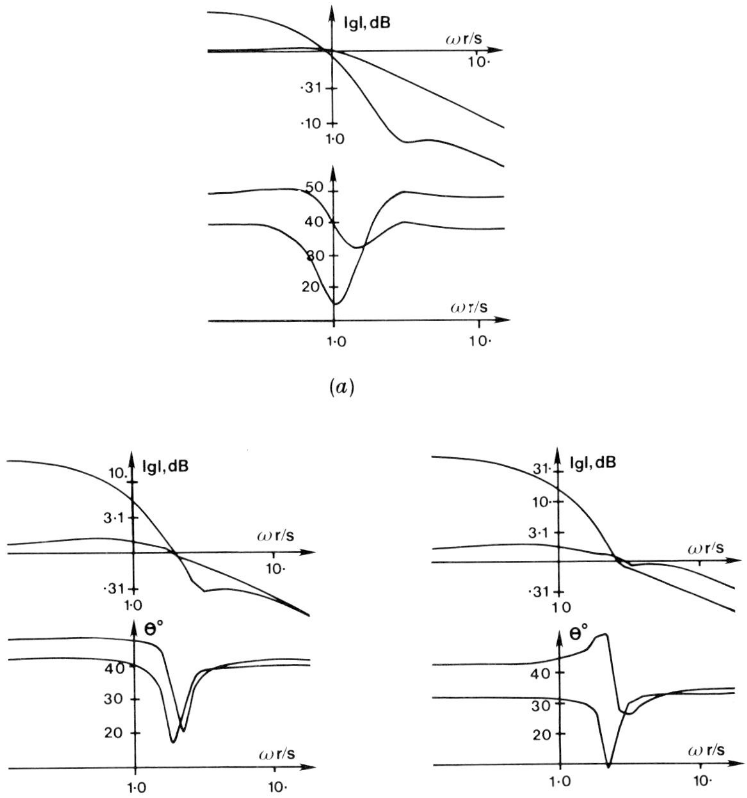

Figure 2

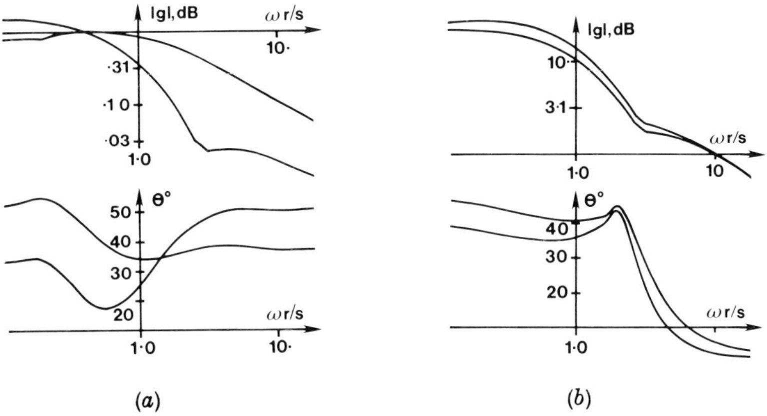

Figure 3

that if $\boldsymbol{\xi}_i$ is parallel with $\mathbf{e}_i$ then $\mathbf{a}_i$ is equal to $\mathbf{w}_i(j\omega)$ at $\omega=\omega_k$, $k=1, 2, \ldots, t$. For this reason we consider the sums of the squares of the moduli of the terms $p_k\mathbf{v}_l^{\mathrm{T}}(j\omega_k)\mathbf{a}_i$ and define $\boldsymbol{\xi}_i$ to be the vector which contains these sums as elements. The proximity of $\boldsymbol{\xi}_i$ to $\mathbf{e}_i$ then gives a measure of the alignment of $\mathbf{a}_i$ with $\mathbf{w}_i(j\omega)$ in the range of frequencies defined by the set of ω_k, $k=1, 2, \ldots, t$. Thus we are led to the problem of maximizing the new index of performance

$$\tilde{\phi}_i=\frac{\sum_{k=1}^{t} p_k^2|(\mathbf{v}_i(j\omega_k), \mathbf{a}_i)|^2}{\sum_{k=1}^{t} p_k^2 \sum_{\substack{l=1\\ l\neq i}}^{m} |(\mathbf{v}_l(j\omega_k), \mathbf{a}_i)|^2} \tag{3.2}$$

Upon substitution of eqn. (2.5) we obtain

$$\tilde{\phi}_i=\frac{\mathbf{a}_i^{\mathrm{T}}\tilde{C}_i\mathbf{a}_i}{\mathbf{a}_i^{\mathrm{T}}\tilde{D}_i\mathbf{a}_i} \tag{3.3}$$

where

$$\tilde{C}_i=\sum_{k=1}^{k} p_k^2\, C_i(j\omega_k) \quad \text{and} \quad \tilde{D}_i=\sum_{k=1}^{t} p_k^2\, D_i(j\omega_k)$$

Note that the matrices C_i and D_i now appear as functions of ω, which indeed they are. The values of C_i and D_i at ω_k may be computed by eqn. (2.6) where $\boldsymbol{\alpha}_i$ and $\boldsymbol{\beta}_i$ now denote the real and imaginary part of the ith row of $G(j\omega_k)$ or $V(j\omega_k)$.

The new objective function $\tilde{\phi}_i$ has the same form as ϕ_i and hence the optimal solution for $\mathbf{a}_i$ may be obtained directly from eqn. (2.7) providing that C_i and D_i are replaced by $\tilde{C}_i$ and $\tilde{D}_i$, respectively.

From the design point of view then, we see that by a simple extension of ALIGN one is able to generate controllers which can be used for the eigenvalue/eigenvector adjustment of $G(s)$ over a range of frequencies. It must be appreciated, however, that normally one may expect the multi-frequency application of ALIGN to result in a deterioration of the overall degree of alignment. For this reason it is essential that the designer has some means of specifying areas in the stipulated frequency range over which he wishes alignment to be most accurate. Fortunately this may easily be accomplished by an appropriate choice of the weighting factors, p_k. Note that in the case of eigenvector adjustments for proper systems, the choice of p_k has a special role to play since with $p_k=1$ for all $k=1, 2, \ldots, t$ the lower frequencies will tend to dominate the calculations ; the p_k then must be chosen so as to compensate for the high frequency attenuation of most physical systems.

To illustrate the effectiveness of the extended ALIGN algorithm we consider the system of eqn. (3.1). ALIGN is applied at two frequencies $\omega=0{\cdot}5$ rad/s and 10 rad/s with the corresponding weights set at $p_1=1$, $p_2=10^{3/2}$. The resulting misalignment angles are shown in Fig. 4 where it is seen that although at $\omega_1=0{\cdot}5$ rad/s and $\omega_2=10$ rad/s θ_1 and θ_2 assume larger values than those obtained by the individual application of ALIGN at 0·5 rad/s or 10 rad/s the multi-frequency application of ALIGN gives the best overall results over the whole frequency range from 0·5 to 10 rad/s.

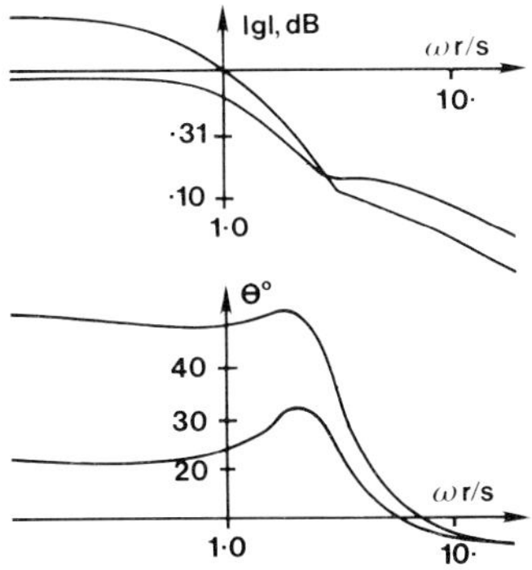

Figure 4

4. A least squares approach to alignment

The discussion thus far indicates that the problem of alignment may be reduced to a generalized eigenvalue/eigenvector problem of the type of eqn. (2.7). The solution of this equation however, depends on the non-singularity of D_i and this may on certain occasions cause difficulties especially when some of the row vectors of $V(j\omega)$ or $G(j\omega)$ are already in the linear manifold of a real vector which tends to reduce the rank of D_i. For this reason here we propose an alternative approach to alignment which is based on a least squares algorithm.

Ideally we wish to choose $\mathbf{a}_i$ such that the vector $V(j\omega)\mathbf{a}_i$ is the ith standard basis vector $\mathbf{e}_i$. This however may not be possible to achieve as $V(j\omega)$ may be complex, in which case $V(j\omega)\mathbf{a}_i$ will be a complex multiple of $\mathbf{e}_i$ plus an error term $\boldsymbol{\epsilon}_i$. Thus after appropriate scaling we may write

$$V\mathbf{a}_i = \exp(j\delta_i)\, \mathbf{e}_i + \boldsymbol{\epsilon}_i \tag{4.1}$$

where δ_i is some angle. Note that for simplicity we drop the argument $(j\omega)$. Both δ_i and $\mathbf{a}_i$ are free parameters which we may choose to minimize the error term. Thus a convenient index of performance is the norm of $\boldsymbol{\epsilon}_i$

$$\phi_i = \|\boldsymbol{\epsilon}_i\| = \bar{\boldsymbol{\epsilon}}_i^{\mathrm{T}}\, \boldsymbol{\epsilon}_i = \boldsymbol{\epsilon}_i^*\, \boldsymbol{\epsilon}_i \tag{4.2}$$

where $\boldsymbol{\epsilon}_i^*$ denotes the transpose of the complex conjugate of $\boldsymbol{\epsilon}_i$, otherwise written as $(\bar{\boldsymbol{\epsilon}}_i)^{\mathrm{T}}$.

Upon substitution of eqn. (4.1) into (4.2) we obtain

$$\phi_i = \mathbf{a}_i^{\mathrm{T}}\, V^* V\mathbf{a}_i - \exp(-j\delta_i)\, \mathbf{v}_i^{\mathrm{T}}\, \mathbf{a}_i - \exp(j\delta_i)\, \mathbf{a}_i^{\mathrm{T}}\, \bar{\mathbf{v}}_i + 1 \tag{4.3}$$

Hence, ϕ_i is a function of both $\mathbf{a}_i$ and δ_i and as such has singular points at

$$\frac{\partial \phi_i}{\partial \mathbf{a}_i} = [V^* V + (V^* V)^{\mathrm{T}}]\mathbf{a}_i - \exp(-j\delta_i)\, \mathbf{v}_i - \exp(j\delta_i)\, \bar{\mathbf{v}}_i = 0 \tag{4.4}$$

and

$$\frac{\partial \phi_i}{\partial \delta_i} = j\exp(-j\delta_i)\, \mathbf{v}_i^{\mathrm{T}}\, \mathbf{a}_i - j\exp(j\delta_i)\, \mathbf{a}_i^{\mathrm{T}}\, \bar{\mathbf{v}}_i = 0 \tag{4.5}$$

Note that using the notation of eqn. (2.6) we may write

$$V^* V + (V^* V)^{\mathrm{T}} = 2C_i + 2D_i = D \tag{4.6}$$

which we substitute into eqns. (4.4) and (4.5) to get

$$\mathbf{a}_i = D^{-1} (\exp(-j\delta_i)\, \mathbf{v}_i + \exp(j\delta_i)\, \bar{\mathbf{v}}_i) \tag{4.7}$$

and

$$\exp(-j2\delta_i)\, \mathbf{v}_i^{\mathrm{T}} D^{-1}\mathbf{v}_i + \mathbf{v}_i^{\mathrm{T}} D^{-1}\bar{\mathbf{v}}_i - \mathbf{v}_i^{\mathrm{T}} D^{-1}\bar{\mathbf{v}}_i - \exp(j2\delta_i)\, \mathbf{v}_i^* D^{-1}\bar{\mathbf{v}}_i = 0 \tag{4.8}$$

Performing all the necessary cancellations and making the substitution $\rho_i \exp(j\psi_i) = \mathbf{v}_i^{\mathrm{T}} D^{-1}\mathbf{v}_i$ into the above, we deduce that

$$\rho_i \exp(j(2\delta_i - \psi_i)) - \rho_i \exp(-j(2\delta_i - \psi_i)) = j2\rho_i \sin(2\delta_i - \psi_i) = 0$$

or

$$\delta_i = \frac{\psi_i}{2} \tag{4.9}$$

This completes the frame alignment algorithm since the solutions $\mathbf{a}_i$ which minimize the index of performance ϕ_i are given by

$$A = [\mathbf{a}_1, \mathbf{a}_2, \ldots, \mathbf{a}_m] = D^{-1}[V^{\mathrm{T}} \exp(-j\tfrac{1}{2}\psi) + V^* \exp(-j\tfrac{1}{2}\psi)] \tag{4.10}$$

where $\psi = \operatorname{diag}\{\psi_i\}$ and ψ_i denotes the phase of diagonal elements of $VD^{-1}V^{\mathrm{T}}$.

As with ALIGN the least squares method will also solve the dual problem of looking for a real frame B which approximates V in the sense that it minimizes the error term in

$$BW = \exp(j \operatorname{diag}\{\delta_i\}) + E \tag{4.11}$$

A mere transposition of the above reduces the problem to the form of that posed at the beginning of the section.

5. The problem of frame scaling

The scaling of the eigenvectors and dual eigenvectors is, of course, arbitrary so that the frames for alignment of W and V may be substituted by $W \operatorname{diag}\{1/k_i\}$ and $\operatorname{diag}\{k_i\}\, V$ where k_i, $i = 1, 2, \ldots, m$, are some scaling factors. The alignment problem then becomes that of finding appropriate diagonal matrices θ_A, θ_B, frames A, B and scaling factors k_i such that the error terms E_A and E_B in

$$\operatorname{diag}\{k_i\}\, VA = \exp(j\theta_A) + E_A, \quad BW \operatorname{diag}\left\{\frac{1}{k_i}\right\} = \exp(j\theta_B) + E_B \tag{5.1}$$

are minimized. The simultaneous minimization of both E_A and E_B is important for the design of the approximately commutative controller of eqn. (2.2), since the effectiveness of the controller in modifying the characteristic loci of $G(s)$ in accordance with eqn. (2.3) depends on the good alignment of both W and V by A and B respectively.

An inspection of eqns. (2.6) and (4.6) establishes that in general different scaling factors will lead to a different matrix D and hence by eqn. (4.10) a different matrix A (and similarly a different matrix B). The dependence of A and B on k_i makes the direct solution of the minimization problem cumbersome.

For this reason an iterative indirect approach may be adopted. Thus for a given set of k_i we use the method of § 4 to obtain the optimum A and B. These are introduced into eqn. (5.1) and the error terms E_A and E_B are minimized subject to appropriate choice of θ_A, θ_B and k_i. The new set of scaling factors now may be used to improve on the solution for A and B. We proceed in this fashion until E_A and E_B cannot be minimized any further.

The solution for θ_A, θ_B and k_j, $j = 1, 2, \ldots, m$ is easy to obtain. First we note that we may choose the ith diagonal element of θ_A and θ_B say δ_A and δ_B and the ith scaling factor, say k, independently for each value of i. Then if $\mathbf{m}$ and $\mathbf{n}$ denote the ith column of VA and $W^{\mathrm{T}}B^{\mathrm{T}}$ we choose δ_A, δ_B and k to minimize the error term given below

$$\begin{aligned}\phi = \boldsymbol{\epsilon}^*\boldsymbol{\epsilon} &= (k\mathbf{m} - \exp(j\delta_A)\,\mathbf{e}_i)^*(k\mathbf{m} - \exp(j\delta_A)\,\mathbf{e}_i) \\ &\quad + \left(\frac{1}{k}\mathbf{n} - \exp(j\delta_B)\,\mathbf{e}_i\right)^*\left(\frac{1}{k}\mathbf{n} - \exp(j\delta_B)\,\mathbf{e}_i\right) \\ &= k^2\mathbf{m}^*\mathbf{m} - k\,(\exp(-j\delta_A)\,\mathbf{e}_i^{\mathrm{T}}\,\mathbf{m} + \exp(j\delta_A)\,\mathbf{m}^*\mathbf{e}_i) \\ &\quad + 2 - \frac{1}{k}\,(\exp(-j\delta_B)\,\mathbf{e}_i^{\mathrm{T}}\,\mathbf{n} + \exp(j\delta_B)\,\mathbf{n}^*\mathbf{e}_i) + \frac{1}{k^2}\,\mathbf{n}^*\mathbf{n} \qquad (5.2)\end{aligned}$$

Differentiation with respect to δ_A, δ_B and k yields

$$\begin{aligned}&\exp(-j\delta_A)\,\mathbf{e}_i^{\mathrm{T}}\,\mathbf{m} - \exp(j\delta_A)\,\mathbf{m}^*\mathbf{e}_i = 0, \\ &\qquad \exp(-j\delta_B)\,\mathbf{e}_i^{\mathrm{T}}\,\mathbf{n} - \exp(j\delta_B)\,\mathbf{n}^*\mathbf{e}_i = 0 \qquad (5.3)\end{aligned}$$

and

$$\begin{aligned}&(2\mathbf{m}^*\mathbf{m})k^4 - (\exp(-j\delta_A)\,\mathbf{e}_i^{\mathrm{T}}\,\mathbf{m} + \exp(j\delta_A)\,\mathbf{m}^*\mathbf{e}_i)k^3 \\ &\qquad + (\exp(-j\delta_B)\,\mathbf{e}_i^{\mathrm{T}}\,\mathbf{n} + \exp(j\delta_B)\,\mathbf{n}^*\mathbf{e}_i)k - 2\mathbf{n}^*\mathbf{n} = 0 \qquad (5.4)\end{aligned}$$

Substituting $\mathbf{e}_i^{\mathrm{T}}\,\mathbf{m} = \mathbf{v}_i^{\mathrm{T}}\,\mathbf{a}_i = \rho_A \exp(j\psi_A)$ and $\mathbf{e}_i^{\mathrm{T}}\,\mathbf{n} = \mathbf{w}_i^{\mathrm{T}}\,\mathbf{b}_i = \rho_B \exp(j\psi_B)$ into the above we get

$$\delta_A = \psi_A, \quad \delta_B = \psi_B \qquad (5.5)$$

and

$$\rho_A^{\,2}\,k^4 - \rho_A\,k^3 + \rho_B k - \rho_B^{\,2} = 0 \qquad (5.6)$$

The real solutions of this equation may be easily obtained by a digital computer and of these the solution that yields the minimum value for ϕ may be selected as the appropriate scaling factor.

6. The duality between the frames aligning the eigenframe and dual eigenframe

A further constraint for the successful application of the approximately commutative controller of eqn. (2.2) is that the frames A and B should be dual to each other, namely that $AB = BA = I$. We consider here the possibility of scaling the columns of A and the rows of B in order to meet this constraint. Thus A and B are replaced by AK_A and K_BB, where K_A and K_B are two diagonal matrices and we seek the optimal choice for K_A and K_B such that the error term E in

$$I = AK_AK_BB + E = AKB + E \qquad (6.1)$$

is minimized. Since the effect of K_A and K_B is accumulated in the product $K = K_A K_B$ we look first at the optimal solution for $K = \text{diag}\,\{k_i\}$. A simple rearrangement of eqn. (6.1) will now yield

$$\begin{bmatrix} a_{11}b_{11} & a_{12}b_{12} \cdots a_{1m}b_{m1} \\ a_{11}b_{12} & a_{12}b_{22} \cdots a_{1m}b_{m2} \\ \vdots & \vdots \qquad \vdots \\ a_{11}b_{1m} & a_{12}b_{2m} \cdots a_{1m}b_{mm} \\ a_{21}b_{11} & a_{22}b_{21} \cdots a_{2m}b_{m1} \\ \vdots & \vdots \qquad \vdots \\ a_{m1}b_{1m} & a_{m2}b_{2m} \cdots a_{mm}b_{mm} \end{bmatrix} \begin{bmatrix} k_1 \\ k_2 \\ \vdots \\ k_m \end{bmatrix} = \begin{bmatrix} \mathbf{e}_1 \\ \mathbf{e}_2 \\ \vdots \\ \vdots \\ \mathbf{e}_m \end{bmatrix} + \begin{bmatrix} \boldsymbol{\epsilon}_1 \\ \boldsymbol{\epsilon}_2 \\ \vdots \\ \vdots \\ \boldsymbol{\epsilon}_m \end{bmatrix} \tag{6.2}$$

where a_{ij} and b_{ij} denote the elements of A and B, and $\mathbf{e}_i$, $\boldsymbol{\epsilon}_i$ denote the ith column of the identity and error matrices I and E respectively.

Ideally we would like all $\boldsymbol{\epsilon}_i$, $i = 1, 2, \ldots, m$ to be zero but since eqn. (6.2) has m^2 scalar relations and there are only m unknowns (the values of k_i) such a solution, in general, will be unobtainable. We may, however, select the k_is so as to minimize the error term $[\boldsymbol{\epsilon}_1, \boldsymbol{\epsilon}_2, \ldots, \boldsymbol{\epsilon}_m]^{\mathrm{T}}$ by the standard least squares technique. Thus if P denotes the coefficient matrix of $[k_1, k_2, \ldots, k_m]^{\mathrm{T}}$ in eqn. (6.2) then the least squares solution will be given by

$$[k_1, k_2, \ldots, k_m]^{\mathrm{T}} = [P^{\mathrm{T}}P]^{-1}P[\mathbf{e}_1, \mathbf{e}_2, \ldots, \mathbf{e}_m] \tag{6.3}$$

Having obtained the optimum values for the scaling factors we may now address the problem of how can these be best distributed in A and B so that the resulting frames give optimum alignment in the sense defined in § 4. Suppose then that we replace A by AK_A in which case, by eqn. (6.1), we have that B must be replaced by $K_A^{-1}\,KB = K_A^{-1}\,\tilde{B}$ where $\tilde{B} = KB$. The question that arises now is how to choose K_A so that the error terms E_A and E_B in

$$VAK_A = \exp\,(j\theta_A) + E_A \quad \text{and} \quad K_A^{-1}\,\tilde{B}W = \exp\,(j\theta_B) + E_B \tag{6.4}$$

for some θ_A and θ_B are minimized. A transposition of both these conditions reduces eqn. (6.4) to the same form as that of eqn. (5.1) the solution of which has already been discussed in § 5. Thus the choice of the scaling factors for the columns of A and the rows of B may be completed using techniques already developed in the previous section.

References

Hawkins, D. J., 1972, *Proc. Instn elect. Engrs*, **119,** 337.

Kouvaritakis, B., 1974, Doctoral Thesis, University of Manchester, England.

Kouvaritakis, B., and Edmunds, J., 1978, National Engineering Consortium Conference, Chicago, U.S.A.

MacFarlane, A. G. J., 1972, *Automatica*, **8,** 455.

MacFarlane, A. G. J., and Kouvaritakis, B., 1977, *Int. J. Control*, **25,** 837.

MacFarlane, A. G. J., and Postlethwaite, I., 1977, *Int. J. Control*, **25,** 81.

Control system design and analysis using closed-loop Nyquist and Bode arrays

J. M. EDMUNDS

In this paper a method is described for designing linear multivariable control schemes which have a closed-loop frequency response as close as possible, in a least squares sense, to a desired response. After using characteristic gain loci to ensure system stability, the closed-loop Bode array gives easily understood information about the controlled system in terms of bandwidth, speed of response, resonance and interaction. The closed-loop Nyquist array indicates the robustness of the control scheme for sensor failures; it also indicates the extent to which state and input noise will be suppressed, since the feedback just multiplies the open-loop disturbances by a unit matrix minus the closed-loop frequency response. Bands of Gershgorin and Ostrowski circles are used to indicate the behaviour for changes in the characteristics of more than one sensor at a time. A similar frequency-response array, obtained by breaking the feedback loops next to the actuators instead of next to the sensors, can be used to predict the behaviour of the controlled system for actuator failures. Parameter sensitivity can be investigated by determining the rate of change of the closed-loop frequency response with changes in the parameter concerned.

Three examples are used to illustrate these closed-loop array methods. Linear multivariable control schemes are designed for a 5-input, 5-output, 33-state jet engine; and for a 3-input, 2-output, 29-state chemical reactor. A parameter-dependent compensator is designed for a 2-input, 2-output, 8-state missile in which the control scheme was required to work for a wide range of values for one of the system parameters.

1. Closed-loop frequency responses

The design method in this paper calculates dynamic multivariable pre-compensators for the feedback configuration shown in Fig. 1, where the outputs **y** are compared with the reference inputs **r**, and the resulting errors **e** used to calculate the control inputs **u**. The aim of this compensator is to allow rapid independent control of the outputs, i.e. to ensure that when one of the reference inputs is changed the time response of the corresponding output rapidly settles to the new reference value, while the disturbance on the other outputs is kept to a minimum. In the frequency domain these requirements become that the bandwidth of each channel of the controlled system should be sufficiently wide, and that each output should not have any

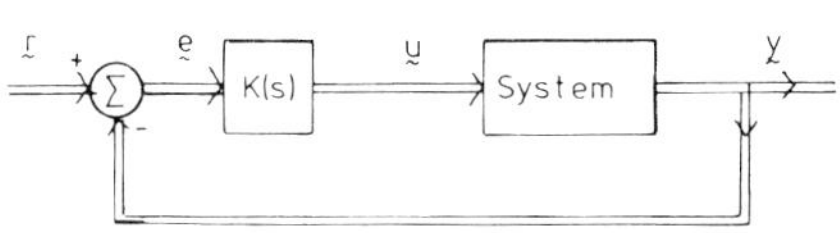

Figure 1. Multivariable feedback system.

significant response to the other reference inputs. This interaction can be rapidly assessed from the closed-loop Nyquist array, in which element i, j is the frequency response of the controlled system from the reference input j to the output i. The bandwidths are clearly displayed by a closed-loop Bode array, that is by an array of magnitude plots of the elements of the closed-loop frequency response.

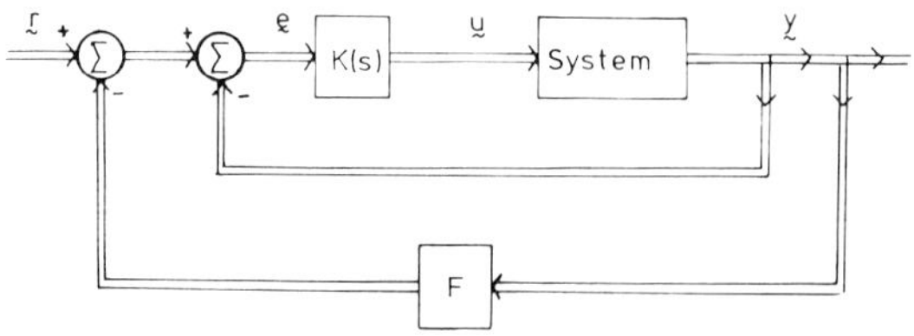

Figure 2. Multivariable feedback system with an extra diagonal feedback to investigate the effects of sensor variations.

The closed-loop Nyquist arrays can also be used to investigate the stability of the system for changes in sensor characteristics, or for changes in feedback gains. To see how to do this suppose that an extra diagonal feedback matrix F is added in parallel with the previous feedback (Fig. 2). The results derived by Rosenbrock (1974) for the direct Nyquist array then indicate the limits on these extra feedback gains, and so give limits to the maximum changes that can be tolerated in the sensor or feedback gains. Usually these closed-loop arrays will give accurate information for parameter changes, since the control scheme has been designed to make the closed-loop frequency response as near diagonal as possible.

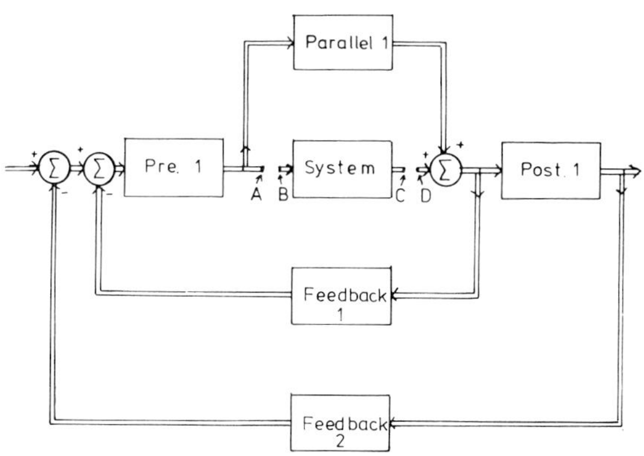

Figure 3. Multivariable system with several feedback loops, showing points at which to find the frequency response to investigate actuator variations.

These closed-loop analysis methods are described in § 3; they are primarily for the control configuration shown in Fig. 1, but they can be applied to systems with several minor feedback loops (Fig. 3).

The design method suggested in § 2 should be considered as one tool for use in a complete control system design, allowing a computer to be used for the tedious job of optimizing parameter values. In § 4 the design method is used to calculate a dynamic pre-compensator which is dependent on a variable parameter in the model. This approach can be used to keep a control scheme tuned, despite system parameter variations, in cases when the model of the system depends on some slowly changing measurable variable. Two other worked examples are given in this paper; a jet engine with 5-inputs, 5-outputs and 33-states, and a 29-state, 3-input, 2-output, two-bed chemical reactor. The design studies for these systems were carried out using the Cambridge Linear Analysis and Design programs, the documentation of which (Edmunds 1978) contains many of the ideas presented here.

2. Closed-loop design

In this section a frequency response method of tuning dynamic multivariable pre-compensators is presented. This parameter tuning is usually preceded by the use of open-loop analysis tools, such as Nyquist arrays and characteristic gain loci, which indicate the difficult features of the design and indicate the response that can be expected from the system. For difficult problems it is sometimes useful to first do a complete design study, using for example the characteristic locus design method (MacFarlane and Kouvaritakis 1977), in order to get a better understanding of the problem. The computer would then be used to optimize the parameters in the compensator; and finally the complete control scheme should be analysed, using for example the characteristic gain loci and the closed-loop Nyquist array techniques of § 3.

2.1. *Basic closed-loop design method*

This design method consists of four main steps. First the desired closed-loop transfer function $H'(s)$ is specified, bearing in mind the guide lines from § 2.2. Next the complexity of the controller $K(s)$ is specified, by choosing the denominators of the controller's transfer function and choosing the orders of its numerators. When the controller is being calculated in state space form its complexity is specified by fixing the A and B matrices and deciding which of the elements in the C and D matrices are to be tuned. The tuning process is then done in two stages; a linear least-squares minimization procedure is used to obtain a first estimate of the best parameter values, then a similar least squares procedure is used iteratively to obtain better estimates.

It has been found that large control schemes can be rapidly tuned using this approach; for instance the 50-parameter proportional-plus-integral control scheme given in the example in § 2.4 took about 45 s to compute on a multi-user minicomputer, while the more complicated controller mentioned in the same section with 175 parameters took about 5 min on the same computer. The computation to obtain the iterative improvement to the parameter values takes longer; for example, the 27-parameter control scheme, in § 2.5, took about 5 min to converge on a PDP11/45. An unexpected feature of this design method is that the optimization sometimes allows the use of a simpler controller form than is obtained by methods such as the characteristic locus design method, and at times leads to lower controller gains. However, the designer is still left with sufficient flexibility in the choice of the desired response and the controller form to be able to avoid difficult features of the system. The other area where this design approach seems to have an advantage is in its ability to remove interaction in the critical frequency range where the characteristic gains pass the critical point.

There are some difficulties with this approach of using the computer to tune the controller parameters ; the main one is how to choose the desired closed-loop response and this is tackled in § 2.2. Also, care needs to be exercised in the choice of the controller structure, since the parameters can wander to large values if too much freedom is allowed ; however, the wandering is restricted by the denominators being fixed. The other main problem is that instability sometimes occurs when integral action is being used and the controller structure is not sufficiently complicated to remove most of the interaction. This situation is prone to cause instability since then the fitting errors at high and intermediate frequencies dominate the errors at low frequencies, allowing the diagonal elements of the closed-loop frequency response matrix to leave the point at $+1$ in different directions to those of the desired response. The characteristic gains of the system depend greatly on these diagonal elements and so this initial difference in direction can lead to a different number of encirclements of the critical point by the open-loop characteristic gain loci of the desired and actual system, and hence lead to instability. Therefore good analysis methods are required to investigate the controllers produced by this parameter tuning.

2.2. *Choice of desired closed-loop response*

The desired closed-loop response $H'(s)$ should be chosen to have a sufficiently rapid response, while keeping the high frequency gain of the controller small to avoid saturation in the actuators. The response speed is also limited by the fact that that on high order systems the optimization of the frequency response can lead to instability if the demanded response is too rapid.

There are several ways of deciding the desired response speed ; for example, it may be that the designer knows enough about the system from past experience to be able to tell how fast a response to expect. In other cases, where not so much is known, it is useful to observe the uncontrolled responses and initially just attempt to speed these up by a small factor. Alternatively, if the bandwidth of the system is known, the required bandwidth can be specified to be slightly wider and then translated into an appropriate transfer function. The positions of the dominant poles also indicate the maximum speed of response which can be easily obtained with a proportional-plus-integral control scheme ; in this case a pole is placed at the origin for each loop and the dominant pole is approximately cancelled, so the expected speed of response is found from the second largest time constant in each loop. On certain problems it has been found advantageous to first do a brief design study (with, for example, the characteristic locus design method) in order to get a better feel for the system.

From the desired closed-loop response $H'(s)$ the desired compensated open-loop response $G'(s)$ can be calculated.

$$G'(s)=(I-H'(s))^{-1}H'(s) \tag{1}$$

For closed-loop stability it is necessary to match the number of poles, at the origin and in the right-half plane, of the desired open-loop response and the actual system, $G(s)K(s)$, since otherwise the parameter estimation will try to give the wrong number of encirclements of the critical point. Also, it is often found necessary to match the number of infinite zeros and right-half-plane zeros in the system and the desired closed-loop model, since these determine the number of origin encirclements by the characteristic gains. These pole and zero conditions are transferred to the desired transfer function as follows. To ensure that the desired open-loop transfer function has an nth order infinite zero the order of the numerator should be n less than the order of the denominator. To ensure that the open-loop transfer function has n poles at the origin in a particular element, the coefficients of s up to that of s^{n-1} in the numerator of the corresponding closed-loop element should be the same as those in the denominator.

Another useful computational tool to force the open-loop characteristic loci to behave in a satisfactory manner at low frequencies is to increase the weighting on the fitting errors at the lowest frequency.

2.3. *Least squares method*

Once the denominators of the controller have been specified, a linear least squares curve fitting procedure can be used to determine the best values of the numerators to make the compensated open-loop transfer function $G(s)K(s)$ as close as possible to the desired compensated open-loop transfer function $G'(s)$; that is, to minimize $E_o(s)$ where

$$G'(s)=G(s)K(s)+E_o(s) \tag{2}$$

(From here on the (s)s in the equations have been omitted for clarity, since all the matrices except for the I depend on frequency.) The left-hand side of this equation is known from eqn. (1), and on the right-hand side the effects of the fixed denominators of $K(s)$ can be amalgamated with the known values of $G(s)$, so leaving the problem in the usual least squares form. The error function to be minimized is the sum of the elements of E_o multiplied by their complex conjugates, since the errors are complex.

Unfortunately the closed-loop problem cannot be reduced to a linear problem since the closed-loop error $E_c(s)$ to be minimized is given by

$$H'=H+E_c \tag{3}$$

$$=(I+GK)^{-1}GK+E_c \tag{4}$$

The occurrence of $K(s)$ in two places in eqn. (4) means that the equation cannot be rearranged to give the correct form for a linear least squares problem. However, an approximation to the solution can be obtained in the following manner. In Appendix A, it is shown that eqn. (4) implies

$$E_c=(I-H)(G'-GK)(I-H') \tag{5}$$

Rewriting, this becomes

$$(I-H)G'(I-H')=(I-H)GK(I-H')+E_c \tag{6}$$

Assuming that the final closed-loop error is going to be small, $H(s)$ in eqn. (6) can be approximated by the desired closed-loop response $H'(s)$, reducing eqn. (6) to a similar form to the open-loop eqn. (2). The most significant difference is the term $(I-H'(s))$ post-multiplying the $K(s)$ on the right-hand side of the eqn. (6). When the $(I-H'(s))$ term is diagonal the columns $\mathbf{k}_m(s)$ of the compensator can be calculated separately, and the $(I-H'(s))$ term is accounted for by multiplying each of the elements in $G(s)$ by $(1-H'_{m,m}(s))$. The separate calculation of the columns of $K(s)$ is particularly useful on systems with many inputs and outputs, since much less computation is required to find several small sets of parameters than one large one. In the unusual case of $(I-H'(s))$ not being diagonal the problem unfortunately does not simplify in such a nice manner, and all of the columns of the controller have to be calculated simultaneously, as follows. Equation (6) has the form

$$Y=AKB+E \tag{7}$$

which can be turned into the usual least squares form

$$\mathbf{y}=A'\mathbf{k}+\mathbf{e} \tag{8}$$

where $\mathbf{k}$ is the vector of all the coefficients to be calculated and $\mathbf{y}$ has n times as many rows as Y. The rows of the matrix A' are formed by taking a set of copies of the relevant row of A and multiplying each of these copies by an element from B (Glover 1973).

The approach (of estimating all the parameters simultaneously) can be extended to give an iterative method of improving the choice of parameters for the cases where the initial remaining errors are too large. Usually the method has converged to a local minimum on the error surface in less than 10 iterations, even with the problem in § 2.5 with 27 parameters to be fitted and some large off-diagonal errors. In the cases investigated these local minima are almost certainly global minima, since they are obtained from a wide variety of starting positions in the parameter space.

The iterative approach used is to choose a control scheme, $K_f(s)$, then use a linear least squares calculation to obtain the changes to the parameter values which will give the global minimum of the local approximation to the error surface $E_e(s)$ and then repeat using the new controller values. Substituting in eqns. (2) and (4) gives that the open- and closed-loop errors with this controller, $K_f(s)$, are $E_o(s)$ and $E_c(s)$ respectively, where

$$E_o=G'-GK_f \quad \text{and} \quad E_c=H'-(I+GK_f)^{-1}GK_f \tag{9}$$

In Appendix B it is shown that for small variations in the controller parameters $K_e(s)$

$$K=K_f+K_e \tag{10}$$

the closed-loop error $E_e(s)$ is related to $K_e(s)$ by

$$E_c=(I-H)GK_e(I+(I-H)E_o)(I-H')+E_e \tag{11}$$

If $H(s)$ is now approximated by the closed-loop response with the controller which was used to calculated E_c, eqn. (11) has the same form as eqn. (8), and so can be solved in the same linear least squares manner, to give the changes which should be made to the controller parameters.

This parameter estimation approach can be extended to at least two further useful cases. One simple modification, to allow the different elements of the error matrix (E_c) to be weighted separately, is to multiply the rows of **y** and A' in eqn. (7) by the weighting factors of the elements to which they correspond. This weighting facility is used in the example in § 2.5; here the third output was just one of the control inputs which has to be returned to its nominal value fairly rapidly, so the corresponding elements in the error matrix were given a very small weighting to allow the use of this control input.

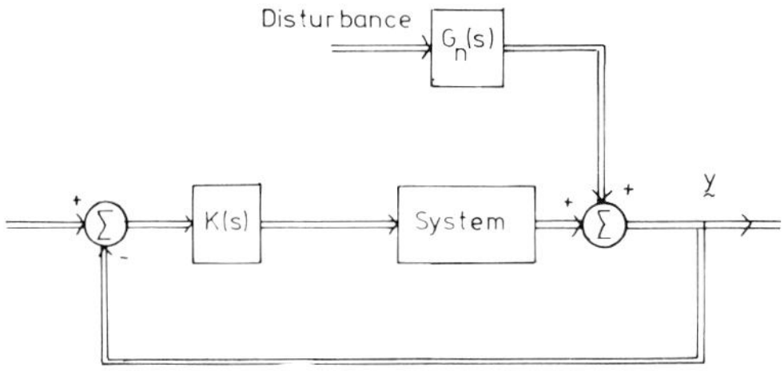

Figure 4. Multivariable feedback system with a disturbance.

The example in § 2.5 also uses a simple extension of the method to allow the response to a disturbance input to be minimized. Suppose a system has a disturbance input, as shown in Fig. 4, and that the Laplace transform of the uncontrolled disturbance on the outputs is $G_n(s)$. The Laplace transform of the closed-loop disturbance, $E_n(s)$, is given by

$$E_n = (I - H)G_n \tag{12}$$

An approximation to the contribution of this disturbance to the error function E_e can be calculated in a similar manner to that used for the iterative version of the method (Appendix C) and is given by

$$E_n = (I - H)GK_e(I - H)G_n + E_e \tag{13}$$

This has a similar form to eqn. (11), and so can be directly included in the iterative method for the parameter estimation.

2.4. *Example 1: a 5-input, 33-state, jet engine model*

This example demonstrates the use of the design method, including the choice of the desired closed-loop response, the initial calculation of the best controller parameter values and some of the analysis of the resulting controller to check for stability and system response. Further analysis is carried out in § 3 to test its robustness under parameter variations. The model of the F100 turbofan jet engine used is the one which was the theme problem for the

International Forum on Alternatives for Linear Multivariable Control, sponsored by the National Engineering Consortium in 1977 in Chicago, Illinois (Sain *et al.* 1978). The complete model is used here including all the sensor and corrected actuator dynamics.

The five control inputs on this engine were as follows: the main burner fuel flow, the nozzle jet area, the inlet guide vane position, the high variable stator position and the customer compressor bleed flow. There are also five variables to be controlled: the engine net thrust level, the total engine airflow, the turbine inlet temperature, the fan stall margin and the compressor stall margin. Good control was wanted of the first three outputs, but the last two outputs only had to stay positive, and so changes would only occasionally be made to these outputs in order to move the engine to a safer operating point. Unfortunately none of the outputs could be measured directly, and so they had to be estimated from the five sensed states of the system. The slowest modes of the system occurred in one of the sensors, so the first step in the design of the controller was to use some phase advance, C_1, on this sensor to improve its response time. Next a filter, F_1, was designed to estimate the outputs by finding the linear combinations of the sensed variables which gave the best approximations to the outputs for steps on each of the control inputs (Kouvaritakis and Edmunds 1978). In order to improve the approximation to the fifth output the fourth input as well as the sensed variables was used in the filter, hence the link F_2 in Fig. 5. The coefficients of this filter

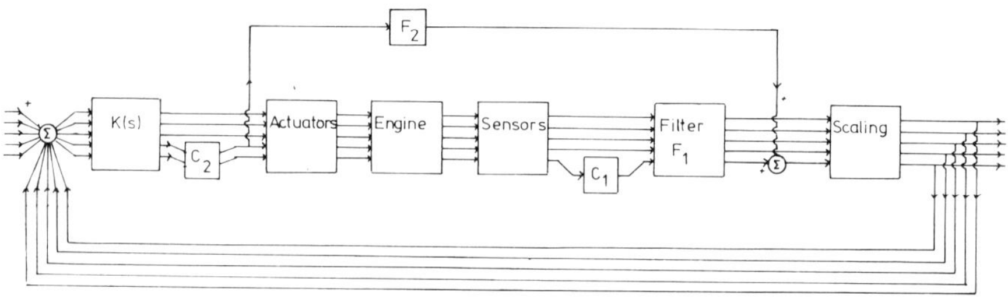

Figure 5. Diagram of jet engine and control scheme.

and of the rest of the control scheme are given in Appendix D. After calculating the filter some initial compensation, C_2, was used to slow down the fourth and fifth actuators since the response to these inputs was much faster than to the other inputs; in particular, the response to the fourth input was very rapid as this was being fed directly to the estimated outputs. Then a suitable relative scaling for the outputs was chosen, since the responses of the outputs had different orders of magnitude owing to the different units used in their measurement. The scaling was chosen so that the steady-state gains had similar magnitudes for all the outputs. A brief characteristic locus design was then carried out, giving the controlled time responses shown in Fig. 6. These responses could have been improved by a longer study, but since they were only to be used as a guide to the possible system response this was not necessary.

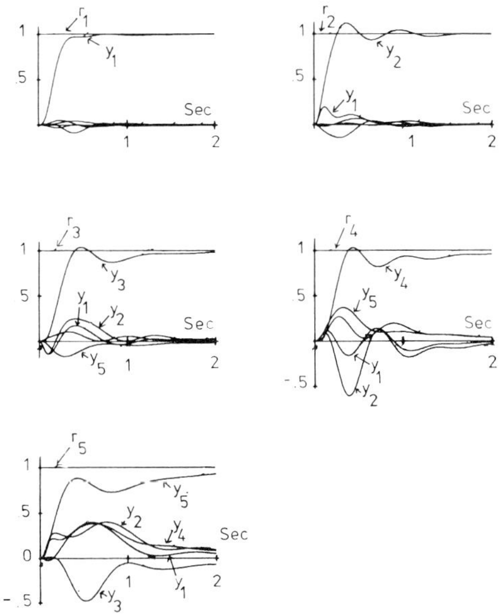

Figure 6. Closed-loop step responses using the controller resulting from the brief characteristic locus design study. Notation: y_1 = thrust, $y_2 = 100 \times$ (total engine airflow), $y_3 = 5 \times$ (turbine inlet temperature), $y_4 = 2500 \times$ (fan stall margin), and $y_5 = 1500 \times$ (compressor stall margin), r_i = demand on y_i.

Choice of desired closed-loop response for the jet engine

All of the infinite zeros of this system are of high order (one pattern of order 3, three patterns of order 4 and one of order 5), so the diagonal elements of the desired closed-loop transfer function were chosen to have high-order infinite zeros. The closed-loop responses obtained with the initial characteristic locus design indicated that the system can respond in about half a second or less. However, it had been observed during the designing of the filter that the estimates of the outputs had fairly large errors at times less than a quarter of a second so it would not be worthwhile trying to make the system respond in less than half a second. These initial time responses also show a tendency to resonate at about 10 rad s^{-1}, so the desired closed-loop transfer function matrix for the parameter optimization was chosen to be a diagonal with all its matrix elements equal:

$$H'(s) = \frac{5000}{s^3 + 50s^2 + 900s + 5000} I \tag{14}$$

$$= \frac{5000}{(s + 20 - 10j)(s + 20 + 10j)(s + 10)} I \tag{15}$$

The characteristic locus design only used a proportional-plus-integral control scheme, so the structure for the compensator was chosen to be $K_p + K_i/s$ with all of the elements of K_p and K_i variable. The initial estimates for the

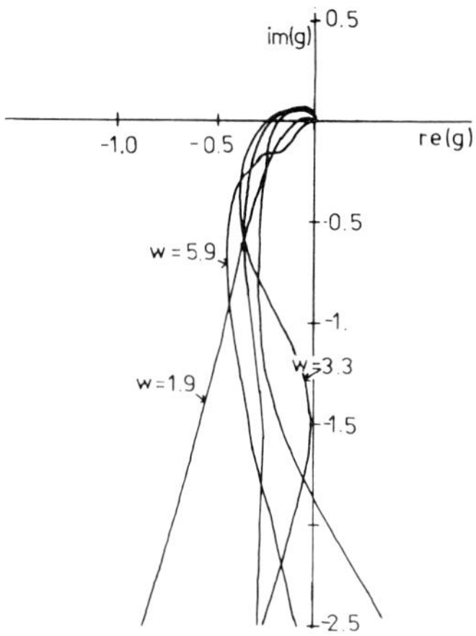

Figure 7. Characteristic gain loci for the jet engine with P.I. controller designed using the least squares approach, showing gain margins of 4.

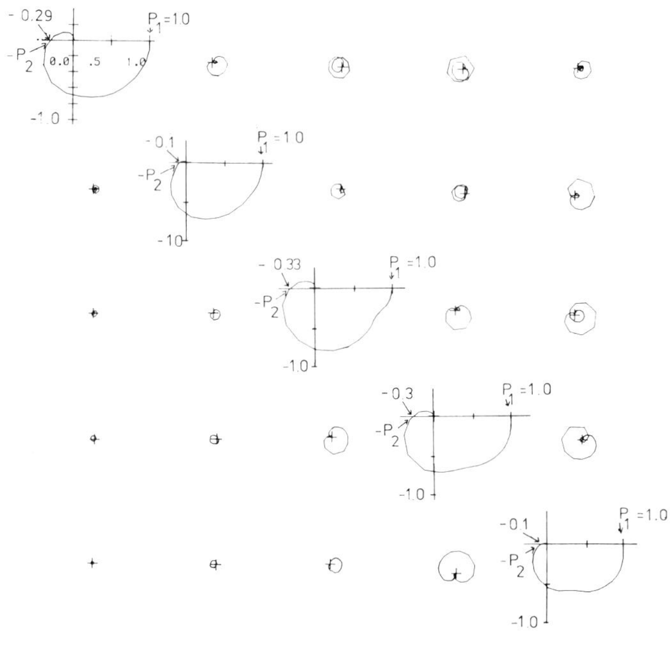

Figure 8. Closed-loop Nyquist array for the jet engine with P.I. controller, showing the remaining interaction to be less than 30%. All the elements are drawn to the same scale.

best parameter values were obtained using the least squares approach, and are given in Appendix D. This computation took about 45 s on a small time-sharing computer. The average error between the desired and actual closed-loop responses in the frequency range from 0·01 to 100 rad s^{-1}, was approximately 0·7%, so it was not considered worthwhile to use the iterative refinement of the least squares design method, particularly as this would be a lengthy procedure with 50 parameters to estimate. The characteristic loci of the system and compensator in Fig. 7 show that the system has a gain margin of about 4, for simultaneous changes in gains of all the feedback loops.

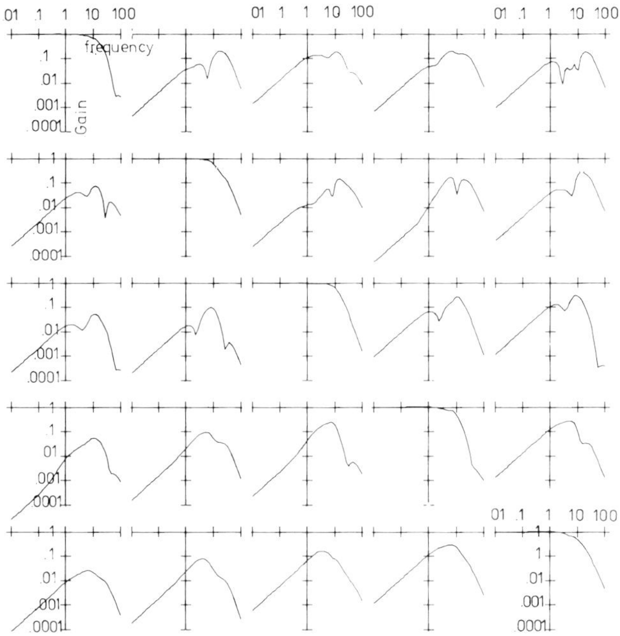

Figure 9. Bode array of magnitude plots for the closed-loop frequency response of the jet engine, with P.I. controller, showing bandwidths of about 10 rad s^{-1} and interaction peaks at 10–20 rad s^{-1}.

The closed-loop Nyquist array in Fig. 8 shows that the closed-loop system will have no excessive resonances, and very little interaction for changes in the first three reference inputs. The last two inputs have more interaction, about 30% in the frequency domain. The closed-loop magnitude array shown in Fig. 9 shows that the interaction occurs mainly in the frequency range from 5 to 20 rad s^{-1}, which as usual corresponds with the approximate cut-off frequency on the system. The time responses for this controller are shown in Fig. 10, on the same scales as those for the initial characteristic locus design in Fig. 6, and it can be seen that the interaction has been considerably reduced, the response speed has been increased slightly and the resonances have been reduced.

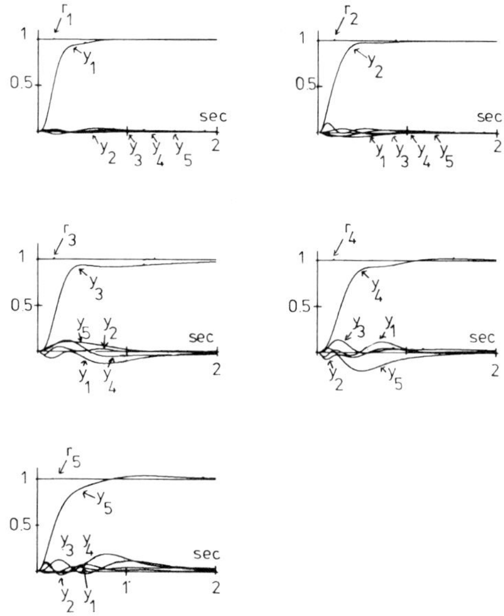

Figure 10. Closed-loop unit step response of the jet engine with the P.I. controller, showing rapid response and less than 20% interaction. The same scales and notation are used as for Fig. 6.

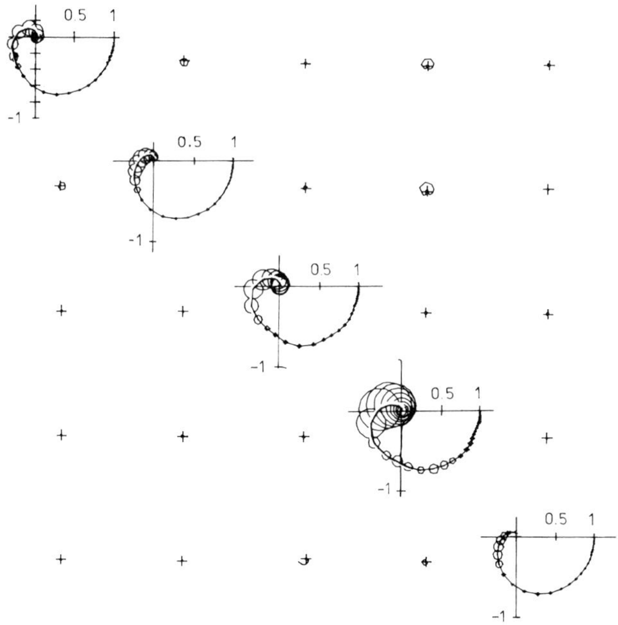

Figure 11. Closed-loop Nyquist array of the jet engine with the high-order controller of § 2.4, showing decreased interaction and hence small Gershgorin column circles.

Further improvements in the performance can be obtained by using a more complicated controller structure ; for example the interaction was almost removed by allowing each of the elements of the controller to have the form $(as^6+bs^5+cs^4+ds^3+es^2+fs+g)/(s(s+1)(s+2)(s+4)(s+10)(s+20))$, as is shown by Fig. 11 which is the closed-loop Nyquist array with the resulting 175-parameter controller. However, it is doubtful that the improvements would justify the extra controller complexity because the inaccuracies in the reconstruction of the outputs from the sensed variables mean that interaction will still occur in the actual outputs.

2.5. *Example 2 : a 29-state, 3-input, 2-output, two-bed chemical reactor*

This section illustrates the use of the iterative improvement to the least squares parameter tuning ; a use of different weighting for different elements of the frequency-response matrix ; and the use of the modification for taking into account noise suppression.

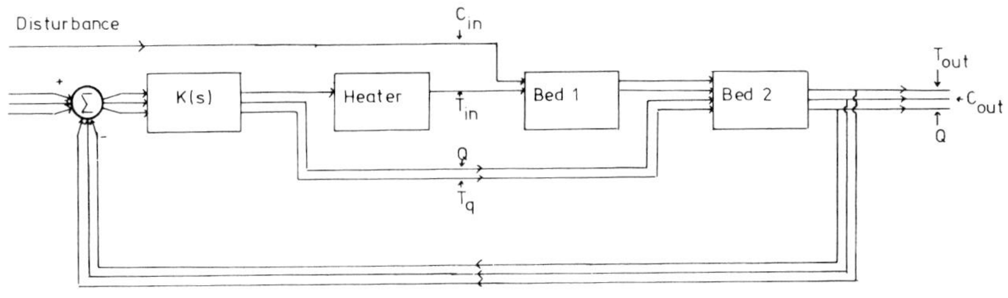

Figure 12. Diagram of the two-bed chemical reactor with control scheme.

The aim of this design is to control the product concentration and temperature of an experimental two-bed exothermic catalytic reactor. Two of the inputs are the flow rate and temperature of a ' quench ' stream injected between the beds, and the third input is the feed temperature, Fig. 12. The control system has been developed to reset the quench flow rate to its nominal value upon sustained system disturbances, since this input can easily saturate. Details of the non-linear partial differential equations describing the system, together with the reduction to linear state space form are given by Silva *et al.* (1979). Details of a characteristic locus controller design together with a reduced order model of the system (Wallman 1977) have been recently submitted for publication (Foss *et al.* 1980).

The time responses of the system with the controller obtained using the characteristic locus design method are shown in Fig. 13, and it can be seen that the system responds to set point changes after about 2 normalized units of time, but does not settle down until about 10. The effects of concentration disturbances in the main input stream are considerably reduced by the control scheme, but unfortunately rather too much use is made of the quench flow rate. Hence the aims of this design study are to reduce the interaction and settling time, while making less use of the quench flow rate.

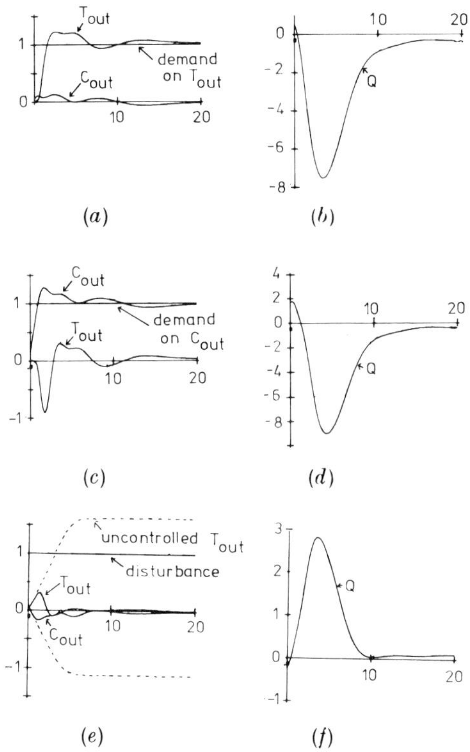

Figure 13. Closed-loop step responses of two-bed chemical reactor with controller designed by characteristic locus method. The units of time for these graphs have been normalized. (*a*) Step on reference for T_{out}. (*b*) Control action for (*a*). (*c*) Step on reference for C_{out}. (*d*) Control action for (*c*). (*e*) Controlled and uncontrolled disturbances resulting from a unit change of feed concentration. (*f*) Control action for (*e*).

Choice of desired closed-loop response for the chemical reactor

The system has two right-half-plane zeros, one first-order infinite zero and one pattern of third-order infinite zeros. The Nyquist array of the open-loop system indicates that the high-order infinite zeros and the right-half-plane zeros are associated with the first output, so these were put in the first diagonal element of the desired closed-loop transfer function. The second output responds immediately to the control inputs, so this element was given no infinite zeros, leaving the first-order infinite zero for the third element. In practice this third output (which is the quench flow rate) also responds immediately ; however, a rapid response is not desirable on this output. Since it is not a real output of the system, but is one of the control inputs, a rapid response would make this input stay near to its set-point so preventing it from being used by the control scheme. The chosen closed-loop response was specified by

$$H'(s)=\operatorname{diag}\left[\frac{64(s-11)(s-65)}{(s+11)(s+65)(s+4)^3}, \frac{10+0{\cdot}5s}{s+10}, \frac{0{\cdot}3}{s+0{\cdot}3}\right] \tag{16}$$

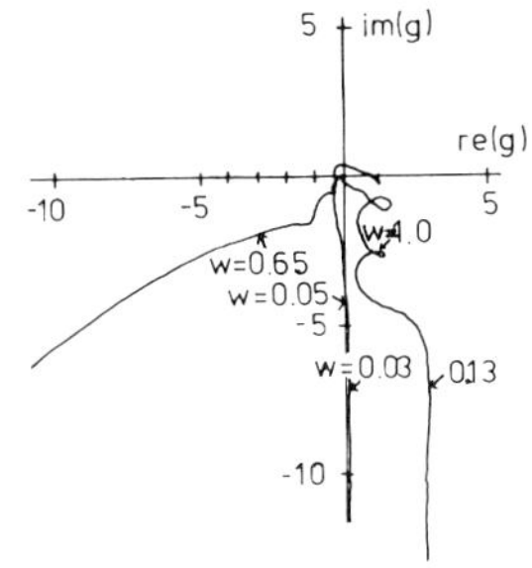

Figure 14. Characteristic loci for the two-bed chemical reactor with control scheme designed using the least squares approach.

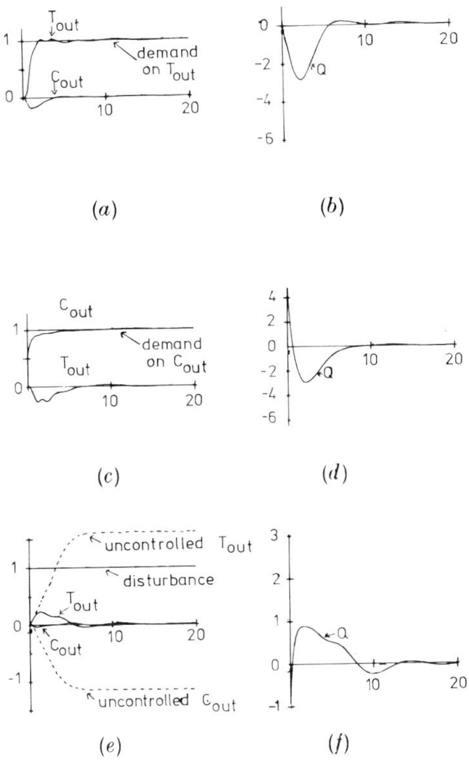

Figure 15. Closed-loop step responses of two-bed chemical reactor, with controller designed using the least squares approach, on the same scales as Fig. 13. The units of time for these graphs have been normalized. (*a*) Step on reference for T_{out}. (*b*) Control action for (*a*). (*c*) Step on reference for C_{out}. (*d*) Control action for (*c*). (*e*) Controlled and uncontrolled disturbances resulting from a unit change of feed concentration. (*f*) Control action for (*e*).

The approach used to reset the quench flow rate to its nominal value was to include it as a system output, so comparatively large initial interactions should be allowed on this output. The matrix, W, of relative weightings of the errors in the closed-loop frequency response was chosen to allow the quench flow rate to be used as a control input, by making the elements 3,1 3,2 and 3,4 small. Dynamic weighting was used on the elements corresponding to the disturbance input to reflect the fact that the control scheme cannot be expected to remove the high-frequency disturbance, but should remove low-frequency disturbances. Thus w was chosen as :

$$W = \begin{bmatrix} 1 & 1 & 1 & 0{\cdot}5/(s+0{\cdot}1) \\ 1 & 1 & 1 & 0{\cdot}5/(s+0{\cdot}1) \\ 0{\cdot}005 & 0{\cdot}005 & 1 & 0{\cdot}0025/(s+0{\cdot}1) \end{bmatrix} \tag{17}$$

Figure 14 shows the characteristic loci with this controller, indicating that the closed-loop system will be stable. The closed-loop time responses are shown in Fig. 15, and it can be seen that the objectives of decreasing interaction and settling time have been achieved, together with a threefold decrease in the use of the quench flow. Also, it can be seen that the noise rejection properties have been improved ; in particular, the disturbance on the output concentration has been reduced.

3. Closed-loop analysis

It is assumed that the stability of the closed-loop system has been checked either by calculating the closed-loop pole positions, or by using the characteristic gain loci before using the analysis methods described in this section.

The Nyquist array for the closed-loop system $(1+G(s))^{-1}G(s)$ can be used to indicate closed-loop response in terms of interaction, resonance, speed of response and the degree to which disturbances will be suppressed by a particular feedback control scheme. The Gershgorin bands (which are bands of circles centred on the diagonal elements of the array, with each circle having a radius equal to the sum of the moduli of the other elements in the row or column at the chosen frequency) on this array indicate how robust the control will be for changes in the characteristics of the sensors, and the Gershgorin bands on a similar array indicate the robustness for actuator variations (Edmunds 1978). The rate at which the closed-loop frequency response array varies with the change of a parameter within the system model indicates the sensitivity of the closed-loop response to uncertainties or changes in that system parameter.

3.1. *Interaction*

The closed-loop interaction in the frequency domain can be readily seen from the off-diagonal elements of the closed-loop Nyquist array. It has been observed that in many cases the interaction on the closed-loop step response is about 0·5 to 0·7 times the maximum interaction on the frequency response. This multiplication rule is illustrated by Fig. 16 (a) which shows two frequency responses with typical shapes for closed-loop off-diagonal elements of first

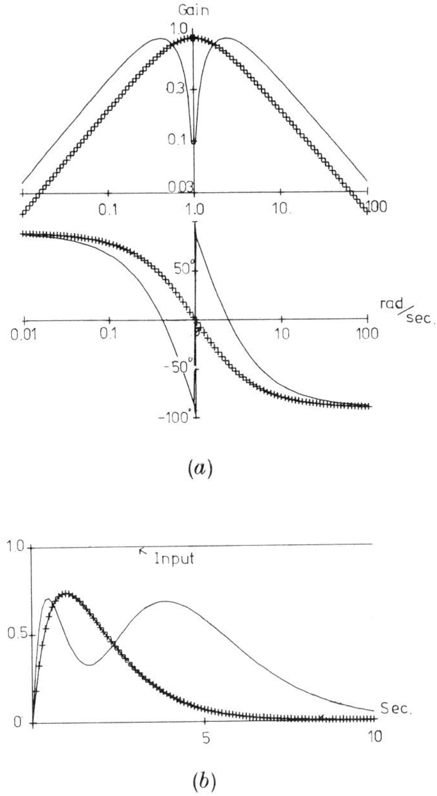

Figure 16. Typical frequency and step responses for the interaction between elements. (*a*) Bode plot; (*b*) step response.

order systems with integral action. One is for the transfer function $2s/(s+1)(s+1)$ and the other is for $4(s+j)(s-j)/(s+1)(s+1)(s+1)(s+1)$, both of which have peaks in their frequency response of magnitude 1·0. The corresponding time responses are shown in Fig. 16 (*b*) and both have peaks of about 0·7. The integral action causes the differentiator in the off-diagonal elements, and the first-order infinite zeros cause them to decrease as a/s for large values of s. The second example has been given since the least squares algorithm of § 2 often seems to put zeros in these off-diagonal elements near to the imaginary axis. The multiplication factor required seems to be smaller for higher order systems where the frequency response decreases more rapidly at high frequencies.

3.2. *Bandwidth, speed of response*

The bandwidth of the closed-loop system can be seen directly from the diagonal elements of the closed-loop Bode array. The speed of response of the closed-loop system can be estimated by approximating the frequency response of these diagonal elements by a transfer function with a single pole or some other transfer function for which the time response can be easily calculated. Resonances can also be easily seen from the closed-loop arrays, since magnitudes of more than 1 indicate that there are resonances.

3.3. *Stability for a single sensor failure*

The diagonal elements of the closed-loop Nyquist array give exact information about system stability under changes of any single sensor. For example, to see how large a change of gain can be tolerated in the sensor i, find the points at which the locus for element i, i crosses the real axis. Suppose that the largest positive crossing is at $+P_1$, and the most negative crossing is at $-P_2$, as shown on Fig. 8. Then use of the Nyquist criterion indicates that feedback gains f with $-1/P_2 < f < 1/P_1$ could be tolerated from output i to input i, Fig. 2. Hence the system will be stable for negative feedback gains f' for $1-1/P_1 < f' < 1+1/P_2$, and so the system will remain stable if the gain of the sensor i is multiplied by any constant in this range.

Using this transformation it can be seen from Fig. 8 that the controlled jet engine would remain stable if the ith sensor gain was multiplied by f'_i, so long as

$$\left.\begin{array}{l} f'_1 < 4\cdot 4 \\ f'_2 < 11\cdot 0 \\ f'_3 < 4\cdot 0 \\ f'_4 < 4\cdot 3 \\ f'_5 < 11\cdot 0 \end{array}\right\} \quad (18)$$

Note that this assumes that only one of the sensors changes.

3.4. *Stability for multiple sensor failures*

The theorems to indicate stability or instability for closed-loop systems by using bands of Gershgorin circles (Rosenbrock 1974), or bands of Ostrowski circles on the direct Nyquist array (Rosenbrock 1974) can be applied to the closed-loop Nyquist array to investigate the effect of adding an extra feedback loop as shown in Fig. 2. The limits on the stable ranges of feedback gains then indicate the allowable sensor changes by using the same transformation as for the variation of a single sensor in § 3.3. Since the controller has been designed to make the closed-loop as diagonal as possible, the closed-loop array will usually be more diagonal than the open-loop array and so the Gershgorin bands will tend to be narrow, giving fairly clear information about the stable and unstable parts of the gain space (MacFarlane 1972, Rosenbrock 1974).

It is interesting to note that this analysis indicates that control systems which have been designed to give low interaction will have good gain margins for changes of gains of the sensors. For example, suppose that a control scheme has been designed for a 3-input system to give no resonances of above 1·4 and to give interaction of less than 0·15. Then in the worst possible case the Gershgorin bands are bound to be in a circle of radius 2, and so will

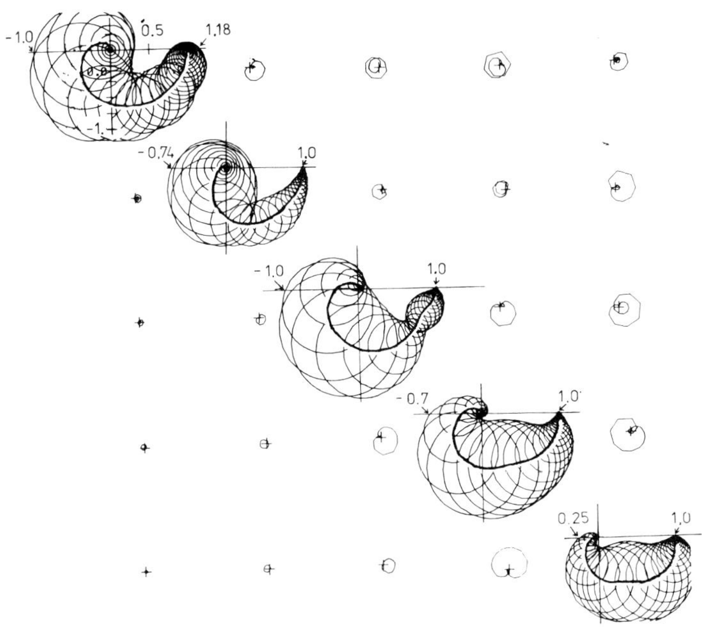

Figure 17. Closed-loop Gershgorin row circles for the jet engine with P.I. controller designed in § 2.4 using the least squares approach.

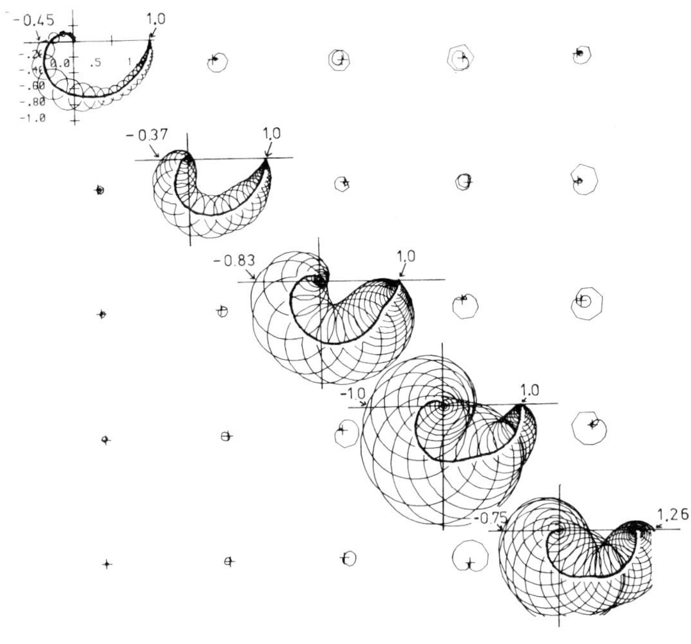

Figure 18. Closed-loop Gershgorin column circles for the jet engine and P.I. controller designed in § 2.4 using the least squares approach.

tolerate at least 50% changes in any combination of sensor gains without losing stability. Figure 17, which shows the bands of Gershgorin row circles for the jet engine example in § 2.4, would be more typical ; stability will be retained if the sensor gains are multiplied by any combination f'_i where

$$\left.\begin{array}{ll} & 0{\cdot}15 < f'_1 < 2{\cdot}0 \\ \text{and} & \\ & 0{\cdot}0 < f'_2 < 2{\cdot}3 \\ \text{and} & \\ & 0{\cdot}0 < f'_3 < 2{\cdot}0 \\ \text{and} & \\ & 0{\cdot}0 < f'_4 < 2{\cdot}3 \\ \text{and} & \\ & 0{\cdot}0 < f'_5 < 5{\cdot}0 \end{array}\right\} \tag{19}$$

By similar arguments the bands of Gershgorin column circles shown in Fig. 18 indicate that stability is assured if

$$\left.\begin{array}{ll} & 0{\cdot}0 < f'_1 < 3{\cdot}2 \\ \text{and} & \\ & 0{\cdot}0 < f'_2 < 3{\cdot}7 \\ \text{and} & \\ & 0{\cdot}0 < f'_3 < 2{\cdot}2 \\ \text{and} & \\ & 0{\cdot}0 < f'_4 < 2{\cdot}0 \\ \text{and} & \\ & 0{\cdot}2 < f'_5 < 2{\cdot}3 \end{array}\right\} \tag{20}$$

Note that the stable ranges indicated by the two sets of circles are different. Figure 11 shows the bands of the Gershgorin row circles for the more complicated 30-state, 175-parameter, control scheme for the jet engine and gives accurate stability and instability information for that design. These two figures also give ranges of gains for which it is known that the closed-loop system is unstable.

Generally the closed-loop Nyquist array from the point D to the point C on Fig. 3 should be used to investigate the effects of changes in sensor gains. This array will be different from the closed-loop Nyquist array if any dynamic or non-diagonal post-compensators have been employed in the control scheme, or if more than one feedback loop has been used.

3.5. *Actuator failures*

The closed-loop Nyquist array from the point B to the point A on Fig. 3 should be used to investigate the effect of changes in the actuators. This can be seen by rearranging Fig. 3 as shown in Fig. 19. This gives the same form as previously, but now the extra feedback gains introduced in the manner shown in Fig. 2 correspond to changes in the actuators. Generally the loops have to be split next to that block of the system which is to be varied. Often these actuator closed-loop arrays are not as diagonal as the closed-loop frequency response ; for example, for the jet engine of § 2.4, element 1,5 of the

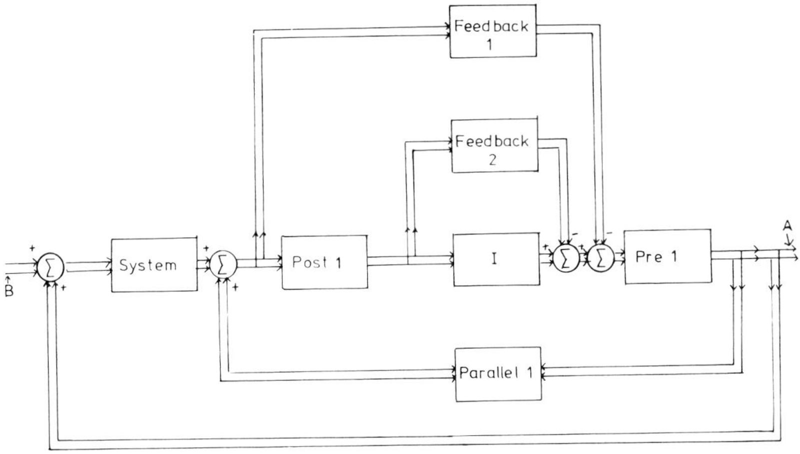

Figure 19. Rearrangement of multivariable feedback system shown in Fig. 3 to investigate the effects of actuator variations.

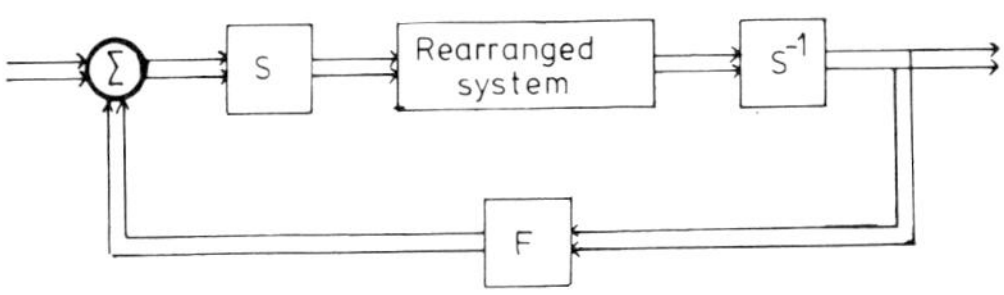

Figure 20. Rearranged system of Fig. 19 with diagonal scaling matrix S and its inverse used to decrease the size of the Gershgorin bands prior to the introduction of an extra diagonal feedback matrix to investigate parameter sensitivity.

actuator array is about 4000 times larger than the diagonal elements. However, it should be noted that sizes of the off-diagonal elements can be traded off against each other by introducing a diagonal scaling matrix S, shown in Fig. 20 together with its inverse, without altering the stability information, since these scaling matrices commute with the diagonal feedback matrix. Figure 21 shows the scaled actuator array for the jet engine with the proportional plus integral control scheme and

$$S = \text{diag}\,(1 \cdot 0,\ 0 \cdot 001,\ 0 \cdot 06,\ 0 \cdot 03,\ 0 \cdot 00007) \tag{21}$$

The diagonal elements of this array indicate that if just one of the actuators is multiplied by f'_i stability will be retained provided that

$$\left.\begin{aligned} f'_1 &< 4 \cdot 3 \\ f'_2 &< 4 \cdot 3 \\ f'_3 &< 4 \cdot 3 \\ -1 \cdot 2 < f'_4 &< 10 \cdot 0 \\ f'_5 &< 10 \cdot 0 \end{aligned}\right\} \tag{22}$$

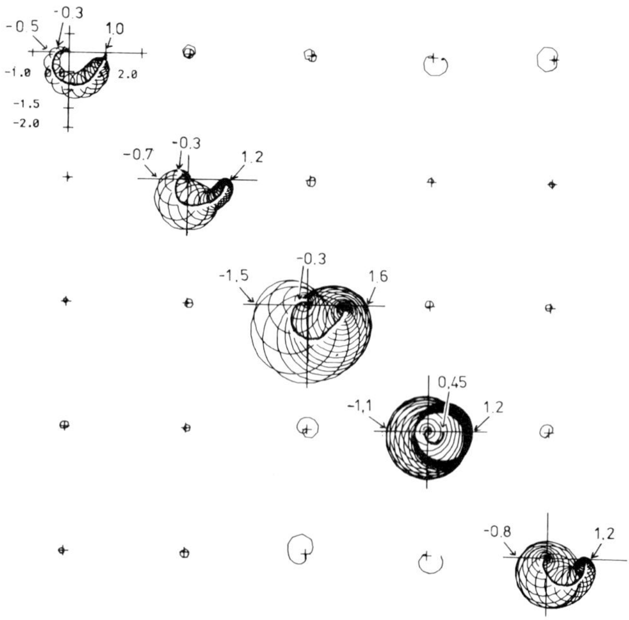

Figure 21. Actuator array with Gershgorin column circles for the jet engine and P.I. controller.

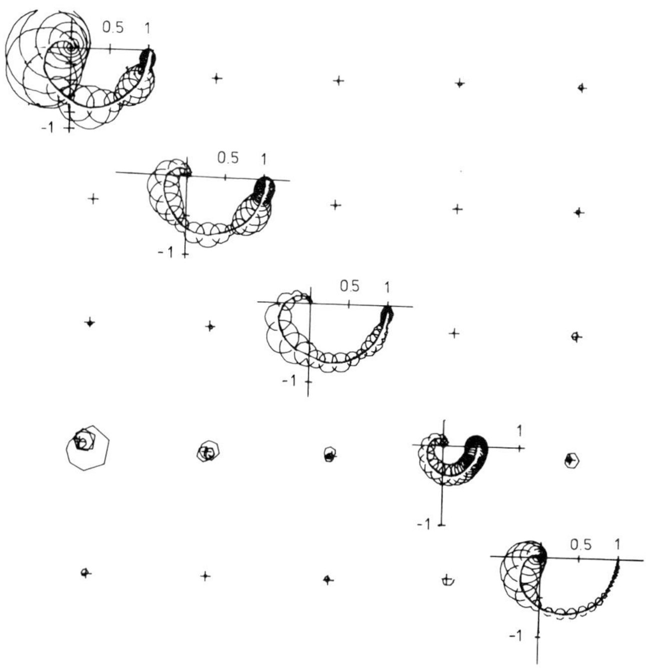

Figure 22. Actuator array with Gershgorin column circles for the jet engine and high-order controller.

It will be seen that element 4,4 of this figure does not start at +1 for low frequencies, despite the integral action in the control scheme. This is due to the way that some of the fourth input is used in the filter to help reconstruct one of the outputs, and will generally happen when the inputs are used to help reconstruct the outputs to be controlled.

The bands of Gershgorin column circles shown on Fig. 21 indicate that stability is assured if

$$\left.\begin{aligned} &0{\cdot}0 < f'_1 < 3{\cdot}0 \\ \text{and}\quad &0{\cdot}16 < f'_2 < 2{\cdot}4 \\ \text{and}\quad &0{\cdot}38 < f'_3 < 1{\cdot}6 \\ \text{and}\quad &0{\cdot}16 < f'_4 < 1{\cdot}9 \\ \text{and}\quad &0{\cdot}16 < f'_5 < 2{\cdot}2 \end{aligned}\right\} \tag{23}$$

Figure 22 shows the scaled actuator array for the same jet engine with the more complicated control scheme. It can be seen that the Gershgorin bands are much narrower, so giving more precise stability and instability information. It is interesting to note that in this case the actuator array becomes more diagonal as the closed-loop array does. This may be due to the fact that the closed-loop system is approaching a diagonal matrix with all its diagonal elements equal, and if equality was actually achieved the actuator array would also be diagonal and identical to the closed-loop array.

3.6. *Disturbance suppression*

State and actuator disturbances can usually be modelled by a disturbance input with a Laplace transform matrix $G_n(s)$ being added to the output of the system, as shown in Fig. 4. The closed-loop disturbance on the output will just be the open-loop disturbance multiplied by $(I+G(s)K(s))^{-1}$, i.e. the final disturbance will be $(I-H(s))G_n(s)$, where $H(s)$ is the closed-loop transfer function matrix of the system. Therefore the degree of noise suppression by the control scheme can be seen from the array $(H(s)-I)$, which is the same as the closed-loop Nyquist array with the origins of the diagonal elements moved to +1. The closed-loop noise will be small if the elements of $(I-H(s))$ are small at all the frequencies with significant open-loop noise.

3.7. *Parameter variations*

The closed-loop frequency response array gives a complete description of the closed-loop behaviour of the system and so it can be used as an aid in the investigation of parameter sensitivity. The closed-loop frequency response at a given frequency may be regarded as a function of the parameter which is varying. In many cases this function will be continuous and smooth in the range of parameter values being considered, and will only become infinite when the system is on the verge of instability since an infinite response would imply that there is some sinusoidal input which will produce an infinite output. The converse is also true that when the system is on the verge of becoming unstable, owing to a change of a parameter in the system, one of the closed-loop characteristic loci will be infinite at some frequency, and so the closed-loop frequency response matrix will have at least one infinite element. Hence a Taylor series expansion about the expected value for the variable parameter may be used to describe the closed-loop frequency response within the range of parameter values which leave the system stable. An approximation to the first term of this series can be easily found by calculating the frequency response at some parameter value close to the original value and so obtaining the rate of change. In particular this matrix of gradients will indicate how parameter variations will effect interaction on the closed-loop system. It would be expected that the most significant variations will occur in the frequency range near the critical point on the characteristic loci, since in this frequency range interaction is only suppressed by making the product of the system and controller approximately diagonal.

4. Parameter dependent control schemes

Parameter-dependent control schemes may be used if the system is known to depend on a measurable parameter. One way of calculating a variable compensator is to design control schemes at two or more operating points for the parameter in question, and then use some method for interpolating to obtain compensators for values between the design points. This interpolation is relatively easy if the states of the compensator are chosen to be independent of the unknown parameter value, since then the past behaviour of the system does not have to be remembered. In order to make the compensator states independent of the variable parameter the denominators of its transfer function should be fixed, or in the case of a state space compensator the A and B matrices should be fixed.

The design method in § 2 is ideally suited to calculating such a parameter-dependent controller, since the denominators are fixed by the designer and so can be made the same for each of the design points. The numerators of the transfer function, or the C and D matrices of the state space matrices of the compensator, are then made to be functions of the variable parameter so that they take the chosen values at the design points. In the example in § 4.1 the C and D matrices are obtained by linear interpolation from their values at the two design points.

4.1. *Example 3: an autopilot for a 2-input, 2-output, 8-state missile*

This example demonstrates the design of a variable compensator for an autopilot for a missile, the model of which contains a variable parameter (p = missile roll rate) which has to be selected. The system can be described by a state space model with the A and C matrices dependent on p; $A = A_f + pA_v$ and $C = C_f + pC_v$, the values of which are given in Appendix E. A peak interaction of 1 or 2% is allowable on the closed-loop step response. A variable compensator, which has approximately P.I.D. form, was designed for this system, with state space matrices A_c, B_c, $C_{cf} + pC_{cv}$, and $D_{cf} + D_{cv}$. The approach used was as follows:

(1) A control scheme was designed for $p = 12{\cdot}5$ such that the diagonal elements of the closed-loop response were as near as possible to (25/

$(s+25))^2$, by using a linear least squares fitting technique. The desired closed-loop response was chosen to be second order since the system was second order. The closed-loop pole was chosen to avoid problems in the frequency range around 120, since the system had zeros at about $-6 \pm 120j$ and $-6 \pm 170j$. These zeros were slightly dependent on the choice of p but seemed to stay in the same vicinity.

(2) A control scheme was designed for $p=17{\cdot}5$ such that the diagonal elements of the closed-loop response were as near as possible to $(25/(s+25))^2$.

(3) The controller was formed by using a linear interpolation on the C and D matrices of these two compensators giving a compensator of the form mentioned above.

This compensator has an unusual symmetric form

$$K(s) = \begin{bmatrix} k_1(s) & k_2(s) \\ -k_2(s) & k_1(s) \end{bmatrix} \tag{24}$$

since the original system also has this form. Figure 23 (a) shows the frequency response of the closed-loop system from the first reference input to the first output for values of p at the extremes required for this variable, and at the central value. Figure 23 (b) shows the frequency response from reference input one to the second output, and it can be seen that the interaction does not exceed 3% for any frequency. Figure 24 (a) shows the response of the first output to a unit step on the first input, and Fig. 24 (b) shows the response of the second output to a unit step on the first reference input. Owing to the symmetric nature of the system and compensator the responses to the second input are the same as those for the first, except that

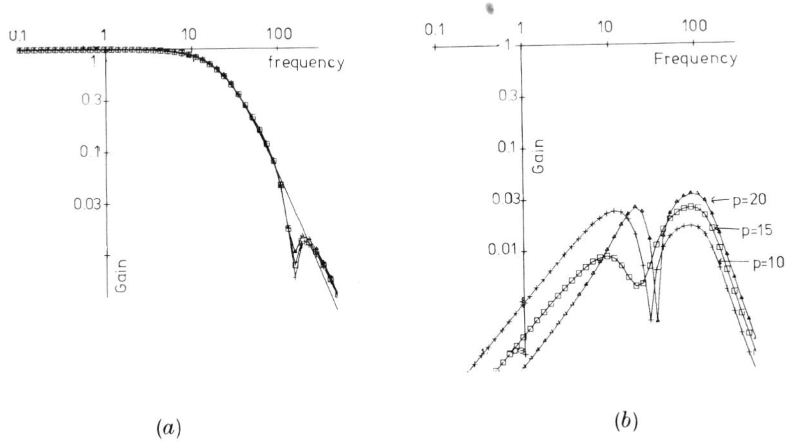

(a) (b)

Figure 23. Closed-loop frequency response of missile with variable control scheme for three parameter values. Normalized units have been used for these displays. (a) Diagonal elements ; (b) off-diagonal elements.

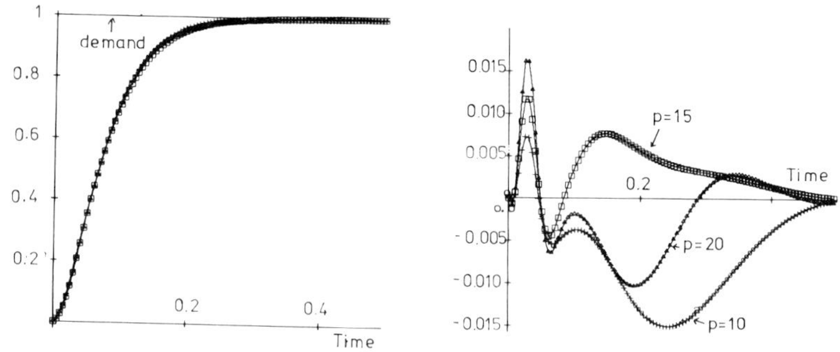

Figure 24. Closed-loop step response of missile with variable control scheme for three parameter values. Normalized units have been used for these displays. (*a*) Diagonal elements ; (*b*) off-diagonal elements.

the direction of the interaction is reversed. It can be seen that the compensator satisfies the stipulated design specification that the maximum transient interaction should not be more than 2%.

5. Conclusions

The design method presented in § 2 provides a way of rapidly tuning linear multivariable control schemes, and has been shown to work on two large problems. The approach sometimes allows simpler compensator structures because of the optimization, and sometimes results in smaller gains. However, care has to be taken to check the stability and robustness of the resulting control scheme.

The closed-loop Bode array gives an easily comprehensible display of closed-loop response and bandwidth, while the closed-loop Nyquist array shows the interaction and robustness of the system. Since the compensator has been designed to give little interaction the Gershgorin bands on the closed-loop arrays will be narrow, and thus provide fairly precise information about the stable and unstable parts of the gain space. This indicates that designing a control scheme to give little interaction and resonance ensures that the resulting scheme will be robust for sensor variations.

The actuator array provides information about the stability of the system for actuator changes, while the sensitivity of the system for general parameter changes can be seen from the appropriate Taylor series expansions, in terms of the variable parameter of the elements of the closed-loop frequency response. In cases where the variable parameter in the system can be measured, a parameter dependent compensator can be used to give improved control.

Appendix A

The relationship between open- and closed-loop errors

Definitions

$G(s)$ = Open-loop frequency response matrix.

$H(s)$ = Closed-loop frequency response matrix.

$K(s)$ = Pre-compensator frequency response matrix.

$G'(s)$ = Desired open-loop frequency response matrix.

$H'(s)$ = Desired closed-loop frequency response matrix $G'(s)(I+G'(s))^{-1}$.

$E_o(s)$ = Error in the open-loop frequency response equal to $G'(s)-G(s)K(s)$.

$E_c(s)$ = Error in the closed-loop frequency response equal to $H'(s)-H(s)$.

Identities

$$H=(I+GK)^{-1}GK=GK(I+GK)^{-1} \quad \text{(A 1)}$$

$$GK=H(I-H)^{-1}=(I-H)^{-1}H \quad \text{(A 2)}$$

$$(1+GK)^{-1}=(I-H) \quad \text{(A 3)}$$

Equation (3) of § 2.3 gives

$$\begin{aligned}
E_c &= H'-H && \text{(A 4)}\\
&= H'-(I+GK)^{-1}GK && \text{(A 5)}\\
&= (I+GK)^{-1}((I+GK)H'-GK) && \text{(A 6)}\\
&= (I-H)(H'+GK(H'-I)) && \text{(A 7)}\\
&= (I-H)(H'(H'-I)^{-1}+GK)(H'-I) && \text{(A 8)}\\
&= (I-H)(G'-GK)(I-H') && \text{(A 9)}\\
&= (I-H)E_o(I-H') && \text{(A 10)}
\end{aligned}$$

Appendix B

The local approximation to the error surface

Suppose that the compensator $K_f(s)$ gives open- and closed-loop errors $E_o(s)$ and $E_c(s)$ respectively. Then from eqn. (A 9) the closed-loop error $E_e(s)$ with a compensator $K_f(s)+K_e(s)$ is given by the following :

$$\begin{aligned}
E_e &= [I-(I+GK_f+GK_e)^{-1}(GK_f+GK_e)][G'-GK_f-GK_e](I-H') && \text{(B 1)}\\
&= [I-(I+GK_f)^{-1}(I+GK_e(I+GK_f)^{-1})^{-1}(GK_f+GK_e)]\\
&\qquad \times [G'-GK_f-GK_e](I-H') && \text{(B 2)}
\end{aligned}$$

for small GK_e this becomes

$$\begin{aligned}
E_e &\simeq [I-(I+GK_f)^{-1}(I-GK_e(I+GK_f)^{-1})(GK_f+GK_e)]\\
&\qquad \times [G'-GK_f-GK_e](I-H') && \text{(B 3)}\\
&\simeq [I-(I+GK_f)^{-1}(GK_f+GK_e-GK_e(I+GK_f)^{-1}GK_f)]\\
&\qquad \times [G'-GK_f-GK_e](I-H') && \text{(B 4)}\\
&= [I-(I+GK_f)^{-1}GK_f+(I+GK_f)^{-1}GK_e((I+GK_f)^{-1}GK_f-I)]\\
&\qquad \times [G'-GK_f-GK_e](I-H') && \text{(B 5)}\\
&\simeq (I-H)(G'-GK_f)(I-H')-(I-H)GK_e(I-H')\\
&\qquad +(I+GK_f)^{-1}GK_e((I+GK_f)^{-1}GK_f-I)(G'-GK_f)(I-H') && \text{(B 6)}\\
&= E_c-(I-H)GK_e[I+(I-(I+GK_f)^{-1}GK_f)(G'-GK_f)](I-H') && \text{(B 7)}\\
&= E_c-(I-H)GK_e[I+(I-H)E_o](I-H') && \text{(B 8)}
\end{aligned}$$

Therefore

$$E_c \simeq (I-H)GK_e(I+(I-H)E_o)(I-H')+E_e \tag{B 9}$$

Appendix C

Noise suppression

Suppose that the compensator $K_f(s)$ leaves a closed-loop disturbance $E_f(s)$, then from eqn. (12) the closed-loop error E_e with a compensator $K_f(s)+K_e(s)$ is given by

$$E_e = (I+GK_f+GK_e)^{-1}G_n \tag{C 1}$$

$$= (I+GK_f)^{-1}(I+GK_e(I+GK_f)^{-1})^{-1}G_n \tag{C 2}$$

For small G_e this is approximately

$$\simeq (I-H)(I-GK_e(I-H))G_n \tag{C 3}$$

therefore

$$E_f \simeq (I-H)GK_e(I-H)G_n + E_e \tag{C 4}$$

Appendix D

Details of the control scheme for the jet engine

The initial compensation on the fifth sensor which detected the fan turbine inlet temperature was

$$C_1 = \frac{(0{\cdot}595s+1)(5{\cdot}49s+1)}{(2{\cdot}10755s+1)(0{\cdot}2s+1)} \tag{D 1}$$

The filter to reconstruct the outputs from the sensed variables was

$$F_1 = \begin{bmatrix} -0{\cdot}562 & -0{\cdot}307 & 45{\cdot}0 & -24{\cdot}9 & 4{\cdot}43 \\ -0{\cdot}0141 & 0{\cdot}00389 & 0{\cdot}505 & -3{\cdot}26 & 0{\cdot}0578 \\ 0{\cdot}00528 & -0{\cdot}0183 & 0{\cdot}873 & -6{\cdot}49 & 1{\cdot}26 \\ 0{\cdot}000314 & -0{\cdot}0000297 & -0{\cdot}00433 & 0{\cdot}0137 & -0{\cdot}000474 \\ -0{\cdot}0000199 & 0{\cdot}000165 & -0{\cdot}00322 & 0{\cdot}0166 & -0{\cdot}00007 \end{bmatrix} \tag{D 2}$$

The feedforward term F_2 from the fourth input to the fifth output to help in the reconstruction of the outputs was $-0{\cdot}00459$.

The scaling matrix to balance the output sizes was

$$D = \text{diag}\,(1,\ 100,\ 5,\ 2500,\ 1500) \tag{D 3}$$

The initial compensation, used for the characteristic locus design and the proportional-plus-integral design, to slow down the fourth and fifth inputs was

$$C_2 = \text{diag}\,(1,\ 1,\ 1,\ 1/(0{\cdot}00009s_3+0{\cdot}0069s^2+0{\cdot}16s+1),\ (0{\cdot}02s+1)/(0{\cdot}1s+1)) \tag{D 4}$$

The proportional-plus-integral controller designed using the parameter tuning of § 2 was :

$$K_p = \begin{bmatrix} 0{\cdot}590 & 0{\cdot}0145 & 2{\cdot}36 & 0{\cdot}976 & 0{\cdot}991 \\ -0{\cdot}00024 & 0{\cdot}000253 & 0{\cdot}000144 & -0{\cdot}000105 & 0{\cdot}000116 \\ -0{\cdot}00962 & 0{\cdot}00461 & -0{\cdot}00830 & -0{\cdot}0646 & 0{\cdot}0138 \\ -0{\cdot}00522 & -0{\cdot}00343 & -0{\cdot}0114 & -0{\cdot}0258 & -0{\cdot}0739 \\ -0{\cdot}0000104 & -0{\cdot}0000154 & 0{\cdot}0000236 & 0{\cdot}0000218 & -0{\cdot}0000222 \end{bmatrix} \tag{D 5}$$

$$K_i = \begin{bmatrix} 2\cdot82 & -1\cdot62 & 3\cdot75 & -3\cdot22 & 7\cdot79 \\ -0\cdot000667 & 0\cdot00194 & -0\cdot000170 & 0\cdot00161 & -0\cdot000971 \\ -0\cdot0180 & 0\cdot0565 & -0\cdot00928 & -0\cdot0919 & -0\cdot103 \\ 0\cdot00175 & 0\cdot0210 & -0\cdot0563 & 0\cdot102 & -0\cdot369 \\ -0\cdot0000890 & -0\cdot000220 & 0\cdot000318 & -0\cdot000431 & 0\cdot000484 \end{bmatrix} \qquad \text{(D 6)}$$

Appendix E

Missile autopilot

The state space matrices of the model of the missile are as follows :

$$A = \begin{bmatrix} 3\cdot23 & p & -476\cdot0 & 0 & 228\cdot0 & 0 & 0 & 0 \\ -p & -3\cdot23 & 0 & 476\cdot0 & 0 & -228\cdot0 & 0 & 0 \\ 0\cdot39 & 0 & -1\cdot93 & -0\cdot8p & -415\cdot0 & 0 & 0 & 0 \\ 0 & -0\cdot39 & 0\cdot8p & -1\cdot93 & 0 & -415\cdot0 & 0 & 0 \\ 0 & 0 & 0 & 0 & 0 & 0 & 75\cdot0 & 0 \\ 0 & 0 & 0 & 0 & 0 & 0 & 0 & -75\cdot0 \\ 0 & 0 & 22\cdot4 & 0 & -300\cdot0 & 0 & -150\cdot0 & 0 \\ 0 & 0 & 0 & -22\cdot4 & 0 & 300\cdot0 & 0 & -150\cdot0 \end{bmatrix}$$

$$B = \begin{bmatrix} 0 & 0 & 0 & 0 & 0 & 0 & -1\cdot0 & 0 \\ 0 & 0 & 0 & 0 & 0 & 0 & 0 & -1\cdot0 \end{bmatrix}$$

$$C = \begin{bmatrix} -2\cdot99 & 0 & -1\cdot19 & 0\cdot123p & -27\cdot64 & 0 & 0 & 0 \\ 0 & -2\cdot99 & 0\cdot123p & 1\cdot19 & 0 & 27\cdot64 & 0 & 0 \end{bmatrix}$$

The resulting controller matrices were

$$A_c = \begin{bmatrix} 0 & 1 & 0 & 0 \\ 0 & -100 & 0 & 0 \\ 0 & 0 & 0 & 0 \\ 0 & 0 & 0 & -100 \end{bmatrix}, \quad B_c = \begin{bmatrix} 0 & 0 \\ 1 & 0 \\ 0 & 0 \\ 0 & 1 \end{bmatrix}$$

$$C_{cf} = \begin{bmatrix} 293\cdot698 & -21\cdot827 & 24\cdot698 & 1\cdot547 \\ -24\cdot847 & -1\cdot547 & 293\cdot698 & -21\cdot827 \end{bmatrix}$$

$$D_{cf} = \begin{bmatrix} 0\cdot407802 & -0\cdot004208 \\ 0\cdot004207 & 0\cdot407802 \end{bmatrix}$$

$$C_{cv} = \begin{bmatrix} -15\cdot325 & 0\cdot028 & -24\cdot418 & -1\cdot612 \\ 24\cdot418 & 1\cdot612 & -15\cdot325 & 0\cdot028 \end{bmatrix}$$

$$D_{cv} = \begin{bmatrix} 0\cdot002789 & 0\cdot006112 \\ -0\cdot006112 & 0\cdot002789 \end{bmatrix}$$

References

EDMUNDS, J. M., 1978, Cambridge Linear Analysis and Design Programs, Report CUED/F–CAMS/TR 170, Engineering Department, University of Cambridge, England.

FOSS, A. S., EDMUNDS, J. M., and KOUVARITAKIS, B., 1980, *Ind. Engng Chem. Fundam.*, (to be published).

GLOVER, K., 1973, Ph.D. Dissertation ESL–R–516, M.I.T., Cambridge, Massachusetts, U.S.A.

KOUVARITAKIS, B., and EDMUNDS, J. M., 1978, *Alternatives for Linear Multivariable Control* (Illinois : National Engineering Consortium), pp. 229–246.

MACFARLANE, A. G. J., 1972, *Automatica*, **18,** 455.

MACFARLANE, A. G. J., and KOUVARITAKIS, B., 1977, *Int. J. Control*, **25,** 837.

ROSENBROCK, H. H., 1974, *Computer Aided Control System Design* (London : Academic Press).

SAIN, M. K., PECZKOWSKI, J. L., and MELSA, J. L., 1978, *Alternatives for Linear Multivariable Control* (Illinois : National Engineering Consortium).

SILVA, J. M., WALLMAN, P. A., and FOSS, A. S., 1979, *Ind. Engng Chem. Fundam.*, **18**.

WALLMAN, P. A., 1977, Ph.D. Thesis, University of California, Berkeley, U.S.A.

V. BIBLIOGRAPHY

ALLWRIGHT, J. C., 1970, Commutative controllers. *Electronics Lett.*, **6,** 276.

AMARA, R. C., 1959, Application of matrix methods to the linear least squares synthesis of multivariable systems. *J. Franklin Inst.*, **268,** 1–6.

AMARA, R. C., 1959, The linear least squares synthesis of multivariable control systems. *Trans. Am. Inst. elect. Engrs*, **78,** 115–119.

ASHON, S. I., and NICHOLSON, H., 1976, Improvement of turbo-alternator response using the inverse Nyquist array method. *Int. J. Control*, **23,** 657–672.

BARMAN, J. F., and KATZENELSON, J., 1974, A generalized Nyquist-type stability criterion for multivariable feedback systems. *Int. J. Control*, **20,** 593–622.

BARNETT, S., 1971, *Matrices in Control Theory* (London : Van Nostrand Reinhold).

BELL, D. J. (editor), 1973, *Recent Mathematical Developments in Control* (London : Academic Press).

BELLETRUTTI, J. J., and MACFARLANE, A. G. J., 1971, Characteristic loci techniques in multivariable control system design. *Proc. Instn elect. Engrs*, **118,** 1291–1297.

BENGTSSON, G., 1973, A theory for control of linear multivariable systems. Lund Institute Technical Report 7341, Lund, Sweden.

BENGTSSON, G., 1977, Feedback realizations in linear multivariable systems. *I.E.E.E. Trans. autom. Control*, **22,** 576–585.

BENGTSSON, G., 1977, Output regulation and internal models—a frequency domain approach. *Automatica*, **13,** 333–345.

BLACK, H. S., 1934, ' Stabilized Feedback Amplifiers '. *Bell Syst. tech. J.*, **13,** 1–18.

BLACK, H. S., 1977, Inventing the negative feedback amplifier. *I.E.E.E. Spectrum*, **14,** 54–60.

BLISS, G. A., 1933, *Algebraic Functions* (Providence, Rhode Island : American Mathematical Society).

BODE, H. W., 1940, ' Relations between attenuation and phase in feedback amplifier design '. *Bell Syst. tech. J.*, **19,** 421–454.

BODE, H. W., 1945, *Network Analysis and Feedback Amplifier Design* (Princeton, N.J. : Van Nostrand).

BODE, H. W., 1960, Feedback—the history of an idea. *Proceedings of the Symposium on Active Networks and Feedback Systems*, Polytechnic Institute of Brooklyn (Polytechnic Press), pp. 1–17.

BOHN, E. V., 1960, Stabilization of linear multivariable feedback control systems. *I.R.E. Trans. autom. Control*, **5,** 321–327.

BOHN, E. V., 1962, Design and synthesis methods for a class of multivariable feedback control systems based on single-variable methods. *Trans. Am. Inst. elect. Engrs*, **81,** 109–116.

BOHN, E. V., and KASVAND, T., 1963, Use of matrix transformations and system eigenvalues in the design of linear multivariable control systems. *Proc. Inst. elect. Engrs*, **110,** 989–996.

BOKSENBOM, A. S., and HOOD, R., 1949, General algebraic method applied to control analysis of complex engine types. National Advisory Committee for Aeronautics, Report NCA-TR-980, Washington, D.C.

BONGIORNO, J. J., JR., and YOULA, D. C., 1977, On the design of single-loop single-input-output feedback control systems in the complex-frequency domain. *I.E.E.E. Trans. autom. Control*, **22,** 416–423.

BRISTOL, E. H., 1966, On a new measure of interaction for multivariable process control. *I.E.E.E. Trans. Autom. Control*, **11(1)**, 133–134.

BRISTOL, E. H., 1967, A philosophy for single loop controllers in a multiloop world. *Proceedings of the 8th National Symposium on Instrumentation in the Chemical and Process Industries*, St. Louis, Mo., Vol. 4, 18–29.

BRISTOL, E. H., 1976, Industrial needs and requirements for multivariable control. *Chemical Process Control*, AIChE Symposium Series No. 159, **72,** 88–98.

BROCKETT, R. W., 1965, Poles, zeros and feedback: state space interpretation. *I.E.E.E. Trans. autom. Control*, **10,** 129–135.

CALLIER, F. M., and DESOER, C. A., 1972, A graphical test for checking the stability of a linear time-invariant feedback system. *I.E.E.E. Trans. autom. Control*, **17,** 773–780.

CALLIER, F. M., and DESOER, C. A., 1976, On simplifying a graphical stability criterion for linear distributed feedback systems. *I.E.E.E. Trans. autom. Control*, **21,** 128–129.

CHEN, C.-T., 1968, Stability of linear multivariable feedback systems, *Proc. Inst. electron. Engrs*, **56,** 821–828.

COOK, P. A., 1973, Stability of linear constant multivariable systems. *Proc. Instn elect. Engrs*, **120,** 1557.

COOK, P. A., 1974, Nyquist-plot methods of investigating multivariable feedback systems. *Proceedings of the Third IFAC Symposium on Multivariable Technological Systems*, Manchester (Institute of Measurement and Control: London), Paper S-28.

DAVIS, J. H., 1972, Fredholm operators, encirclements, and stability criteria. *SIAM J. Control*, **10,** 608–622.

DAVIS, J. H., 1972, Encirclement conditions for stability and instability of feedback systems with delays. *Int. J. Control*, **15,** 793–799.

DAVISON, E. J., 1969, A nonminimum phase index and its application to interacting multivariable control systems. *Automatica*, **5,** 791–799.

DAVISON, E. J., and WANG, S. H., 1974, Properties and calculation of transmission zeros of linear multivariable systems. *Automatica*, **10,** 643–658.

DAVISON, E. J., 1976, The robust decentralized control of a general servomechanism problem. *I.E.E.E. Trans. autom. Control*, **21,** 529–534.

DAVISON, E. J., and WANG, S. H., 1977, Properties and calculation of transmission zeros of linear multivariable feedback systems. *Control System Design by Pole–Zero Assignment*, edited by F. Fallside (London: Academic Press), pp. 16–42.

DAVISON, E. J., and WANG, S. H., 1978, An algorithm for the calculation of transmission zeros of the system (C, A, B, D) using high gain output feedback. *I.E.E.E. Trans. autom. Control*, **23,** 738–741.

DECARLO, R., and SAEKS, R., 1977, The encirclement condition: An approach using algebraic topology. *Int. J. Control*, **26,** 279–287.

DESOER, C. A., 1965, ' A general formulation of the Nyquist criterion '. *I.E.E.E. Trans. Circuit Theory*, **12,** 230–234.

DESOER, C. A. 1974, Multivariable systems: factorization of det $[I+FG]$ as a product of measurable gains. *I.E.E.E. Trans. autom. Control*, **19,** 606.

DESOER, C. A., and SCHULMAN, J. D., 1973, Cancellations in multivariable continuous-time and discrete-time feedback systems treated by greatest common divisor extraction. *I.E.E.E. Trans. autom. Control*, **18,** 401–402.

DESOER, C. A., and SCHULMAN, J. D., 1974, Zeros and poles of matrix transfer functions and their dynamical interpretation. *I.E.E.E. Trans. Circuits Syst.*, **21,** 3–8.

DESOER, C. A., and VIDYASAGAR, M., 1975, *Feedback Systems: Input–Output Properties* (New York: Academic Press).

DESOER, C. A., and WANG, Y. T., 1978, On the generalized Nyquist stability criterion. Memorandum UCB/ERL M79/11, Electronics Research Laboratory, University of California, Berkeley.

DESOER, C. A., and WU, M. Y., 1968, Stability of linear time-invariant systems. *I.E.E.E. Trans. Circuit Theory*, **15,** 245–250.

DESOER, C. A., and WU, M. Y., 1968, Stability of multiple-loop feedback linear time-invariant systems. *J. math. Analysis Applic.*, **23,** 121–130.

DOYLE, J. C., 1978, Guaranteed margins for LQG regulators. *I.E.E.E. Trans. autom. Control*, **23,** 756–757.

EDMUNDS, J. M., 1978, Cambridge Linear Analysis and Design Programs. Report No. CUED/F-CAMS/TR170, Cambridge University Engineering Department.

EDMUNDS, J. M., 1979, Cambridge Linear Analysis and Design Programs. *Pro-*

ceedings of the IFAC Symposium on Computer-Aided Control System Design, Zurich.

EDMUNDS, J. M., 1979, Characteristic gains, characteristic frequencies and stability. *Int. J. Control*, **29,** 669–706.

EDMUNDS, J. M., 1979, Control system design and analysis using closed-loop Nyquist, and Bode Arrays. *Int. J. Control*, **30,** 773–802.

EDMUNDS, J. M., and KOUVARITAKIS, B., 1979, Extensions of the frame alignment technique and their use in the characteristic locus design method. *Int. J. Control*, **29,** 787–796.

EVANS, W. R., 1948, Graphical analysis of control systems. *Trans. Am. Inst. elect. Engrs*, **67,** 547–551.

EVANS, W. R., 1950, Control system synthesis by root locus method. *Trans. Am. Inst. elect. Engrs*, **69,** 1–4.

EVANS, W., 1953, *Control System Dynamics* (New York : McGraw-Hill).

FALB, P. L., and FREEDMAN, M. I., 1969, A generalized transform theory for causal operators. *SIAM. J. Control*, **7,** 452–471.

FALLSIDE, F. (editor), 1977, *Control System Design by Pole-Zero Assignment* (London : Academic Press).

FOSS, A. S., 1973, Critique of chemical process control theory. *I.E.E.E. Trans. autom. Control*, **18,** 646–652.

FOSS, A. S., EDMUNDS, J. M., and KOUVARITAKIS, B., 1979, Multivariable control system for two-bed reactors by characteristic locus method. To be published in *Ind. Eng. Chem. Fund.* during 1980.

FOSSARD, A. J., 1977. *Multivariable System Control* (Amsterdam : North-Holland).

FRANCIS, B. A., 1975, The linear multivariable regulator problem. Memorandum ERL-M560, Electronics Research Laboratory, University of California, Berkeley.

FRANCIS, B. A., 1977, The multivariable servomechanism from the input-output viewpoint. *I.E.E.E. Trans. autom. Control*, **22,** 322–328.

FRANCIS, B. A., and WONHAM, W. M., 1975, The role of transmission zeros in linear multivariable regulators. *Int. J. Control*, **22,** 657–681.

FREEDMAN, M. I., FALB, P. L., and ZAMES, G., 1969, A Hilbert space stability theory over locally compact Abelian groups. *SIAM J. Control*, **7,** 479–495.

FREEMAN, E. A., 1973, Stability of linear constant multivariable systems. *Proc. Instn elect. Engrs*, **120,** 379–384.

FREEMAN, E. A., 1974, Relation between some criteria for the stability of constant linear multivariable systems. *Proc. Instn elect. Engrs*, **121,** 725–729.

FREEMAN, H., 1957, A synthesis method for multipole control systems. *Trans. Am. Inst. elect. Engrs*, **76,** 28–31.

FREEMAN, H., 1958, Stability and physical realizability considerations in the synthesis of multipole control systems. *Trans. Am. Inst. elect. Engrs*, Part 2, **77,** 1–15.

GOLOMB, M., and USDIN, E., 1952, A theory of multidimensional servo systems. *J. Franklin Inst.*, **253(1),** 28–57.

GREGSON, M. J. (editor), 1978, *Recent Theoretical Developments in Control* (London : Academic Press).

HARVEY, C. H., and STEIN, G., 1978, Quadratic weights for asymptotic regulator properties. *I.E.E.E. Trans. autom. Control*, **23,** 378–387.

HAWKINS, D. J., 1973, Determination of stability regions with the inverse Nyquist array. *Proc. Instn elect. Engrs*, **120,** 1445–1448.

HOROWITZ, I., 1960, Synthesis of linear, multivariable feedback control systems. *I.R.E. Trans. autom. Control*, **5,** 94–105.

HOROWITZ, I. M., 1963, *Synthesis of Feedback Systems* (New York : Academic Press).

HOROWITZ, I., and SHAKED, U., 1975, Superiority of transfer function over state-variable methods in linear time-invariant feedback system design. *I.E.E.E. Trans. autom. Control*, **20,** 84–97.

HUGHES, F. M., 1977, A simplified frequency-response design approach for power-station control. *Int. J. Control*, **25,** 575–587.

IBRAHIM, T. A. S., and MUNRO, N., 1975, Design of sampled-data multivariable control systems using the inverse Nyquist array. *Int. J. Control*, **22,** 297–311.

KALMAN, R. E., 1964, When is a linear control system optimal? *J. bas. Engng*, **86,** 51–60.

KALMAN, R. E., 1965, Irreducible realizations and the degree of a rational matrix. *SIAM Jl Control*, **13,** 520–544.

KARCANIAS, N., 1978, The use of Grassman products in the computation of the invariant zero polynomial of square linear proper systems. *Int. J. Control*, **28,** 361–381.

KAVANAGH, R. J., 1957, Noninteraction in linear multivariable systems. *Trans. Am. Inst. elect. Engrs*, **76,** 95–100.

KAVANAGH, R. J., 1957, The application of matrix methods to multivariable control systems. *J. Franklin Inst.*, **262,** 349–367.

KAVANAGH, R. J., 1958, Multivariable control system synthesis. *Trans. Am. Inst. elect. Engrs*, Part 2, **77,** 425–429.

KAVANAGH, R. J., 1966, The multivariable problem. *Prog. Control Engng*, **3,** 94–129.

KONTAKOS, T., 1973, Ph.D. Thesis, University of Manchester.

KOUVARITAKIS, B., 1976, A geometric approach to the inversion of multivariable systems. *Int. J. Control*, **24,** 593–622.

KOUVARITAKIS, B., 1978, The optimal root loci of linear multivariable systems. *Int. J. Control*, **28,** 33–62.

KOUVARITAKIS, B., 1978, Gain margins and root locus asymptotic behaviour in multivariable design, Parts I and II. *Int. J. Control*, **27,** 705–724 and 725–751.

KOUVARITAKIS, B., 1978, Asymptotic multivariable root locus behaviour for non-proper systems. *Int. J. Control*, **28,** 419–440.

KOUVARITAKIS, B., and EDMUNDS, J. M., 1978, The characteristic frequency and characteristic gain design method for multivariable feedback systems. *Alternatives for Linear Multivariable Control* (Chicago: National Engineering Consortium), pp. 229–246.

KOUVARITAKIS, B., and EDMUNDS, J. M., 1979, Multivariable root loci: a unified approach to finite and infinite zeros. *Int. J. Control*, **29,** 393–428.

KOUVARITAKIS, B., and MACFARLANE, A. G. J., 1976, Geometric approach to analysis and synthesis of system zeros. Part 1: square systems. *Int. J. Control*, **23,** 149–166.

KOUVARITAKIS, B., and MACFARLANE, A. G. J., 1976, Geometric approach to analysis and synthesis of system zeros. Part 2: non-square systems. *Int. J. Control*, **23,** 167–181.

KOUVARITAKIS, B., MURRAY, W., and MACFARLANE, A. G. J., 1979, Characteristic frequency/gain design study of an automatic flight control system. *Int. J. Control*, **29,** 325–358.

KOUVARITAKIS, B., and SHAKED, U., 1976, Asymptotic behaviour of root loci of linear multivariable systems. *Int. J. Control*, **23,** 297–340.

KWAKERNAAK, H., 1976, Asymptotic root loci of multivariable linear optimal regulators. *I.E.E.E. Trans. autom. Control*, **21,** 378–382.

LAUB, A. J., and MOORE, B. C., 1976, Calculation of transmission zeros using QZ techniques, Systems Control Report No. 7618, Department of Electrical Engineering, University of Toronto.

LAYTON, J. M., 1976, *Multivariable Control Theory* (Stevenage: Peter Peregrinus).

MACFARLANE, A. G. J., 1970, The return-difference and return-ratio matrices and their use in the analysis and design of multivariable feedback control systems. *Proc. Inst. elect. Engrs*, **117,** 2037–2049.

MACFARLANE, A. G. J., 1970, Commutative controller. *Electronics Lett.*, **6,** 121–123.

MACFARLANE, A. G. J., 1971, Return-difference matrix properties for optimal stationary Kalman–Bucy filter. *Proc. Inst. elect. Engrs*, **118,** 373–376.

MACFARLANE, A. G. J., 1976, Feedback. *J. Inst. Measmt Control*, **9,** 449–462.

MACFARLANE, A. G. J., 1979, The development of frequency-response methods in automatic control. *I.E.E.E. Trans. autom. Control*, **24,** 250–265.

MacFarlane, A. G. J. (editor), 1979, *Frequency Response Methods in Control Systems*, I.E.E.E. Reprint Series (New York : I.E.E.E.).

MacFarlane, A. G. J., and Belletrutti, J. J., 1973, The characteristic locus design method. *Automatica*, **9,** 575–588.

MacFarlane, A. G. J., and Karcanias, N., 1976, Poles and zeros of linear multivariable systems : a survey of the algebraic, geometric and complex-variable theory. *Int. J. Control*, **24,** 33–74.

MacFarlane, A. G. J., and Karcanias, N., 1978, Relationships between state-space and frequency-response concepts. *Preprints of Seventh IFAC World Congress*, Vol. 3, pp. 1771–1779.

MacFarlane, A. G. J., and Kouvaritakis, B., 1977, A design technique for linear multivariable feedback systems. *Int. J. Control*, **25,** 837–874.

MacFarlane, A. G. J., Kouvaritakis, B., and Edmunds, J. M., 1978, Complex variable methods for multivariable feedback systems analysis and design. *Alternatives for Linear Multivariable Control* (Chicago : National Engineering Consortium), pp. 189–228.

MacFarlane, A. G. J., and Postlethwaite, I., 1977, The generalized Nyquist stability criterion and multivariable root loci. *Int. J. Control*, **25,** 81–127.

MacFarlane, A. G. J., and Postlethwaite, I., 1977, Characteristic frequency functions and characteristic gain functions. *Int. J. Control*, **26,** 265–278.

MacFarlane, A. G. J., and Postlethwaite, I., 1978, Extended Principle of the Argument. *Int. J. Control*, **27,** 49–55.

MacFarlane, A. G. J., and Rosenbrock, H. H., 1970, New vector space structure for dynamical systems. *Electronics Lett.*, **6,** 162–163.

MacFarlane, A. G. J., and Scott-Jones, D. F. A., 1979, Vector gain. *Int. J. Control*, **29,** 65–91.

Marshall, S. A., and Owens, D. H., 1976, Feed-forward-feedback control of multivariable systems. *Int. J. Control*, **23,** 693–701.

Mayne, D. Q., 1973, The design of linear multivariable systems. *Automatica*, **9,** 201–207.

Mayne, D. Q., 1974, The effect of feedback on linear multivariable systems. *Automatica*, **10,** 405–412.

McMillan, B., 1952, Introduction to formal realizability theory. *Bell Syst. tech. J.*, **31,** 217–279, 541–600.

McMorran, P. D., 1970, Design of gas-turbine controller using inverse Nyquist method. *Proc. Instn elect. Engrs*, **117,** 2050–2056.

McMorran, P. D., 1970, Extension of the inverse Nyquist method. *Electronics Lett.*, **6,** 800–801.

McMorran, P. D., 1971, Parameter sensitivity and inverse Nyquist method. *Proc. Instn elect. Engrs*, **118,** 802–804.

Munro, N., 1972, Design of controllers for open-loop unstable multivariable system using inverse Nyquist array. *Proc. Instn elect. Engrs*, **119,** 1377–1382.

Nyquist, H., 1932, ‘ Regeneration Theory ’. *Bell Syst. tech. J.*, **11,** 126–147.

Newton, G. C., Jr., Gould, L. A., and Kaiser, J., 1957, *Analytical Design of Linear Feedback Controls* (New York : Wiley).

Niederlinski, A., 1970, Analysis and design of 2-variable interacting control systems using inverse polar plots. *Proc. Instn elect. Engrs*, **117,** 2056–2059.

Niederlinski, A., 1971, A heuristic approach to the design of linear multivariable interacting control systems. *Automatica*, **7,** 691–701.

Nwokah, O., 1975, Stability and the eigenvalues of $G(s)$. *Int. J. Control*, **22,** 125–128.

Ostrowski, A. M., 1952, Note on bounds for determinants with dominant principal diagonal. *Proc. Am. math. Soc.*, **3,** 26–30.

Owens, D. H., 1973, Dyadic approximation method for multivariable-control systems analysis with a nuclear-reactor application. *Proc. Instn elect. Engrs*, **120,** 801–809.

Owens, D. H., 1973, Multivariable-control-system design concepts in failure analysis of a class of nuclear-reactor spatial-control systems. *Proc. Instn elect. Engrs*, **120,** 119–125.

Owens, D. H., 1974, Sequential design of linear multivariable systems, retaining full output feedback. *Electronics Lett.*, **10,** 79–80.

Owens, D. H., 1974, Dyadic expansion for the analysis of linear multivariable systems. *Proc. Instn elect. Engrs*, **121,** 713–716.

Owens, D. H., 1974, Dyadic modification to sequential technique for multivariable control system design. *Electronics Lett.*, **10,** 25–26.

Owens, D. H., 1975, Dyadic approximation about a general frequency point. *Electronics Lett.*, **11,** 331–332.

Owens, D. H., 1975, Dyadic expansion, characteristic loci and multivariable-control-systems design, *Proc. Instn elect. Engrs*, **122,** 315–320.

Owens, D. H., 1975, First- and second-order-like structures in linear multivariable-control-systems design. *Proc. Instn elect. Engrs*, **122,** 935–941.

Owens, D. H., 1975, Compensation of multivariable 1st-order-type systems. *Electronics Lett.*, **11,** 511–512.

Owens, D. H., 1976, Modal decoupling and dyadic transfer function matrices. *Int. J. Control*, **23,** 63–65.

Owens, D. H., 1976, Integrity of multivariable first-order-type systems. *Int. J. Control*, **23,** 827–835.

Owens, D. H., 1976, Cascade canonical form for linear multivariable systems. *Int. J. Control*, **23,** 837–850.

Owens, D. H., 1977, Invariant zeros of multivariable systems : a geometric analysis. *Int. J. Control*, **26,** 537–548.

Owens, D. H., 1977, A note on series expansions for multivariable root loci. *Int. J. Control*, **26,** 549–557.

Owens, D. H., 1978, Multivariable root loci and the inverse transfer-function matrix. *Int. J. Control*, **28,** 345–351.

Owens, D. H., 1978, Dynamic transformations and the calculation of multivariable root loci. *Int. J. Control*, **28,** 333–343.

Owens, D. H., 1978, *Feedback and Multivariable Systems* (Hitchin : Peter Peregrinus).

Owens, D. H., 1979, Some unifying concepts in multivariable feedback design. *Proceedings of the IFAC Symposium on Computer-aided Control System Design*, Zurich.

Patel, R. V., 1978, On transmission zeros and dynamic output feedback. *I.E.E.E. Trans. autom. Control*, **23,** 741–742.

Pernebo, L., 1978, Algebraic control theory for linear multivariable systems. Lund Institute Technical Report LUTFD2/TFRT–1016/1–307, Lund, Sweden.

Postlethwaite, I., 1977, The asymptotic behaviour, the angles of departure, and the angles of approach of the characteristic frequency loci. *Int. J. Control*, **25,** 677–695.

Postlethwaite, I., 1977, A generalized inverse Nyquist stability criterion. *Int. J. Control*, **26,** 325–340.

Postlethwaite, I., 1978, A note of the characteristic frequency loci of multivariable linear optimal regulators. *I.E.E.E. Trans. autom. Control*, **23,** 757–760.

Postlethwaite, I., 1979, Association between branch points and stationary points on the gain and frequency surfaces. *Int. J. Control*, **29,** 255–260.

Postlethwaite, I., 1979, A note on parametric stability. *Int. J. Control*, **29,** 261–266.

Postlethwaite, I., 1979, Methods for assessing the robustness properties of a linear multivariable feedback system. *Proceedings of the ONR/MIT–L.I.D.S. Workshop on Recent Advances in the Robustness Theory of Multivariable Systems*, M.I.T., Cambridge, Mass.

Postlethwaite, I., 1979, Algebraic function theory in the analysis of multivariable feedback systems. *Proc. Instn elect. Engrs*, **126,** 555–562.

Postlethwaite, I., and MacFarlane, A. G. J., 1979, *A Complex Variable Approach to the Analysis of Linear Multivariable Feedback Systems*, Lecture Notes in Control and Information Sciences, Vol. 12 (Berlin : Springer-Verlag).

Ramani, N., and Atherton, D. P., 1975, A note on the stability and eigenvalues of $G(s)$. *Int. J. Control*, **22,** 701–704.

RAYMOND, F. H., 1953, Introduction a l'etude des asservissements multiples simultanes. *Bull. Soc. fr. Mec.*, **7,** 18–25.

RETALLACK, D. G., 1970, Extended root-locus technique for design of linear multivariable feedback systems. *Proc. Instn elect. Engrs*, **117,** 618–622.

RETALLACK, D. G., 1970, Transfer-function matrix approach to observer design. *Proc. Instn elect. Engrs*, **117,** 1153–1155.

ROSENBROCK, H. H., 1966, On the design of the linear multivariable control systems. *Proceedings of the Third IFAC Congress*, London, I, 1–16.

ROSENBROCK, H. H., 1969, ' Design of multivariable control systems using the inverse Nyquist array '. *Proc. Instn elect. Engrs*, **116,** 1929–1936.

ROSENBROCK, H. H., 1970, *State Space and Multivariable Theory* (London : Nelson).

ROSENBROCK, H. H., 1971, Progress in the design of multivariable control systems. *Trans. Inst. Measmt Control*, **4,** 9–11.

ROSENBROCK, H. H., 1972, The stability of multivariable systems. *I.E.E.E. Trans. autom. Control*, **17,** 105–107.

ROSENBROCK, H. H., 1973, The zeros of a system. *Int. J. Control*, **18,** 297–299.

ROSENBROCK, H. H., 1973, Bounds for transfer functions in multivariable systems. *I.E.E.E. Trans. autom. Control*, **18,** 54–56.

ROSENBROCK, H. H., 1974, Correction to the zeros of a system. *Int. J. Control*, **20,** 525–527.

ROSENBROCK, H. H., 1974, *Computer-aided Control System Design* (New York : Academic Press).

ROSENBROCK, H. H., and COOK, P. A., 1975, Stability and the eigenvalues of $G(s)$. *Int. J. Control*, **21,** 99–104.

ROSENBROCK, H. H., and MCMORRAN, P. D., 1971, Good, bad or optimal ? *I.E.E.E. Trans. autom. Control*, **16,** 552–554.

ROSENBROCK, H. H., and MUNRO, N., 1978, The inverse Nyquist array method, *Proceedings of the International Forum on Alternatives for Linear Multivariable Control* (Chicago : National Engineering Consortium), pp. 101–137.

ROSENBROCK, H. H., and ROWE, A., 1970, Allocation of poles and zeros. *Proc. Instn elect. Engrs*, **117,** 1879–1886.

SAEKS, R., 1975, On the encirclement condition and its generalization. *Trans. I.E.E.E. Circuits Syst.*, **22,** 780–785.

SAFANOV, M. E., and ATHANS, M., 1977, Gain and phase margins for multiloop LQG regulators. *I.E.E.E. Trans. autom. Control*, **22,** 173–179.

SAIN, M. K., PECZKOWSKI, J. L., and MELSA, J. L (editors), 1978, *Alternatives for Linear Multivariable Control* (Chicago : National Engineering Consortium).

SANDBERG, I. W., 1963, On the theory of linear multiloop feedback systems. *Bell Syst. tech. J.*, **42,** 355–382.

SHAKED, U., 1976, Design techniques for high feedback gain stability. *Int. J. Control*, **24,** 137–144.

SHAKED, U., 1976, The angles of departure and approach of the root loci in linear multivariable systems. *Int. J. Control*, **23,** 445–457.

SHAKED, U., 1977, The zero properties of linear passive systems. *I.E.E.E. Trans. autom. Control*, **22,** 973–974.

SHAKED, U., 1978, The asymptotic behaviour of the root loci of multivariable optimal regulators. *I.E.E.E. Trans. autom. Control*, **23,** 425–430.

SHAKED, U., 1979, The design of multivariable systems having zero sensitive poles. *I.E.E.E. Trans. autom. Control*, **24,** 117–119.

SHAKED, U., and DIXON, M., 1977, Generalized minimal realization of transfer-function matrices. *Int. J. Control*, **25,** 785–803.

SHAKED, U., and KARCANIAS, N., 1976, The use of zeros and zero-directions in model reduction. *Int. J. Control*, **23,** 113–135.

SHAKED, U., and KOUVARITAKIS, B., 1977, The zeros of linear optimal control systems and their role in high feedback gain stability design. *I.E.E.E. Trans. autom. Control*, **22,** 597–599.

SILVERMAN, L. M., 1969, Inversion of multivariable linear systems. *I.E.E.E. Trans. autom. Control*, **14,** 270–276.

Simon, J. D., and Mitter, S. K., 1969, Synthesis of transfer function matrices with invariant zeros. *I.E.E.E. Trans. autom. Control*, **14,** 420–421.

Springer, G., 1957, *Introduction to Riemann Surfaces* (Reading, Mass. : Addison–Wesley).

Truxal, J. G., 1955, *Automatic Feedback Control System Synthesis* (New York : McGraw-Hill).

Valenca, J. M. E., and Harris, C. J., 1979, A Nyquist type criterion for the stability of multivariable linear systems. *Proceedings of the 17th I.E.E.E. Conference on Decision and Control*, San Diego.

Vesty, P., 1975, Geometric interpretation of complex vectors arising in the frequency-domain analysis of multivariable systems. *I.E.E.E. Trans. autom. Control*, **20,** 157–158.

Vidyasagar, M., 1972, Input–output stability of a broad class of linear time-invariant multivariable feedback systems. *SIAM J. Control*, **10,** 203–209.

Vidyasagar, M., 1975, Coprime factorizations and stability of multivariable distributed feedback systems. *SIAM J. Control*, **13,** 1144–1155.

Weyl, H., 1955, The Concept of a Riemann Surface, third edition (Reading, Mass. : Addison–Wesley).

Willems, J. C., 1971, The Analysis of Feedback Systems (Cambridge, Mass. : M.I.T. Press).

Wolovich, W. A., 1974, *Linear Multivariable Systems* (New York : Springer-Verlag).

Wolovich, W. A., 1977, Multivariable system zeros. *Control System Design by Pole-Zero Assignment*, edited by F. Fallside (London : Academic Press), pp. 226–236.

Wolovich, W. A., and Falb, P. L., 1969, On the structure of multivariable systems. *SIAM J. Control*, **7,** 437–451.

Wonham, W. M., 1974, *Linear Multivariable Control: A Geometric Approach* (Berlin : Springer-Verlag).

Woon, S. K., and Marshall, S. A., 1975, Design of multivariable control systems using reduced-order models. *Electronics Lett.*, **11,** 341–342.

Youla, D. C., Bongiorno, J. J., Jr., and Jabr, H. A., 1976, Modern Wiener–Hopf design of optimal controllers. Part I: The single-input-output case. *I.E.E.E. Trans. autom. Control*, **21,** 2–13.

Youla, D. C., Bongiorno, J. J., Jr., and Lu, C. N., 1974, Single-loop feedback stabilization of linear multivariable dynamical plants. *Automatica*, **10,** 159–173.

Youla, D. C., Jabr, H. A., and Bongiorno, J. J., Jr., 1976, Modern Wiener–Hopf design of optimal controllers. Part II: The multivariable case. *I.E.E.E. Trans. autom. Control*, **21,** 319–338.

INDEX